The Prosciutto Sundial

The *orologio di prosciutto* from the Villa dei Papiri in Herculaneum. Silver coated cast bronze. 11.3 cm high × 7.8 cm wide × 1.9 cm thick. MANN 25494. Author's photo. Su concessione del Ministero della Cultura—Museo Archeologico Nazionale di Napoli.

The Prosciutto Sundial

Casting Light on an
Ancient Roman Timepiece
from the Villa dei Papiri
in Herculaneum

CHRISTOPHER PARSLOW

OXFORD
UNIVERSITY PRESS

Oxford University Press is a department of the University of Oxford.
It furthers the University's objective of excellence in research, scholarship,
and education by publishing worldwide. Oxford is a registered trade mark of
Oxford University Press in the UK and in certain other countries.

Published in the United States of America by Oxford University Press
198 Madison Avenue, New York, NY 10016, United States of America.

© Oxford University Press 2024

All rights reserved. No part of this publication may be reproduced,
stored in a retrieval system, or transmitted, in any form or by any means,
without the prior permission in writing of Oxford University Press,
or as expressly permitted by law, by license or under terms agreed with
the appropriate reprographics rights organization. Inquiries concerning
reproduction outside the scope of the above should be sent to
the Rights Department, Oxford University Press, at the address above.

You must not circulate this work in any other form and you must impose this same condition on any acquirer

Library of Congress Cataloging-in-Publication Data
Names: Parslow, Christopher Charles, author.
Title: The Prosciutto Sundial : casting light on an ancient Roman timepiece
from the Villa dei Papiri in Herculaneum / Christopher Parslow.
Other titles: Casting light on an ancient Roman timepiece from the
Villa dei Papiri in Herculaneum
Description: New York, NY : Oxford University Press, [2024]
Identifiers: LCCN 2024012839 | ISBN 9780197749388 (hardback) |
ISBN 9780197749401 (epub)
Subjects: LCSH: Sundials—Italy—Herculaneum (Extinct city) | Villa of the
Papyri (Herculaneum) | Excavations (Archaeology)—Italy—Herculaneum
(Extinct city)
Classification: LCC QB215.P37 2024 | DDC 681.1/112—dc23/eng/20240618
LC record available at https://lccn.loc.gov/2024012839

DOI: 10.1093/9780197749418.001.0001

Integrated Books International, United States of America

For Christina and Olivia

Brief Contents

Preface and Acknowledgments	xiii
Abbreviations	xvii
1. Discovery	1
2. Analysis	81
3. Enlightenment	154
4. Observations	263
Appendix A: Comparison of Grid Dimensions	301
Appendix B: Tables of Horary Readings	304
Bibliography	317
Index	339

Detailed Contents

Preface and Acknowledgments   xiii

Abbreviations   xvii

1. Discovery   1
   - 1.a) Una cosa la quale deve far strepito fra i letterati: *Paderni's Letter to the Royal Society*   4
   - 1.b) *Identifying the Prosciutto's Findspot in the Villa dei Papiri*   7
   - 1.c) *The Prosciutto and the "English Gentleman" in the* Memorie   17
   - 1.d) Autant de fautes que de mots: *D'Alembert's* boustrophedon *and the* Encyclopédie   28
   - 1.e) Un' altra collera antiquaria: *Paciaudi's "Carafe," the* Monumenta Peloponnesia, *and Tanucci*   32
   - 1.f) Les tablettes vides et la mémoire pleine: *The Accounts of Barthélemy and La Condamine*   49
   - 1.g) *Prosciutto Mania: Mazzocchi, Martorelli, and Piaggio's* L'Orologio Solare   57

2. Analysis   81
   - 2.a) Una Cosa Inedita: *The Prosciutto vs. the Egyptianizing Base in* Le Antichità d'Ercolano   83
   - 2.b) *Caylus and the Oedipuses of the Sphinxes of Herculaneum*   99
   - 2.c) *The Accademia Strikes Back: The Preface to Volume III of* Le Antichità d'Ercolano   106
   - 2.d) *Chaerephon's Twelve-Foot Dinner: Deciphering Greek and Roman Time*   116
   - 2.e) *Carcani's Empirical Research and Analysis*   124
   - 2.f) *The Transit of Venus, the Obliquity of the Ecliptic, and Dating the Prosciutto*   129

xii   *Detailed Contents*

2.g) *The Accademia's Engraving of the Prosciutto*	139
2.h) *Paderni and the Sundials of Pompeii*	144

3. Enlightenment   154
    3.a) *Lalande's "Moveable" Gnomon and the Follies of the*
          Encyclopédie*'s Successors*   154
    3.b) Totum ex aere nitidissime elaboratum:
          Calkoen's Analysis of the Prosciutto   184
    3.c) *"Nice Invention, for Not Forgetting Dinner":*
          Montucla's Histoire des Mathématiques   194
    3.d) Ingénieuse et assez simple: *Delambre's Trigonometric Analysis*   198
    3.e) *Woepcke's* Disquisitiones *and the Temporal Hour*   204
    3.f) *The Dawn of "Dialling"*   212
    3.g) *The Prosciutto's Peregrinations and the Horologists*
          of the Nineteenth Century   217
    3.h) *Comparetti and De Petra and the Villa dei Papiri*   235
    3.i) *Drecker's "Little Clock"*   240
    3.j) *De Solla Price's "Mass of Bronze"*   246
    3.k) *Nothing New Under the Sun: Prosciutto Studies in the*
          Late Twentieth Century   252

4. Observations   263
    4.a) *The Prosciutto and Philodemus in the Villa dei Papiri*   264
    4.b) Hoc eius erat matutinum: *Piso Pontifex and the Augustan*
          Calendrical Reforms   268
    4.c) *Epicurus' Pig?*   270
    4.d) *Designing and Casting the Prosciutto Sundial*   278
    4.e) *"The Craft Is Closely Related to the Science"*   283
    4.f) *Slicing and Dicing: Recreating Carcani's Empirical*
          Research in 3D   295

Appendix A: Comparison of Grid Dimensions   301

Appendix B: Tables of Horary Readings   304

Bibliography   317

Index   339

Preface and Acknowledgments

SINCE ITS DISCOVERY in 1755, scores of travelers to the Bay of Naples have laid eyes on the so-called *orologio di prosciutto*, the portable sundial in the shape of an Italian cured ham from the Villa dei Papiri in Herculaneum currently displayed in the Museo Archeologico Nazionale in Naples (MANN). They have scrutinized the irregular hour grid and minuscule inscription carved into its bronze surface and credulously accepted its identification as a sundial, though its shape resembles no sundial they have seen before. Many have documented their encounters with the prosciutto in letters and travel journals or simply listed it among the antiquities they observed in the museum, each bringing a slightly different perspective to their description of it. Others have undertaken more detailed assessments, by probing the ancient literary sources for parallels and explications of how time was reckoned in the ancient world, by comparing it to the designs of other ancient sundials, or by analyzing it on the basis of geometrical and astronomical principles. Early treatments of it were hamstrung by the strangle hold placed by the Bourbon court on the dissemination of information about the antiquities recovered in the cities buried by Vesuvius. The resulting obligation to record their observations largely from memory limited the quality and quantity of treatments of this sundial. The consequences of this prohibition were long lasting and manifested in recurring odd notions, misconceptions, or butchered graphic representations of the prosciutto, often recycled from the same published sources long after they should have been recognized for the follies they were. Many who wrote about it did so without seeing the original; few recent studies have offered new approaches to analyzing it. Its droll appearance and unique method of marking the hours with its tail endear it to all but seemingly render it difficult to take seriously.

This monographic study reconstructs the history of the prosciutto from its discovery in the early excavations at Herculaneum down to the modern day and surveys the attempts to describe, depict, and analyze it over that period. It is the first study since the original 1762 publication to examine the prosciutto based on first-hand analysis of it coupled with an assessment of its ability to keep time based on a 3D scaled model. The documentation surrounding its discovery and the initial attempts at its decipherment are surprisingly rich and detailed, overshadowed only by that available for the eponymous papyri also found in the Villa at roughly the same time. Like the papyri its story plays out against the backdrop of the other remarkable discoveries from these excavations, with a cast of characters comprising not only the colorful figures overseeing that work but also intellectuals in the arts and sciences from the leading international academic institutions of the day, all hungry for details that could satisfy their appetite for anything that might shed new light on the ancient world. The prosciutto's story provides a unique window into the workings of the Bourbon court in dealing with the antiquities and highlights the increasing recognition during the Age of Enlightenment of the inherent value of material objects to illustrate aspects of the ancient world. Its classification as a scientific device forced the court to completely alter their approach to publishing the antiquities—at least in this case—from one based exclusively on artistic aesthetics and literary exegesis to one that applied the most recent astronomical discoveries. It is the earliest known Roman portable sundial, found at a time when other ancient sundials were first being recovered archaeologically and analyzed scientifically, and coinciding with important developments in the history of astronomy. Its uniquely hybrid quality as a work of art, an astronomical device, the representation of a quadruped, and a practical tool helped distinguish it from other antiquities from Pompeii and Herculaneum and served to attract the attention not only of scholars and archaeologists of Classical antiquity, but astronomers and mathematicians, Grand Tour travelers, industrialists, horologists, even a veterinarian. As an archaeological object it has largely been overlooked, its provenance in the richly appointed Villa dei Papiri having been forgotten for a full century after its discovery and the significance of its relationship to that context left unexplored. It is the ability to draw on so wide an array of source material from such varied perspectives and fields of inquiry as a means of piecing together its place in the history of the excavations, tracing the modes of assessment and the avenues by which information about it was transmitted, and supplementing that story with new analysis that makes the prosciutto an ideal object for monographic treatment.

While ancient sundials and the means of reckoning time in antiquity figure throughout, this study is not a history of sundials or even a history of

Preface and Acknowledgments xv

ancient portable sundials, for which excellent sources already are available. Moreover, to paraphrase Mrs. Gatty from her *Book of Sundials*, anyone expecting to find an astronomically sophisticated treatment of the prosciutto sundial likely will be disappointed. While it is a remarkable document to the state of astronomical knowledge in its time, the geometrical principles underlying its design are relatively basic, and the goal of this study was to reconstruct how the prosciutto marked the hours of the day over the course of the year and who may have used it and make this information comprehensible to a general audience. The detailed information set forth here for the first time should facilitate more complex scientific analysis.

As one of the familiar objects brought to light by Karl Weber, the Swiss military engineer overseeing the early excavations in Herculaneum and the subject of my first book, I had stood before the prosciutto dozens of times with that same vague sense of confidence that I understood how it worked, but I did not realize its complexities much less know of its long history in the modern era. My learning curve began when my wife, Christina Trier, a Roman archaeologist in her own right, challenged me to explain what I knew about it. I subsequently used it as an exercise in 3D modelling for a digital humanities course I team taught with my colleague in Greek archaeology at Wesleyan, Kate Birney, entitled *Visualizing the Classical*. This set me on the path of not just learning how it worked but actually getting it to work. I am grateful to Dott. Paolo Giulierini, former director of the MANN, for granting me permission to consult, photograph, model, and publish the prosciutto and for enthusiastically supporting my research. His staff at the MANN in the Archivio e Laboratorio Fotografico and Biblioteca provided technical support and cheerful assistance without which this study would not have been possible. Christopher Chenier, professor of the practice in integrative science and IDEAS at Wesleyan, worked wonders nursing the 3D printers over many hours and several iterations of the prosciutto model. I was able to consult, photograph, and measure the original 1762 publication of the prosciutto in *Le Antichità d'Ercolano* in Olin Library's Special Collections and Archives thanks to the help of Suzy Taraba, Dietrich Family director *emerita*, and her staff. Michaelle Biddle, former curator of the Olin Library Book Conservation Laboratory, and her student volunteers undertook the herculean task of cleaning and rebinding all ten volumes of *Le Antichità di Ercolano*—one of Wesleyan's great treasures—leaving me a crisp text with which to work. My colleague, Michael Roberts, Robert Rich Professor of Latin *emeritus*, graciously reviewed my translation of Paciaudi's tortured Latin discussing the prosciutto, helping clarify several aspects of his analysis. I could not ask for better support for my scholarship than I have received from my other colleagues in the Department of Classical Studies, Andy Szegedy-Maszak and Eirene Visvardi, and from Wesleyan University. A special thanks to Bill Herbst, professor of astronomy *emeritus*, who read preliminary drafts of some sections, fiddled with my 3D models, had his students test an early version of it in one of his classes, and patiently listened to my attempts to wrap my head around the astronomical principles. Any errors and misconceptions in that regard are entirely my own. I am indebted to Kate Wolfe of Olin's Office of Interlibrary Loans for her remarkably swift acquisition of many obscure publications and, in particular, for negotiating with the Biblioteca Nazionale in Naples for the loan of the microfilm containing the text of Antonio Piaggio's *L'Orologio Solare del Museo Ercolanese* and his other letters that enriched this study immeasurably and from which excerpts are translated here for the first time. I am grateful as well to the staff of the Biblioteca Nazionale and the Archivio dello Stato in Naples for facilitating my consultation of other important manuscripts incorporated into this study. Cécile Martini, director of the Bibliotèque of the École Française de Rome, and her staff, kindly provided me with new digital images of Paderni's manuscript. I have benefitted from conversations and assistance related to matters of the prosciutto with Ken Lapatin, Carol Mattusch, Kate Brunson, David Konstan, David Sider, Alexander Jones, Allan Berlind, and Jerry Lalonde. My mom, Michelle Parslow, helped me capture the spirit of the more colloquial French passages; she would have liked to see the final product. Finally, I am most appreciative of the support and encouragement of my editor at Oxford, Stephan Vranka, without whom this work would not have seen the light of day.

For humoring me over the years through yet another of my eureka moments with the prosciutto and for their forbearance through all the times when the "other woman," Miss Piggy, came between us, I dedicate this to my wife, Christina, and my daughter, Olivia, with deepest love.

Abbreviations

ASN	= Archivio dello Stato, Naples
CIL	= *Corpus Inscriptionum Latinarum*
CronErc	= *Cronache Ercolanese*
LSJ	= H. G. Liddell, Robert Scott, H. S. Jones, *A Greek-English Lexicon* (Oxford 1996).
MAAN	= Museo Archeological Nazionale, Naples
Paderni *Diario*	= Camillo Paderni, *Diario de' Monumenti antichi, rinvenuti in Ercolano, Pompei e Stabia dal 1752 al 1799 formata dal Sig.r Camillo Paderni, Custode del Real Museo Ercolanese in Portici, e proseguito dal Signor Francesco La Vega*, Manuscript in the Archivio di Stato in Naples *Maggiordomia Maggiore e Soprintendenza Generale di Casa Reale*, III Inventario, Categorie Diverse, Vol. 461.
Paderni *Monumenti Antichi*	= Camillo Paderni, *Monumenti antichi rinvenuti ne Reali Scavi di Ercolano e Pompei*, Manuscript in the library of the École Française de Rome, Manuscript 26.
PAH	= G. Fiorelli, *Pompeianarum Antiquitatum Historia* (Naples 1860).
PdE	= Reale Accademia ercolanese di archaeologia di Napoli, *Delle Antichità di Ercolano Esposte: Le Pitture di Ercolano e contorni*, 4 vols. (Naples 1757–1765).
PhilTrans	= *Philosophical Transactions, Royal Society of London*
StErc	= M. Ruggiero, *Storia degli Scavi di Ercolano* (Naples 1885).

Abbreviations of the works of Classical authors are generally those of the *Oxford Classical Dictionary*[4] (Oxford 2012).

I

Discovery

AMONG THE LEGIONS of bronze and marble statuary and the mounds of charred papyrus scrolls recovered during the Bourbon excavations in the Villa dei Papiri in Herculaneum emerged a humble yet practical object: a silver-plated bronze portable sundial in the shape of a prosciutto (Figure 1.1). Barely larger than the palm of a hand and easily confused for a simple clump of metal, its silver leaf had corroded to be as black as the solidified pyroclastic flow in which it had been entombed by the eruption of Vesuvius in 79 CE. Fortune had favored the excavators who discerned its form while cutting the dimly lit, narrow excavation tunnels that snaked their way along the walls of the Villa's rooms, twenty-seven meters below ground. They had chanced upon the earliest known example of a Roman portable sundial, a remarkable monument to the state of astronomical knowledge in the era of its manufacture and likely one of the types of sundials identified as *viatoria pensilia*, literally "travelling pendants," by the Roman author Vitruvius (Vitr. 9.8.1) who was writing in the precise period in which the prosciutto had been crafted. The ring attached to the prosciutto's hock for suspending it inverted, as would be an actual prosciutto for purposes of curing and display, helps support this identification even though the design of its dial limited its functionality to a single latitude. The dial is etched into the rind of the ham and consists of seven vertical lines representing the months of the zodiacal calendar, the abbreviations of the corresponding month names inscribed below, crisscrossed by seven undulating horizontal lines that mark the seasonal or "temporal" hours of the Roman day (Figure 1.2). The pig's tail served as the gnomon, defying gravity by projecting up and out to cast its shadow to the right across the dial, though this delicate feature evidently had snapped off at some point prior to the sundial's recovery. That its form was that of a prosciutto, that its markings were those of a sundial, and even that its inscription consisted of Latin characters were all matters that immediately confounded the excavators and scholars who confronted it and quickly sparked heated debate concerning its design

FIGURE 1.1 The *orologio di prosciutto* from the Villa dei Papiri in Herculaneum. Silver-coated cast bronze. 11.3 cm high × 7.8 cm wide × 1.9 cm thick. MANN 25494. Author's photo. Su concessione del Ministero della Cultura – Museo Archeologico Nazionale di Napoli.

and how it worked. They could hardly have imagined that almost every aspect of this remarkable find would remain poorly understood and misinterpreted for centuries.

Details about the sundial's discovery are known to a remarkable degree, thanks largely to documents written by Camillo Paderni, the curator (*custode*) of the royal collection of antiquities at the palace at Portici. Its earliest appearance would have been an entry in the diary he had maintained that doubled as an inventory of all the small finds in bronze, glass, and terracotta transported from the excavation sites to the museum. The entry is dated June 11, 1755, and states simply, "The supervisor of the excavation tunnels consigned to me a small prosciutto of silver, that is, of bronze covered in a thin sheet of silver, this prosciutto served as a sundial."[1] This is clearly a presentation copy

---

1. "Dal Soprastante delle grotte me fu' consegnato un picciolo presciutto d'Arg(ent)o, cioè di Rame rivestito con una piangia sottile di Arg(ent)o, il d(ett)o presciutto serviva di orologio solare." Paderni's inventory was continued by his successor, Francesco La Vega, and the resulting compilation was entitled the *Diario de' Monumenti antichi, rinvenuti in Ercolano, Pompei e Stabia dal 1752 al 1799 formata dal Sig.r Camillo Paderni, Custode del Real Museo Ercolanese*

Discovery

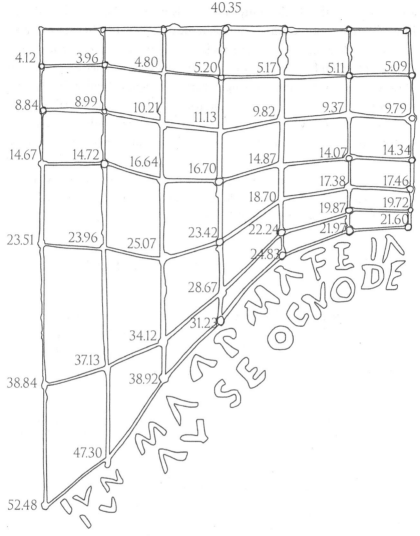

FIGURE 1.2 Close-up of dial's grid and calendrical abbreviations. Digital copy taken from a digital photograph of the original dial. Author's image.

of his *Diario*, however, compiled in 1763, not the working copy he amended as the finds came in. This description therefore already reflects knowledge he

---

in Portici, e proseguito dal Signor Francesco La Vega. The original is in the Archivio di Stato in Naples (ASN, *Maggiordomia Maggiore e Soprintendenza Generale di Casa Reale*, III Inventario, Categorie Diverse, vol. 461), hereafter Paderni *Diario*, with date of entry. The full text was edited and published in M. Forcellino, *Camillo Paderni Romano e l'immagine storica degli scavi di Pompei, Ercolano e Stabia* (Rome 1999). Forcellino (1999, 39) hypothesized that small squares drawn beside some entries, as here, highlighted particularly noteworthy finds.

4   THE PROSCIUTTO SUNDIAL

acquired about the sundial only later, as his own writings would indicate.
More illustrative of the initial confusion the object elicited in the excavators
is seen in its documentation in the weekly inventory of finds from the excavations prepared on June 15, 1755, by Roque Joaquin de Alcubierre, the director
of excavations to the Bourbon king, Charles of Naples. Alcubierre's description offers fewer details and does not acknowledge it as a sundial, perhaps
because he had not yet seen the object and was relying only on the report
supplied by the excavation workmen: "In these same tunnels has also been
discovered a bronze ham about the size of a hand which appears to be covered
in silver leaf and has a ring for suspending it, and some stripes and signs." It
was only later that Alcubierre would acknowledge, somewhat skeptically, that
these "stripes and signs" distinguished the object as a sundial.[2]

1.a) Una cosa la quale deve far strepito fra i letterati: *Paderni's Letter to the Royal Society*

Seventeen days after its discovery, on June 28, 1755, Paderni penned a more
detailed description of the sundial and its discovery in a letter to his English
friend Thomas Hollis.[3] Hollis' subsequent translation of Paderni's letter into
English for the Royal Society of London was published in 1756 in the Society's

---

2. "En las mismas grutas se ha encontrado tambien en pernil de metal que es grande cosi como
una mano, el qual pareze està cubierto de una hoja de plata y tiene su anillito para colgarle y
algunas rayas y señales"; *StErc* 172 (June 15, 1755). Alcubierre edited his excavation diaries for
the period 1738 to 1756 for a presentation copy for the prime minister, Berardo Tanucci, and
emended this entry to reflect its identification as a sundial "en la mismas grutas se encontrò
tambien una pieza de metal, de la figura de un pernil, que es grande casi como una mano, la
qual, parece, està cubierta con una oja de plata y tiene su anillo para estar pendiente, y alcunas
rayas y señales, y se ha reputado por un reloj de sol" ("In these same tunnels has also been discovered a bronze ham about the size of a hand which appears to be covered in silver leaf and has
a ring for suspending it, and some stripes and signs, and it has been identified as a sundial"). In
an index of the principal finds for this period at the end of the manuscript, Alcubierre described
it as "1 small ham which is assumed to be a sundial" (1 pernil pequeno, que se ha supuesto un
reloj de sol). For this later version, see Pannuti 1983, 353 (June 15, 1755), a transcription of the
"Noticia de las alajas antiguas que se han descubierto en las escavaciones de Resina, y otras, en
los diez y ocho años, que han corido desde 22 de octubre de 1738, en que se empezaron, hasta 22
de octubre de 1756, que se van continuado," completed by Alcubierre on April 4, 1757, and
archived in the library in the Società Napoletana di Storia Patria, in Naples (XX.B.19bis).
Finally, the date of June 11, 1755, is confirmed in the *Catalogo di statue di bronzo e parti delle
stesse trovate negli scavi di Ercolano, Pompei, Stabiae, ecc. Dalli…giugno 1750 a tutto li* (27 aprile
1765) compiled by Weber and his successor, Francesco La Vega, published in Comparetti and
De Petra 1883, 254 #66: "11 Giugno 1755. Sotto il bosco di S. Agostino. Si trovò l'orologio
(il giorno segnato sul guscetto)."

3. For additional details about Hollis and his relationship to Paderni, see Chapter 1.g.

Philosophical Transactions. Paderni took the opportunity to embellish the story of the sundial's discovery to give himself credit for identifying it as such, a claim that at least one of his contemporaries would dispute.

> [In Herculaneum, since March]... nothing of merit has been recovered, with the exception of one thing which will make noise among the learned [*una cosa la quale deve far strepito fra i letterati*] and which I believe is unique in the world. This is a small prosciutto of silver-plated bronze, the whole leg being five *oncie* (11 cm) long. On the upper surface of this is inscribed a sundial formed on a quadrant, and since this prosciutto forms a quarter circle, the artisan took the center of this quadrant from the extremity, or rather the shank of the prosciutto, for tracing the hour lines. The lines that indicate the months came to form the usual compartments, or small squares, some larger, some smaller, which have been divided into six by six both in height and in length. Below the lower squares, which are the smallest, are read the names of the months, placed in two lines in retrograde order, in such a way that the month of January came to be the last in the first line which includes the subsequent five months. In the second line are naturally written the other six, in such a way that the month of December came to be below that of January, and in this way the shorter, middle, and longer months, two by two, have in common one compartment for each pair. At almost the extremity of the right [*sic*] side is the animal's tail, somewhat twisted, and this does duty as the gnomon. At the end of the bone, that is of the shank, or rather the center of the quadrant, is a ring for holding the sundial in balance, and one supposes that here would have been its plumb line which in a similar kind of sundial one lets fall on the current month to determine the shadow of the gnomon on the hour lines. One should moreover note, since these sundials are drawn on a flat plane, according to a set and, as it were, material rule, the surface of the prosciutto here being in one place concave and in another raised, one does not know which rule the artisan employed to design a sundial of so difficult a nature upon a surface so irregular and capricious.

This prosciutto was found on the eleventh of this month and was consigned to me around eleven in the evening, and it was not known what it was because the patina that surrounded it made the excavators believe it was a chunk of iron. My curiosity piqued me to see quickly

6 THE PROSCIUTTO SUNDIAL

what it was. I began to discover the shape of a prosciutto, but regardless it seemed to me impossible to believe it was so until then I confirmed it, discovering it was of silver and immediately appeared the lines which formed the compartments and then the characters that denoted the twelve months. I remained so content with such a discovery that I immediately went down to the Royal Garden in which were His Majesty the King and the Queen, and I presented it to them and they were extremely content.[4]

As a renowned painter and skilled draughtsman, Paderni's eye for details should have enabled him to produce an accurate description of the sundial while his role as custodian would have afforded him unfettered access to it and, with that, abundant time to hone a description and puzzle out how it worked.[5] Nevertheless, Paderni had not been up to the challenges posed by this unique scientific instrument. While his summary of its physical characteristics is generally correct, he erred in situating the tail to the right (*sull' estremità del lato destro*) rather than the left of the dial and in identifying the entire lower row of the dial's squares as the "smallest," this being the case only with those on the right, those on the left being among the largest. His artist's instincts are evident in his inference that the designer chose the prosciutto shape because it corresponded with the form of the dial that would be etched onto it, and he recognized the potential for inaccuracies in the dial that might arise from the choice of such an irregular surface, an aspect that later commentators would focus on as well. The description was not entirely his own, however, as was revealed years later when Antonio Piaggio, the Jesuit priest tasked with unrolling the papyrus scrolls from the Villa dei Papiri, claimed

---

4. Paderni 1755–1756, 498–500. This translation is based on the original Italian text, for which see Knight 1997, 41–42. Paderni's claim of having identified it as a sundial and showing it to the royal consorts recalls the story retold by Antonio Piaggio of Paderni's claim that he recognized the carbonized scrolls of papyrus from this same site as written texts: "he ran immediately to the palace and introduced himself to Their Majesties, even though it was inconvenient, and having opened one of the rolls (that is, he cut one with a knife) in their presence, made them conceive the value of the hidden treasure he had uncovered." See Piaggio's "Memorie [autografe] del Padre Ant.o Piaggi [*sic*] impiegato nel R. Museo di Portici relative alle antichità, e Papiri p. anno [*sic*] 1769. 1771," Società Napoletana di Storia Patria, Ms. 31-C-21, fols. 3–4. This was written by Piaggio between 1769 and 1771 for Conte Guglielmo Maurizio Ludolf, who had requested information on the discovery and unraveling of the papyrus scrolls; see also D. Bassi, "Il P. Antonio Piaggio e i primi tentativi per lo svolgimento dei papiri ercolanesi," *Archivio Storico per le Province Napoletane* 32 (1907) 659–66, and Parslow 1995, 104–5. For Piaggio, see Chapter 1.g.

5. On Paderni in general, see De Vos 1993, 99–116; Mansi 1997, 79–108; Forcellino 1993, 49–64; and Forcellino 1999, 9–33.

that he had written a description for Paderni after positively identifying it as a sundial. It was Piaggio who had equated the shape of the quadrant with that of the prosciutto's dial and explained how it functioned.[6]

1.b) Identifying the Prosciutto's Findspot in the Villa dei Papiri

There can be no doubt that the prosciutto was found in the Villa dei Papiri because this was the only site in Herculaneum under investigation at that time. Less certain is its exact findspot within the Villa, since the usually comprehensive documentation provided by the supervisor of the excavations, the Swiss military engineer Karl Jacob Weber (1712–1764), falls short in this regard. In addition to the weekly inventory of finds he filed with Alcubierre and an assortment of scribbled notes included among his personal papers, Weber had carefully recorded his discoveries in the Villa on the large plan he began in 1754 that united a visual record of the progress of the excavations with a plan of the site and an inventory of the finds (Figure 1.3). Although the earliest finds listed in the plan's legend date to August 15, 1750, the latest ones date to July 20, 1754, a full eleven months before the recovery of the prosciutto.[7] Yet Weber had continued to append to his plan the additional tunnels cut in the villa through October 1758. Consequently, while the prosciutto is not listed in the inventory and its findspot is not recorded, the tunnel in which it was found is likely represented on this plan.

It was only in 1883 that Giulio De Petra (1841–1925), in reviewing the Bourbon-era excavation reports from the Villa, concluded that the prosciutto had been recovered from the area of the *atrium* and, specifically, one of the rooms to the east of the atrium.[8] He based this on a reference in the reports to the removal of plaques of marble that he believed had served as the lining of the atrium's *impluvium*. Features associated with the impluvium were first encountered in November 1754, in the course of digging a tunnel running north–south through the center of the atrium that Weber had drawn clearly on his plan (Figure 1.4).[9] In November and December 1754, Paderni noted in

---

6. For Piaggio's version of the sundial's discovery and interpretation, see Chapter 1.g.

7. Parslow 1995, 85–103.

8. Comparetti and De Petra 1883, 177, 233. De Petra was director of excavations at Pompeii, Herculaneum, and Stabiae in this period.

9. De Petra labeled this main tunnel with the numeral 24 in red on his revised edition of Weber's plan of the Villa.

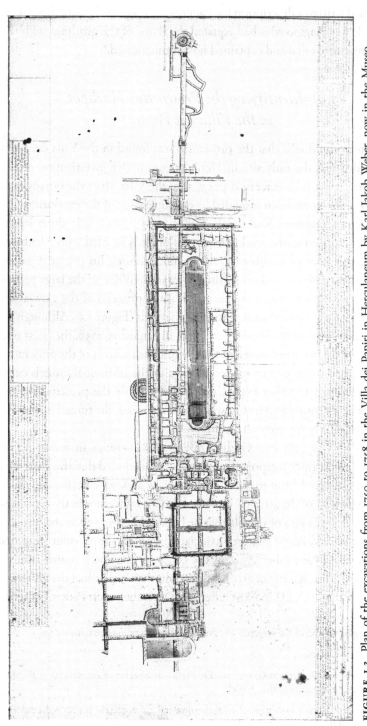

FIGURE 1.3 Plan of the excavations from 1750 to 1758 in the Villa dei Papiri in Herculaneum by Karl Jakob Weber, now in the Museo Archeologico Nazionale, Naples. 0.58 m × 1.21 m. The plan is flipped, putting north toward the bottom. The area of the atrium where the prosciutto was found is on the left. Su concessione del Ministero della Cultura—Museo Archeologico Nazionale di Napoli.

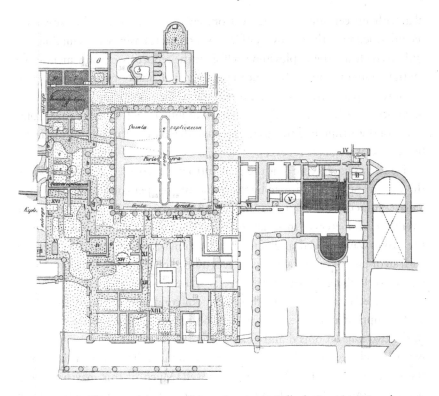

FIGURE 1.4 Close-up of the area of the atrium in the Villa dei Papiri in Herculaneum. The east *ala* (Ala e) is 19; the *impluvium* in the *atrium* is 20. From Comparetti and DePetra's *La Villa Ercolanese dei Pisoni* (Turin 1883).

his inventory that he had descended into the tunnels to recover an assortment of bronze statuettes. These had been fitted with lead pipes to function as fountain jets and evidently adorned the impluvium's rim and interior.[10] In January 1755, Paderni descended again and assisted in the recovery of a bronze bust he specifically stated had been found near the statuettes discovered earlier.[11] In the following period, the entries in Paderni's diary list primarily bronze door fixtures and ornaments, none of which can be situated on the

---

10. Paderni 1755, 110–12; Forcellino 1999, 65–66; *StErc* 163–65 (November 27, 1754–December 18, 1754). For these statuettes, eleven of which were found in all (MANN 5006, 5007, 5011, 5012, 5015, 5020, 5028, 5029, 5030, 5031, 5033), see Mattusch and Lie 2005, 296–315; Wojcik 1986, 232–40.

11. Forcellino 1999, 67: Paderni explicitly states they were excavating "not far from where the 11 figurines consisting of putti and fauns in the month of (Novem)ber"; *StErc* 166–67 (January 10, 1755); Comparetti and De Petra 1883, 264; Mattusch and Lie 2005, 262–63 (= MN 5596: Hellenistic dynast with bushy hair). De Petra had conjectured that the bust came from a niche Weber drew on the atrium's east wall, which De Petra labeled red 27.

plan with any certainty.[12] In the week of May 11, 1755, however, the excavation records document the discovery of "a type of marble basin" with a marble rim, and this is clearly the impluvium itself since the report specifically mentions a central fountain spout. This indicates they had finally cleared the atrium down to its pavement.[13] The actual removal of the impluvium's marble revetment occurred, however, only in the week of June 1, 1755, showing that exploration of the atrium had continued through that date.[14] This was only ten days before the prosciutto's recovery.

De Petra's conjecture that the prosciutto was recovered in a room east of the atrium appears confirmed by the results of the open-air excavations of the Villa conducted between 1996 and 1998 (Figure 1.5). In that period, the twenty-seven meters of material covering the portion of the Villa containing the atrium was removed, and much of the atrium was brought to light together with all or part of fifteen adjacent spaces framed on the south by a three-sided loggia overlooking the sea. While several of the mosaic pavements within this area survived intact, the walls are in very poor condition, the combined effect of the eruption and the extensive Bourbon tunneling. The walls stand tallest in the two *alae* (rooms d and e) at the south end of the atrium where they reach as high as 2.20 and 2.90 m.[15] Virtually all the wall painting, primarily of the so-called Second Style, is absent as well. Much of this may have been deliberate: one of the Bourbon-era tunnels cutting through the southernmost rooms was filled with fragments of wall stucco used by the Bourbons as backfill.[16] Nevertheless, some patches of stucco remaining on the north and east

---

12. Paderni 1755–56, 498 (= Knight 1997, 41), Paderni stated that "nothing of value" had been found between the time of the discovery of the "colossal bronze bust" in the atrium and the sundial. De Simone and Ruffo 2003, 289, likewise note that the majority of finds in the most recent excavations were found on the pavement and consisted of bronze locks and hinges.

13. *StErc* 171 (May 11, 1755). In the interim Weber must have cut the two tunnels flanking the impluvium, which De Petra labeled 25 and 26 in red on his plan. The impluvium's dimensions are given as 12 *palmi* square and 7 *onzas* deep (3.17 × 0.14 m), though in fact the remains in situ appear to be rectangular and somewhat wider and substantially longer; the published findings of the recent excavations provide no details.

14. *StErc* 171 (June 1, 1755): the marble was removed in many pieces. Already in December 1754 Alcubierre had described the extensive excavations here that had resulted in a vast cavern necessitating the construction of several supportive pillars. In March 1755, he describes the site as a "labyrinth of tunnels." These efforts yielded a number of mosaic pavements over the course of this period.

15. De Simone 2010, 9–11; Guidobaldi and Esposito 2010, 25–28; De Simone and Ruffo 2003, 287–89. The maximum height of the walls also is between 2.20 and 2.90 m.

16. Guidobaldi et al. 2009, 128–29.

Discovery   11

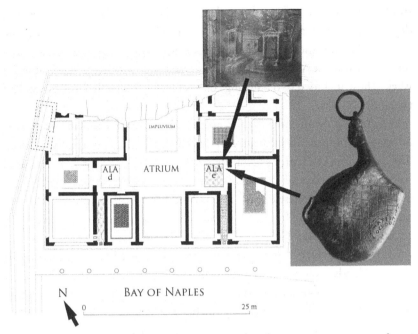

FIGURE 1.5   Plan of the area of the *atrium* exposed in the open-air excavations of 1996 to 1998 showing relationship of the *atrium* with its *impluvium* to the east *ala* (Ala e) with the probable original findspot of the prosciutto and the landscape painting (MANN 9423) likely cut down from the north wall. Adapted from Guidobaldi et al. 2009, 129 Fig. 87. Su concessione del Ministero della Cultura – Museo Archeologico Nazionale di Napoli.

walls of the east ala (room e) may be useful in establishing a more precise provenance for the prosciutto.

The Bourbon excavators removed only a handful of paintings during this period when their efforts appear to have been concentrated in the atrium and surrounding rooms. In the week of December 15, 1754, they cut down two paintings. One, measuring 0.80 m by 0.95 m, was described as representing architecture, while the other, depicting a man, a tree, and a goat, measured 0.50 m square.[17] A small painting of a bird was recovered in the week of January 5, 1755, and a somewhat larger painting of a duck was cut down in the week of January 10.[18] Six small paintings were found in the same week as the

---

17. *StErc* 164 (December 15, 1754); Pannuti 1983, 338. Pagano and Prisciandaro 2006, 212, identified the first with NR 732 (Arditi 1260, Sangiorgio 1137) in the MANN inventories, but were not able to trace the second.

18. *StErc* 166 (January 5, 1754); Pannuti 1983, 339. The painting of the bird measured 0.20 m square while that of the duck was 0.32 m square. Pagano and Prisciandaro 2006, 212, could not locate either of these.

sundial. Three were of small "masks"; another was a landscape; the fifth depicted a pitcher; and the sixth was another landscape populated with six figures.[19] Finally, another five small paintings—the last group of paintings removed from the Villa for some time—were inventoried in the following week. These included two paintings of tigers, one of a pitcher and a cup, and two more "masks."

Of these, only one of the landscapes can be restored with reasonable certainty to its proper role in the decorative scheme of the alae (rooms d and e). The coloration and patterning of the borders on the scant patch of surviving stucco on the north wall of the east ala (room e) compare favorably with those of MANN 9423, a classic sacro-idyllic landscape with architecture but no humans. It is therefore likely that it came from one of the outer panels of the central zones of these alae.[20]

There is no evidence, however, for securely linking MANN 9423 to any of the small landscapes inventoried as recovered in this period. The only one for which dimensions were provided was that cut down in the week of December 15, 1754. This measured 0.80 m by 0.95 m, but MANN 9423 is 0.67 m by 0.87 m, a substantial enough difference to call the identification into question.[21] This would, moreover, be six months before the sundial's discovery, so the association between the two would only make sense if the sundial was encountered as the excavation's focus eventually turned away from the higher wall surfaces toward the floor, where the sundial most likely had come to rest.[22]

---

19. *StErc* 172 (June 15, 1755); Pannuti 1983, 353. No measurements were provided for these paintings. Pagano and Prisciandaro 2006, 212: the masks (actually floating heads) are MANN 8821 = *PdE* 4.17, pl. 4 (though a note on *PdE* 4.355 shows the Accademia believed these came from Civita = Pompeii) = NR 743; and the pitcher is MANN 9891 = NR 744.

20. Moormann 2010, 155–56, a correction of an earlier reconstruction (Moormann 1983, 643–44, fig. 10) that restored this painting to the atrium. The identification of the painting with the Villa was the work of Allroggen-Bedel 1976, 85–88. Pagano and Prisciandaro 2006, 212 n. 225, included MANN 9423 among several paintings believed to have come from this area but did not associate it with any specific find in the excavation inventories.

21. Comparetti and De Petra 1883, 282 #106; De Simone and Ruffo 2003, 292, and Moormann 2010, 67–68. Allroggen-Bedel 1976, 87, ascribes this to the difference between how paintings were measured first and then cut down from the walls. Paintings from the atrium itself, such as a Greek key pattern (MANN 8548 = Weber's "labyrinth" painting) and a still life of suspended geese and two deer (MANN 8759), had been inventoried already, respectively, on March 10, 1754, and June 23, 1754—more than a year before the sundial's discovery. Similarly, on his plan, Weber labeled the northwest corner of the atrium with a "XI" and the southwest corner with a "XIII" and listed the principal finds in the "Quinta explicacion" of his legend. This chronology underscores the length of time devoted to clearing the atrium and the adjacent rooms.

22. This would follow the pattern seen with the impluvium: its fountain statuary was recovered first and then the impluvium itself, sunken into the floor.

Discovery 13

De Petra had conjectured that the excavations had explored the west ala (room d) in September, 1754, but this was only because it fit his chronology of how the tunnels snaked their way through the atrium; Weber had documented no specific finds from this room. Similarly, De Petra's conjectural chronology had situated the excavations in the rooms to the south of the atrium earlier than June 1755, though there are no specific finds to confirm this.[23] Consequently, it is only the coincidence of the discovery of the sundial in the same week as the discovery of a landscape painting (MANN 9423), which fits the room's decorative scheme, that makes the east ala (room e) the prosciutto's likely provenance (see Figure 1.5).

Despite the alacrity with which Paderni penned his account of the sundial's discovery, it was not the earliest published description. That distinction should have fallen to Ottavio Antonio Bayardi to whom Charles of Bourbon had granted exclusive rights in 1747 to oversee the publication of the finds from the royal excavations (Figure 1.6). Bayardi had famously squandered this privilege by taking five years to publish the initial two volumes of his five volume *Prodromo delle antichità di Ercolano* (Naples 1752), a lengthy excursus on the labors of Hercules with barely a reference to the excavations. He followed this with his *Catalogo degli antichi monumenti dissotterrati dalla discoperta città di Ercolano* (Naples 1755), which was essentially an inventory of the first 2,962 archaeological finds, its entries limited, with a few exceptions, to succinct descriptions offering minimal analysis. It had the unique quality, for a catalog of antiquities, of being entirely bereft of illustrations. Neither publication had met scholarly expectations of how the antiquities should be presented to the public.[24]

The prosciutto appears near the end of the *Catalogo*, grouped together with 732 other antiquities recovered after he had already completed the rest of his text. This underscores that the prosciutto was a relatively recent find. Characteristically, the entry provides no dimensions and does not record the

---

23. Comparetti and De Petra 1883, 231–32.

24. According to Castaldi 1840, 32, 92, the first two volumes of the *Prodromo* appeared in 1752 and the last three in 1756. Hollis 1843, 391 (Hollis to Ward, December 25, 1752): Thomas Hollis observed from Genova in a letter to John Ward that, "In confidence, this Work has not answered the expectations of the Public here, as it has entered but slightly into the subject proposed, and on the contrary has dilated itself into foreign and consequently needless digressions. With the same confidence it is added that this prelate, though a very learned person, yet is not believed to be so deep and well skilled in certain parts of Antiquity, as from the post in which he is might be expected." Barthélemy 1801, 120–25: Letter 20 to Caylus, from Rome (April 7, 1756), describes in detail Bayardi's publication record up to that point along with the early work of the Accademia Ercolanese.

FIGURE 1.6 Caricature of Ottavio Antonio Bayardi (1694–1764) in pen and brown ink by Pier Leone Ghezzi (1745). British Museum (1871, 0812.998). ©The Trustees of the British Museum. All rights reserved.

provenance. Yet Bayardi reiterated two facts about the sundial that both Alcubierre and Paderni already had provided but would later be in dispute: that its bronze was covered in silver and that its shape resembled that of a prosciutto. Bayardi was the first to attempt to date it post-Augustan on the basis of the abbreviations IV[lius] and AV[gustus] for the two months renamed for Caesar and the first emperor. Moreover, he was clearly intrigued enough by its underlying science to anticipate dedicating himself to a more detailed astronomical study and blamed his brief treatment of it on the quantity of objects that already filled his catalog:

> DCCXVIII. A sundial of metal covered in silver. It is formed throughout in the manner of our prosciutti with a ring to suspend it. On it one sees the lines and under the lines are indicated the months. It is after the time of Augustus and, as a consequence, after the Julian corrections, and, as a result, indicated among the other months are JULIUS

ET AUGUSTUS. I will give an explication of it in its time. I wanted to provide an account in the present place but I refrained since I had too much in the *Catalog* to talk about it at length.[25]

It is not clear, however, that Bayardi's actually was the first published description, despite his lock on publication rights. He certainly had penned his catalog entry soon after its discovery in June 1755, and his finished manuscript was published by the end of that same year.[26] Though Paderni had written his letter in July, after being translated from Italian into English it was only read before the Royal Society of London on February 17, 1756, and did not appear in press before 1757. Paderni had corresponded with members of the Royal Society about new discoveries in the royal excavations on at least five previous occasions, beginning already in 1739, years before he had any official role in the excavations.[27] Ironically, it was Paderni's first letters describing the earliest discoveries at Herculaneum, along with his sharp criticism of the excavators' poor handling of the paintings, that had led not only to his being barred from the ruins but to the implementation of a strict prohibition against anyone outside the Bourbon court taking notes or making drawings in the excavations or the royal collection. Nevertheless, his reputation as a skilled copier of ancient works of art had allowed him to overcome this initial setback to become *custode*, or director, of the royal collection of antiquities in 1751.[28] Paderni claimed the king himself had awarded him this position "without having made the least application for it, and in opposition to divers

---

25. Bayardi 1755, 423 #718.

26. Bayardi 1755, 420, in the *Catalogo* includes entry 709 for the ithyphallic bronze tripod adorned with satyrs (MANN 27874), which Bayardi specifically states was found "a few days ago"; see Chapter 1.f. This was recovered only two days after the sundial, on June 13, 1755, in the *sacrarium* of Isis in the *Praedia Iuliae Felicis* in Pompeii (Regio II.iv.1–12). This helps establish that he was in the final stages of preparing his manuscript in this period.

27. C. Paderni, "Extracts of two letters from Sig. Camillo Paderni at Rome to Allan Ramsay, painter...concerning some ancient statues, pictures, and other curiosities found in a subterraneous town lately discovered near Naples...Rome, November 20, 1739," *PhilTrans* 41 (1739–1741) 484–89. In a subsequent letter, dated February 20, 1740, and published together with the earlier one, Paderni explicitly stated that he was keen to draw the paintings in an official capacity and noted that he even had left instructions on how the paintings should be properly conserved; see also N. Zanni, "Lettere di Camillo Paderni ad Allan Ramsey: 1739–1740," *Eutopia* 2.2 (1993) 65–77.

28. Knight 1997, 16–53; De Vos 1993, 99–116; Mansi 1997, 79–108. See also C. Piva, "Custodi: Da curatori a guardian, un ruolo professionale per i musei italiani e le sue definizioni storiche," *Venezia Arti* 25 (December 2016) 93.

16   THE PROSCIUTTO SUNDIAL

persons who had great interest with our Princes and other great people."[29] Perhaps because he studiously avoided divulging too many specifics, and because these were personal letters without illustrations rather than formal publications, his subsequent correspondence in his position as *custode* evidently was not viewed as violating the prohibition against publishing the antiquities. The close relationship he developed with Charles of Bourbon doubtless also played a role in the freedom he exercised in divulging this information, a freedom notably curtailed by the King's departure for Spain in 1758.[30] Moreover, by the time Paderni's description of the sundial had appeared, the publication rights already had been transferred from Bayardi to the Reale Accademia Ercolanese, the committee of learned Neapolitans tasked by the prime minister, Berardo Tanucci, with assuming this mantle in December 1755.[31]

The curious nature of the prosciutto sundial as both archaeological discovery and ancient astronomical instrument had quickly piqued scholarly and popular interest in it, yet the tight grip the Bourbon court sought to maintain on the antiquities had consequences for how knowledge of the sundial came to be disseminated and the accuracy of that information. The injunction against taking notes or making sketches on the excavation sites and in the museum helped fuel the appetite for clandestine information about all the finds, whether it was transmitted verbally or in writing, but all now had to be produced from memory. Correspondence like Paderni's had become the principal means for broadcasting news of the discoveries in scholarly circles, but sources lacking Bayardi's or Paderni's degree of access to the finds were inevitably fraught with errors.[32]

---

29. Hollis 1843, 390–91: Paderni made the claim in this 1752 letter to his friend Thomas Hollis, but Hollis observed that it was "known to a few that [Paderni] has been almost the sole means by his own judicious well-timed reiterated insinuations to influence his Master to the before-mentioned right and becoming resolution of forming a Gallery of Antiquities." In a letter from Naples to Joseph Spence dated March 1, 1750, Robert Lowth remarked that Paderni was "very well with the King, has access to him at all times, and frequent conversations with him" (Klima 1975, 437).

30. Knight 1997, 54. Roberts 2015, 68–73. D'Amore 2019, 107–12, identified Paderni as "the Royal Society's main correspondent from the Neapolitan area until the later 1750s." At this point, Charles of Naples assumed the title of King Charles III of Spain. Tanucci, in turn, served as regent to Charles' third son, the new young king who in 1759 took the titles of Ferdinand IV of the Kingdom of Naples and Ferdinand III of the Kingdom of Sicily.

31. Castaldi 1840, 33–36; Choisi 1986, 495–517.

32. In general, see Allroggen-Bedel 2008, 53–72; Gordon 2007, 37–57; Echlin 2014, 145–59.

1.c) The Prosciutto and the "English Gentleman" in the Memorie

Both the inability of the court to prevent information from leaking out and the consequences of limited access to the finds are reflected in what appears to have been the first widely circulated report on the sundial. This is a letter dated July 19, 1755, that appeared in the *Memorie per Servire all' Istoria Letteraria*, a monthly journal of scholarly news and debate published in Venice.[33] The date of the letter—little more than a month after the sundial's discovery—reveals how quickly the initial discussions about it began, while the publication of this fascicule of the *Memorie*, in late August or the end of 1755, shows the speed with which news of it spread. If the fascicule in which the letter appeared was in fact in circulation already in August, as seems likely, it anticipated Bayardi's publication by several months; if it was published only as part of the sixth volume of the *Memorie*, it either appeared contemporaneously with Bayardi's or followed quickly on its heels.[34] Regardless, the *Memorie* would have reached a wider circle of scholarly readers more rapidly than Bayardi's publication, which had a relatively limited publication run and select distribution.[35]

In keeping with the format of the *Memorie*, the information was communicated in the form of a letter, postmarked in Florence, with both the author and the addressee anonymous. The author attributed the information to "an English gentleman who came here from Naples a few days ago," a sufficiently vague attribution that must reflect a desire to protect the identity of the source. The flawed description of the instrument underscores the likelihood it was based, at best, on only a brief, direct observation, if not entirely on

---

33. *Memorie per Servire all' Istoria Letteraria* 6.2 (August 1755) 3–7. The journal was edited by Angelo Calogerà, Zaccaria Serimàn, and Girolamo Zanetti, and published by Pietro Valvasense in Venice.

34. The last letter in the sixth fascicule of the *Memorie*'s sixth volume was dated December 7, 1755, so the volume as a whole could not have been published before late 1755 or, more likely, early 1756. The last letter of the second fascicule, in which this letter appears, is dated August 20, 1755, so the fascicule would not have appeared in public before late August or early September. Publications like the *Memorie* and the *Symbolae Litterariae* generally were produced and distributed as monthly or bi-monthly pamphlets and then bound together into a single annual volume.

35. On Bayardi's *Catalogo*, see Mansi 2008, 115–19; Mattusch and Lie 2005, 65–67. Hollis 1843, 362 (Hollis to Ward, December 25, 1752), noted that most English staying in Naples had received their copies of Bayardi's *Catalogo* from Giuseppe Maria Pancrazi, an antiquarian who had published a two-volume work on the antiquities of Sicily entitled *L'antichità siciliane spiegate colle notizie generali di questo regno* (Naples 1752).

word of mouth, without the benefit of direct explication by Paderni or the opportunity to study it in any detail:

An English gentleman who came here from Naples a few days ago, whom I knew by chance, told me of a new discovery about which I am prepared to give you an account in this letter. It consists of an ancient instrument which for good reason was judged a sundial, as we are accustomed to say, "portable," or, if you wish, "hanging." And so that you may have a clear image without my sending you a drawing, I will offer a detailed description of it. The body of this instrument is circular with a handle that extends from the circle's edge and narrows the closer it reaches its extremity, in precisely the manner of a flask yet not as swollen in its lower part. At the end of the handle, which is almost as long as the diameter of the circle, there is a small hole in which there is a ring that once served to suspend it wherever one wished. The entire instrument is metal, somewhat raised, or we should say bulging, on both parts of the two surfaces. A gnomon is situated on one side, not straight but in the manner of a flame, pointed and somewhat long. It is perhaps one fourth of the diameter. One of the surfaces, which I will call the upper one, and which is entirely covered in silver, is divided by twelve lines, which are parallel and divide and form as many square niches, a few somewhat concave. The last of these niches, or squares, as we want to call them, which have the lower edge of the circle for their boundary, are disposed in this manner and contain the following letters,

| IV | MA | AΓ | MA | FE | IA |
| IV | AV | SE | OC | NO | DE |

which, as you can well see, are the initials of the names of the months. The "P" in the month of April formed in so open a manner, as you perceive, reveals the antiquity of this work, this being precisely the oldest form among the Romans. Notable as well is the manner in which are described these months, which is clearly *boustrophedon*, that is, in

Discovery 19

the manner that oxen plow, a manner not unknown to the Latins and somewhat more familiar to the Greeks.

The letter's author continues with a brief digression on the history of sundials among the Romans drawn from literary sources, beginning with the attempts of Censorinus and Pliny to trace their earliest appearances in Rome. Such discourses were common stock in treatments of ancient time keeping, but this is the earliest published example in the case of the prosciutto sundial. The historical details are limited to Censorinus and Pliny the Elder, with Vitruvius contributing the names of early dialers and their inventions, and the author of the letter quotes these sources more or less verbatim.

> If we believe Censorinus [*de Die Nat.* 23] that the use of sundials among the Romans was most ancient, the precise time of them cannot be established with certainty. "Others," suggests this author, "say that the earliest was at the Temple of Quirinus, others that of Diana on the Aventine, and others on the Capitoline. But without a doubt a sundial was not seen in the forum before M[arcus] Valerius, who placed one brought back from Sicily near the Rostra, which, being made for the climate of that island and not therefore being able to serve exactly the Romans, L[ucius] Philippus the censor placed another in front of it." Pliny [*HN* 7.60], somewhat more believable, or, if you wish, more diligent than Censorinus, claims to provide a detailed history of sundials in Rome, where he writes "that L[ucius] Papirius, eleven years before the war against Pyrrhus (that is, in the year 461 of Rome) placed the first sundial near the Temple of Quirinus, so wrote Fabius Vestalis, of which however he does not explain the type or the design by which it was made, nor the place from which it was transported, or the writer from whom he had this news. M[arcus] Varro recounts that in the First Punic War, M[arcus] Valerius Messala, having conquered Catania in Sicily, brought back from there to Rome the first sundial and placed it near the Rostra. And for that reason, the hours did not correspond (in Rome) with the lines. Q[uintius] Martius Philippus, who was censor with L[ucius] Paullus, placed a larger one and it was very welcome. Scipio Nasica then being advised that the sundials could not indicate the hours or the days made one of water."

Pliny (*HN* 7.60.212–15) had offered a somewhat more detailed historical excursus on the introduction of timepieces into Rome than Censorinus, who wrote two centuries later and largely echoes his predecessor. Pliny reveals that interest in a precise reckoning of the division of the day and night into hours was adopted relatively late and had not even coincided with the erection of the first sundials in the city. The *Twelve Tables*, Rome's earliest written laws traditionally dated to 450 BCE, had referenced only sunrise and sunset. Official state recognition of noon (*meridies*) had come some years later, when the herald (*accensus*) of the consuls began to announce it from the steps of the *Curia* as he observed the sun passing between the *Rostra* and the *Graecostasis*.[36] Similarly, the last hour before sunset was proclaimed when the sun was observed on its course between the *Columna Maenia* and the *Carcer Mamertinus*.[37] Pliny emphasizes that such sightings were naturally limited to clear days. The first sundial in Rome had been erected during the course of the First Punic War by Lucius Papirius Cursor in the Temple of Quirinus on the Quirinal Hill.[38] Pliny's source provided no details about the origins or design of this sundial. That it was the earliest was in dispute since Pliny also records that the antiquarian Marcus Varro, while similarly dating the earliest sundial to the First Punic War, had situated it instead next to the *Rostra*. This may have been the one Censorinus credits to Manlius Valerius Messalla as being the first, which he dedicated during his consulship in 263 BCE.[39] Because this had come to Rome as war plunder from Catania, Sicily, and therefore was designed for a more southerly latitude, it failed to provide the proper time. This evidently did not trouble the Romans for almost a century, and it was not until 164 BCE that the censor Quintus Marcius Philippus erected a more accurate one adjacent to the *Rostra*. But the fact that such dials could only operate when the sun was shining led the next censor, Scipio Nasica, in 159 BCE, to dedicate a water clock (*clepsydra*), a device not dependent on sunlight that could keep the hours day and night. This he had covered in some way in the Basilica Aemilia in the Forum.[40]

---

36. Richardson jr 1992, 102–3, s.v. *Curia Hostilia*, notes that this early version of the Curia was oriented to the cardinal points, allowing the *accensus* to look directly south between two monuments, which may have been different from the *Rostra* and *Graecostasis* that Pliny references.

37. *Lexicon Topographicum Urbis Romae*, vol. 1 (Rome 1993) 301–2, s.v. *Columna Maenia* (M. Torelli); 1.236–37, s.v. *Carcer* (F. Coarelli).

38. Richardson jr 1992, 326–27, s.v. *Quirinus, Aedes*: Papirius Cursor's father had vowed the temple, but he dedicated it during his consulship in 293 BCE. It is unclear why Pliny goes on later in this passage to date the sundial to 264 BCE.

39. *Lexicon Topographicum Urbis Romae*, vol. 4 (Rome 1999) 336, s.v. *solarium* (E. Papi).

40. Varro (*Ling.* 6.4) distinguishes between the *solarium*, used for the object on which the hours were read in the *sol* (sun), and the water clock (the text here is corrupt so the actual term

Discovery

This leads the author of the letter to a catalog of literary references to the Romans' division of the day into twelve parts. This must be an attempt to associate the "twelve lines" etched on the sundial's surface with this division of the day, though he does not explicitly say so.

> Nevertheless, however, it was, and this is something almost certain, only after the introduction of sundials in Rome that both day and night began to be divided into 12 hours, writes Censorinus [*de Die Nat.* 10]: *Hoc*, that is the division in 12 hours, night and day, *credo Romae post inventa horologia observatim* [This, I believe, was observed at Rome after the invention of the sundial]; and comparing this with Pliny [*HN* 7.60], where he writes that before the use of sundials *indiscreta Romano Populo lux fuit* [Daylight was indistinguishable for the Roman people], that the division into 12 hours was unknown. And Vitruvius [9.8] when he reasons about sundials, says *omnium autem figurarum descriptionumque earum effectius unus ut dies aequinoctialis brumalisque, item solstitialis in duodecim partes aequaliter fit divisus* [The effect of all these figures and designs is that the equinoxes and the winter and summer solstices are divided into twelve equal parts (Vitr. 9.7.7)].

The author then uses the reference in Petronius to Trimalchio's possession of a *horologium in triclinio* (Petr. *Sat.* 26: a sundial in his dining room) as evidence for the existence of portable sundials before concluding with an acknowledgment of the difficulty of analyzing such ancient instruments with such limited documentation:

> But we return now to our, as we already said, portable one. A passage of Petronius [26] which narrates the dutiful Trimalchio having had *horologium in Triclinio*, makes me suspect that in those times were possibly in use these hanging and portable sundials. And the strength of this conjecture seems to grow increasingly, joining this passage with the other, where in describing the same room he describes a thick table near the door on which is seen painted the course of the moon, the seven planets and *qui dies boni quique incommode essent, distinguente*

---

is lost), "which Cornelius shaded in the Basilica Aemilia and Fulvia," implying he set it up inside that building on the Roman forum.

bulla, notabantur [which days were lucky and which unlucky were indicated by a decorative knob (Petr. 30)]. But seeing then that the same Trimalchio, with crazy vanity, wanted one of these sundials placed on his tomb, not being able to put there one of these of this design which soon would have been stolen and would need to be stable and firm, I think that rather than looking for examples and authority to prove that ours is hanging, it is appropriate on the contrary to appeal to it in order to establish the use of them among the Romans.

But I did not intend to write a dissertation. Others know how to make good use of this noble piece of Roman antiquity. But it would be worthwhile for them to see it and to contemplate it well up close; I know well from experience how hazardous an enterprise it is for those who undertake to illustrate ancient monuments with only a drawing before their eyes. The celebrated Sign. Canon Alessio Simmachus Mazzocchi could perhaps satisfy the public expectation and give us once and for all a complete series of the rarest and singular things found in Herculaneum, so much and so long up to this time longed for by the entire Republic of Letters.

Despite the comment about the difficulties of working from illustrations alone, it is nowhere stated that the English gentleman had offered a drawing to accompany his description, and the author did not supply one with this letter. The manner in which the grid for the calendrical annotations was rendered in the *Memorie's* text, with the central vertical lines longer than those on the sides, is vaguely reminiscent of the undulating grid of the dial. This could indicate the author of the letter was recalling a diagram presented to him, of at least the dial itself if not the entire object. Yet on the actual instrument the months themselves are listed below the grid and are not framed by it. The specific reference to so fine an aspect as the open form of the letter "P" in the month AP[rilis] also suggests a detail captured in a sketch of the sundial. The paleography of the inscription had clearly been a point of discussion, with the letter form seen as a means for establishing its Roman pedigree and the novelty of the *boustrophedon* ordering of the months inviting debate on textual parallels. Both, however, were features that could have been transmitted orally without the need for a sketch or through a rendering of the inscription alone. Ultimately, the fact that no engraving was offered in the *Memorie*, despite the capacity of the printer to include engraved plates in the publication, favors the conclusion that no illustration of the entire instrument had been made available to the letter's author.[41]

That the English gentleman had only a fleeting look at the sundial, did not have the benefit of an explication by Paderni, had a very poor memory, or some combination of all these, is supported by the fact that he did not recognize its principal characteristic: that it was shaped like a prosciutto. This is clear from the attempt to relate its form to a flask. This, in turn, led him to miss another essential feature of its design: that it had employed the pig's tail as the gnomon. The association of the gnomon's shape with a flame would recur in other treatments of the sundial that likened it to a flask rather than a prosciutto, while the recollection that the gnomon stood upright, one-quarter of the instrument's diameter, which would make it about 2.0 cm high, suggests a conflation of the actual, incomplete remains of the gnomon with the canonical design of the gnomon on a horizontal sundial. The description of the dial itself, as related in the *Memorie*, also ignores a crucial aspect of its design: that while the vertical lines are rectilinear and parallel, five of the six horizontal lines are undulating. The author speaks only of "squares" resulting from this combination of twelve (*sic*) intersecting lines without stating their total number, but of the thirty-six squares on the actual dial, formed in fact by seven horizontal and seven vertical lines, only a handful are equilateral, primarily those along the top of the dial. The *boustrophedon* ordering of the months was sufficiently remarkable to become commonplace in most treatments of the sundial, though the term is applied inaccurately here. The *boustrophedon* writing style technically means that all the letters proceed from right to left in the usual fashion but then reverse to run left to right, like, as the anonymous Englishman remarked, oxen plowing a field. In the case of the prosciutto, the pairs of letters forming the calendrical abbreviations are grouped correctly right to left but the abbreviations in the upper line are in retrograde, or arranged left to right, with the lower line returning right to left

---

41. An elegant rendering of the so-called Colonna Nania from Melos was reproduced for a similarly anonymous letter later in this same fascicule: *Memorie per Servire all'Istoria Letteraria* 6.2 (August 1755) 71–73 (letter dated August 20, 1755, from Venice), the engraving appears on page 73; J. T. Clarke, "A Doric shaft and base found at Assos," *American Journal of Archaeology and of the History of the Fine Arts* 2.3 (July–September 1886) 273, notes that the Colonna Nania was first described and engraved by Girolamo F. Zanetti in his *Due antichissime greche iscrizioni* (Venice 1755), where, however, the column shaft was printed upside down. Zanetti was one of the editors of the *Memorie*. The column formed part of the collection of the Nani brothers of Venice; see G. Nani, *Collezione di tutte le antichità che si conservano nel museo Naniano di Venezia* (Venice 1815) I, fig. 8; and Chapter 1.e.

to correspond to the movement of the sun's shadow across the dial.[42] On the other hand, the scrutiny of the open form of the letter P in Aprilis is the only attempt in the modern literature to date the instrument paleographically.[43] Finally, the author did not attempt to explain how the sundial actually functioned.

The parade of Englishmen who passed through Italy and Naples in the eighteenth century on the Grand Tour makes it virtually impossible to identify with certainty the "English gentleman" credited with providing this description so soon after the prosciutto's discovery. The long-serving British Consul at Naples, Isaac Jamineau, would have been an important first contact for Englishmen passing through the city. Jamineau penned several letters on the eruptions of Vesuvius in 1754 and an earthquake in 1756, extracts of which were published in the *Philosophical Transactions* of the Royal Society of London, though he was not himself ever a member of the Society. His collection of artworks would have drawn many of the same scholars and artists keen to see the archaeological finds, but there is no evidence linking him directly or even indirectly with the excavations.[44]

Another major figure from this period offers an intriguing resumé boasting many of the necessary connections. From 1744 to 1746, Sir James Gray, second baronet, served as secretary to Robert D'Arcy, Fourth Earl of Holdernesse, the British ambassador extraordinary to Venice, and he remained there as British Resident through 1753 (Figure 1.7). As a founding member of the Society of Dilettanti in 1738, Gray not only would have had regular contact with other Englishmen passing through Italy on the Grand Tour, a requisite for admission to the Society, but also recommended several of them for membership in the Society. This included James Stuart and

---

42. Subsequent scholars routinely described the arrangement of the abbreviations as *boustrophedon*, so the term has been retained throughout this study.

43. Scholars continue to view this open, or "unclosed," form, where the curved stroke fails to touch the vertical bar (*hasta*), as a feature of inscriptions dating down to the end of the Julio-Claudian period, though later examples exist; see Salomies 2015, 169; Gordon and Gordon 1957, 109–10, 211, fig. 14.2.

44. I. Jamineau, "An Extract of the Substance of Three Letters from Isaac Jamineau, Esq; His Majesty's Consul at Naples, to Sir Francis Hoskins Eyles Stiles, Bart. and F. R. S. concerning the Late Eruption of Mount Vesuvius," *PhilTrans* 49 (1755–1756) 24–28; the letters date between December 7 and 28, 1754. Jamineau, also spelled Jemineau, was British Consul from 1753 to 1779, and was elected to the American Philosophical Society in 1768 in recognition of the assistance he had provided its members; see W. J. Bell, *Patriot-Improvers: Biographical Sketches of Members of the American Philosophical Society* (American Philosophical Society 1997) 427–28. For his collection of art, primarily paintings, see E. Dodero, *Ancient Marbles in Naples in the Eighteenth Century: Findings, Collections* (Leiden 2019) 52, 197, 266.

FIGURE 1.7   Portrait of Sir James Gray, 2nd Baronet (1708–1773), Envoy-Extraordinary to the Bourbon court at Naples. Pastel on paper from 1744–1745 by Rosalba Carriera (1675–1757). Digital image courtesy of Getty's Open Content Program (2009.80).

Nicholas Revett, whom he met in Venice in 1750 as they were setting off on their expedition to document the antiquities of Athens.[45] His antiquarian interests would have brought him in contact with Venetian intellectuals with similar interests, perhaps including those responsible for the *Memorie*.[46] In late 1753, Gray had been appointed Envoy-Extraordinary to the Bourbon

---

45. *A Dictionary of British and Irish Travellers in Italy, 1701–1800* (New Haven 1997) 424–25, s.v. *Gray, Sir James*. His brother, George, was secretary of the Society and an architect in his own right. For his founding of, and involvement in, the Society of Dilettanti, see L. Cust, *History of the Society of Dilettanti* (London 1914) esp. 76–78. For Stuart and Revett and their *The Antiquities of Athens*, 5 vols. (London 1762–1816), see L. Lawrence, "Stuart and Revett: Their literary and architectural careers," *Journal of the Warburg Institute* 2.2 (1938) 128–46; and J. Kelly, *The Society of Dilettanti: Archaeology and identity in the British enlightenment* (New Haven 2009) 145–71.

46. C. Whistler, "Venezia e l'Inghilterra: Artisti, collezionisti e mercato dell' arte, 1700-1750," in L. Borean and S. Mason, eds., *Il Collezionismo d'Arte a Venezia: Il Settecento* (Vicenza 2009) 94–95. Gray's portrait (see Figure 1.8) was painted by Rosalba Carriera, a Venetian artist with long-standing ties to the British diplomatic and Venetian artistic and commercial circles. See also J. Kelly, "The Portraits of Sir James Gray (c. 1708–1773)," *The British Art Journal* 8.1 (2007) 15–19.

court of Naples and served in this role until 1763. This position allowed him to establish close ties with leading Neapolitan intellectuals and antiquarians. He quickly developed a warm relationship with Charles of Bourbon, advising him on the designs for his new palace at Caserta.[47] In correspondence addressed to Lord Holdernesse, Gray lamented that he could not contribute to the King's discussions of his two passions, dogs and hunting, but that "as architecture and antiquities are equally favorite topics I have my part."[48] In particular his correspondence confirms he had enjoyed first-hand access to the antiquities recovered in the royal excavations at Pompeii and Herculaneum. In another letter he remarked that "when anything curious is discovered, The King usually shews it to us after his Dinner."[49] In a letter dated October 29, 1754, to Sir Thomas Robinson, Gray reported on the discovery of the first papyrus scrolls from the Villa dei Papiri at Herculaneum, noting specifically that Alessio Simmacho Mazzocchi (1684–1771), Canon of the Duomo in Naples and a luminary of the Neapolitan intellectual scene, was in charge of reading them, and this demonstrates that he was conversant in the excavations' current affairs.[50] That the prosciutto had passed from the excited hands

---

47. In 1759, Charles had requested that Gray negotiate the return to Naples of one fragment of the *Tabulae Heracleenses* (MANN 2481), tablets inscribed in Greek and Latin that had been found in Heraclea in 1732, that had found its way to England in 1735 through the antiquarian market. This joined the other tablets housed in the museum at Portici that had been published in 1754 without reference to the English fragment; see Chapter 1.g. P. C. Webb, *An Account of a Copper Table: Containing Two Inscriptions, in the Greek and Latin Tongues, Discovered in the Year 1732 Near Heraclea, in the Bay of Tarentum, in Magna Grecia* (London 1760) 10, who initially thought the absence of the text of the fragment from England in the publication was a printer's error.

48. British Library Manuscripts and Archives, *Egerton Mss.* 3464, ff. 250–71, 303–18, 321–48: *Sir James Gray, 2nd Baronet, diplomatist; Correspondence with Lord Holdernesse, 1749–1759*: (January 29, 1754, and April 9, 1754).

49. British Library Manuscripts and Archives, *Egerton Mss.* 3464, ff. 250–71, 303–18, 321–48: *Sir James Gray, 2nd Baronet, Diplomatist; Correspondence with Lord Holdernesse, 1749–1759* (October 29, 1754).

50. James Gray, "Extract of a letter from Sir James Gray, Bart. His Majesty's Envoy to the King of Naples, to the Right Honorable Sir Thomas Robinson Knight of the Bath...relating to the same discoveries at Herculaneum," *PhilTrans* 48 (1754) 825–26. In this same letter, Gray also reported on the discovery of a bronze bust of Epicurus (probably MANN 5465, but found November 4, 1753) and two bronze heads, one of "Seneca" (MANN 5616, found September 27, 1754, also known as the "Pseudo-Seneca") and the other of "a captive king" (MANN 5590, found on October 23, 1754), and announced the intention of the king to publish *Le Antichità*. Gray's letter was preceded in this volume by one from Paderni on the latest finds from the Villa dei Papiri in which he refers to this portrait of a "captive king," now identified with Ptolemy Soter I, as representing a Syrian king (*re di Siria*), which suggests Gray had learned of this identification from Paderni. For these heads, see Comparetti and De Petra 1883, 264, 265–66; Wojcik 1986, 94–95, 97–99; Mattusch and Lie 2005, 249–53, 260–62.

of Paderni into those of the King in the garden of the palace at Portici to subsequently serve to animate an after-dinner conversation at which Gray was present is easy to imagine. The fragmented, hasty nature of the description provided to the anonymous recipient in Florence, and the singling out of Mazzocchi in the letter, bear all the hallmarks of the recollections from a discussion held under just such conditions, perhaps even with Mazzocchi holding forth, while Gray's need to continue interacting with the court would have made the preservation of his anonymity desirable. Yet it would be incredible if he had not observed the sundial's prosciutto shape.

Further evidence of the manner in which clandestine information moved within Italy and beyond is provided by another letter published in 1755 by the Royal Society. The contents of the letter "from a Learned Gentleman of Naples" was one of the first to provide a detailed description of the scrolls from the Villa dei Papiri, their appearance, literary contents, the manner employed by the Bourbon court in unrolling and deciphering them, and the individuals involved in the process. As had been the case with the Englishman's description of the prosciutto, and as it would be with Gray's own reference to the papyri, Mazzocchi's talents in particular were highlighted, pointing to him as the likely original source of information. In this case the link between Naples and England passed through Gasparo Cerati of Parma, then resident in Pisa, who communicated it to Henry Baker of the Royal Society.[51] Cerati, in turn, was a correspondent with one of the major intellectual figures of the day, Anton Francesco Gori (1691–1767), and was a member of his society of antiquarians, the Società Colombaria.[52]

Gori's role as the unidentified Florentine conduit of the details divulged in the *Memorie* is entirely speculative, but he had a long history of serving as a kind of clearinghouse for passing along details about the discoveries in the Vesuvian region. As a consequence he was looked upon with suspicion by the Bourbon court.[53] A founder of the Società Colombaria in 1735, he maintained

---

51. *Anonymous*, "Copy of a Letter from a Learned Gentleman of Naples, Dated February 25, 1755, Concerning the Books and Antient [*sic*] Writings Dug out of the Ruins of an Edifice Near the Site of the Old City of Herculaneum; To Monsignor Cerati, of Pisa, F. R. S. Sent to Mr. Baker, F. R. S. and by Him Communicated; with a Translation by John Locke, Esq; F. R. S." *PhilTrans* 49 (1754–55) 112–15.

52. Cerati had described some elephant bones recovered in a field near Fucecchio, west of Florence, in a letter to Gori dated January 18, 1754: *Gori Archives* BVII7.833r (Record 11953): Biblioteca Marucelliana in Florence (BMF): http://www.maru.firenze.sbn.it/gori/a.f.gori. htm.

53. E.g., A. F. Gori, *Notizie del memorabile scoprimento dell' antica città di Ercolano* (Florence 1748), a collection of letters primarily addressed to Gori from other scholars, many of them

active correspondence with scholars throughout Italy, including Venice, where he had published the gem collection of the antiquarian Anton-Maria Zanetti.[54] Zanetti's cousin was the archaeologist Girolamo Francesco Zanetti, a member of Gori's Società Colombaria and one of the editors of the *Memorie*, where Gori's name is cited frequently.[55] As an antiquarian Gori would have relished the opportunity to retell the historical and literary evidence concerning Roman sundials. As an epigrapher, he would have been especially interested in the prosciutto's inscription, and this may be why it and the orthography of the character "P" featured so prominently in the transmission of its description, provided in the course of a conversation with the English gentleman and subsequently retold in the letter published in the *Memorie*. If it had not been Gray himself, who does not appear among Gori's many correspondents and with whom there is no specific evidence for a connection or travel through Florence, it may have been one of Gray's dilettanti, relaying a description originating in a conversation with Gray. Even more suspect, however, is Mazzocchi, who numbered among Gori's correspondents, who was an epigrapher, and whose own misconceptions about the sundial could easily have corrupted the early descriptions of it.

1.d) Autant de fautes que de mots: *D'Alembert's* boustrophedon *and the* Encyclopédie

Either the anonymous author, the English gentleman, or this notice in the *Memorie* must have served as the source for the next published description of the sundial. This was incorporated into the entry for *Gnomonique* in Denis Diderot and Jean Le Rond d'Alembert's *Encyclopédie, ou dictionnaire raisonné des sciences, des arts et des métiers* and, because of the popularity and broad distribution of the *Encyclopédie*, may be considered the first widely circulated

---

anonymous "letterati Napoletani," transmitting the early discoveries at Herculaneum. Gori's correspondence, numbering in the thousands, is archived in the Biblioteca Marucelliana in Florence (BMF): http://www.maru.firenze.sbn.it/gori/a.f.gori.htm. Vazquez-Gestal 2019, 41–42, quotes a letter dated September 16, 1752, from Bayardi to Gori (BMF B VII 4 cc. 6r–7v), showing that Gori still maintained his contacts inside the court. For Gori's and other connections between Tuscany and Naples, see A. Castorina and F. Zevi, "*Antiquaria Napoletana* e cultura Toscana nel Settecento," in G. Cafasso et al., eds., *Il Vesuvio e le Città Vesuviane, 1730–1860: In ricordo di Georges Vallet* (Naples 1998) 115–32.

54. A. F. Gori, *Le Gemme Antiche di Anton-Maria Zanetti di Girolamo* (Venice 1750).

55. E.g., *Memorie per Servire all' Istoria letteraria* 1.4 (1753) 53–57: Gori leads a call for scholars to submit contributions to a history of Tuscany. G. F. Zanetti had published the Colonna Nania in the Nani brother's collection in Venice.

FIGURE 1.8 Jean le Rond d'Alembert (1717–1783). Pastel portrait by Maurice Quentin de la Tour (1753). Creative Commons Public Domain https://commons.wikimedia.org/wiki/File:Jean_Le_Rond_d%27Alembert,_by_French_school.jpg.

notice of the sundial's existence. It was also the discussion that would generate the most controversy. D'Alembert himself penned the entry, which appeared in the seventh volume of the series. Since the volume's canonical date of publication is November 1757, more than two years had passed since the sundial's discovery.

As a mathematician and scientist d'Alembert (1717–1783) could provide the first assessment of the sundial as a scientific instrument (Figure 1.8). The similarities between his physical description and that in the *Memorie* are unmistakable, however, and reveal that d'Alembert had not examined the sundial himself:

> A portable sundial has been found in the ruins of Herculaneum. This sundial is round and equipped with a handle at the end of which is a ring which no doubt serves to suspend the sundial wherever one wishes. The entire instrument is bronze and a bit convex on its two

surfaces. It has a somewhat long and jagged style on one side which is roughly a fourth part of the diameter of this instrument. One of the two surfaces, which can be regarded as the upper surface, is entirely covered in silver and divided by twelve parallel lines which form an equal number of small, slightly sunken squares. The final six squares which end in the lower part of the circle's circumference are disposed as one can see and contain the following characters which are the initial letters of each month's name.

| JU. | MA. | AV. | MA. | FE. | JA. |
| JU. | AV. | SE. | OC. | NO. | DE. |

The manner in which these months are arranged is remarkable in that it is in *boustrophedon*. One could believe that this disposition of the months on the dial comes from the fact that in the months that are one below the other, for example in April and September, the sun is somewhat at the same height on certain corresponding days. But in this case, the dial is not quite exact in this regard because this correspondence happens only in the two first halves of each of these months: in the last fifteen days of April, the sun is much higher than in the last fifteen days of September; this also occurs in the other months.[56]

D'Alembert dispensed with likening the shape to a flask and altered only slightly the description of the gnomon. He again highlighted the *boustrophedon* ordering of the calendrical abbreviations, representing them in the accompanying diagram as fully framed in a rectilinear grid that recalls the *Memorie*'s representation while similarly bearing little resemblance to what is actually on the sundial. D'Alembert's diagram introduced several errors into his transcription of the inscription. He wrote "AV" rather than "AP" for the abbreviation of Aprilis; substituted initial "J" for "I" in writing JA[nuarius], JU[nius], and JU[lius]; employed "U" rather than "V" in both Iunius and

---

56. *Encyclopédie, ou Dictionnaire Raisonné des Sciences, des Arts et des Métiers, par un societè de gens de lettres* (Paris 1757) 7.725–726 s.v. *Gnomonique* (D'Alembert). For d'Alembert, see *Biographical Encyclopedia of Astronomers* (New York 2014) 499–502, s.v. *d'Alembert, Jean-Le-Rond* (M. Chapront-Touze).

Discovery 31

Iulius, while retaining the "V" in Augustus; and wrote only "JU" rather than "IVN" for Iunius, the only month actually reduced to three letters on the sundial. As the abbreviations "AV" and "JU" show, however, d'Alembert had transcribed the Latin abbreviations into the French ones for avril and juin and juillet, respectively, for the benefit of his audience but at the expense of accuracy.

In retaining the *Memorie*'s flawed description of the dial's design, d'Alembert revealed that he did not fully comprehend how it functioned. In his defense, however, the information he had in front of him was both inaccurate and incomplete. His main contribution was his attempt to explain the science behind the pairing of the twelve months. These pairings were, in fact, the one aspect of the sundial for which his evidence was sound, notwithstanding his errors of transcription, precisely because of the availability of a reproduction of the inscription. These specific months had been joined, he noted, because the sun's position in the sky was more or less equivalent on certain corresponding days of these months. In other words, the pairings were not just a convenient means of organizing the twelve months on a constricted space but could be ascribed to established astronomical computations of the sun's movement throughout the year, of which he would have been well aware. His clarification that this concordance was valid only in the first half of the paired months is true in that the sun's altitude deviates only between one and five degrees on the first day of the month in each of these pairings, and between two and seven degrees mid-month. D'Alembert was also correct in observing that the variations are wider in the second half of these months, when the altitude of the sun can vary up to eighteen degrees. In fact, his example of April and September is, together with March and October, precisely when this difference in solar altitudes is greatest.[57] But the inexactitude d'Alembert perceived was not due to the ignorance of the sundial's designer but to d'Alembert's failure to recognize that the dial was not based on the lunar calendar but on the zodiacal calendar in which the days and hours align much more closely as the sun moves in and out of the zodiacal signs, the days of entry and departure of which occur closer to the middle of the lunar months.

---

57. The values of solar altitudes at latitude 41.5° (longitude -72.66°) in 2022 illustrate this: April 1 = 53.18° vs. September 1 = 56.52°, April 30 = 63.26° vs. September 30 = 45.28°; March 1 = 39.38° vs. October 1 = 44.89°, March 31 = 52.80° vs. October 31 = 33.92°; but, e.g.: January 1 = 23.77° vs. December 1 = 23.95°, January 31 = 29.85° vs. December 31 = 23.64°; June 1 = 70.44° vs. June 30 = 71.58°, July 1 = 71.51° vs. July 31 = 66.61°.

1.e) Un' altra collera antiquaria: *Paciaudi's* "*Carafe,*" *the* Monumenta Peloponnesia, *and Tanucci*

A more detailed description of the sundial with an attempt to explain its underlying science and mode of operation was offered in 1761 by Paolo Maria Paciaudi (1710–1785) in his *Monumenta Peloponnesia*, a two-volume work cataloging the antiquities collection of Bernardo and Giacomo Nani of Venice (Figure 1.9).[58] An important antiquarian himself, Paciaudi corresponded with an international cast of the period's leading scholars, including Gori, Giacomo Martorelli, the Comte de Caylus, the abbé Jean-Jacques Barthélemy, and Johann Joachim Winckelmann, while his work took him from Venice to Rome, Malta, Florence, Naples, and, ultimately, Parma.[59] His attention had turned to the sundial during the course of his analysis of the so-called Columna Dianiae. This fragmentary votive column in the Nani collection bore an inscription in Greek recording the dedication of a sundial, a column, and a base (*horologium et columnam et basim*) to Diana "Kelkaia" by Colirius Superbus and Claudia Nicephoris. The column had served as a base for the sundial, which was no longer extant. Nevertheless, this had prompted Paciaudi to offer a survey of the history of ancient sundials, and this, in turn, had served as a useful path to discussing the prosciutto sundial.

At the outset of his discussion, and again in his table of contents, his index, and his explication, Paciaudi claimed that while others before him had treated the history of sundials, he was presenting this particular ancient example for the first time: "it was always my task to seek out some unpublished monument and make it public so that at least some dignity for [my] work might be procured through the compensation of novelty."[60] It is not readily clear on

---

58. Paciaudi 1761, I.49–55. This is the same collection containing the Columna Nania that was reproduced in the *Memorie* volume with the anonymous letter describing the prosciutto sundial.

59. Paciaudi et al. 1802, iii–vxiii; Caylus et al. 1877, i–ciii. The Comte de Caylus' full name and titles is Anne-Claude-Philippe de Tubieres de Grimoard de Pestels de Lévis de Caylus, Marquis d'Esternay, Baron de Branzac. Paciaudi became a member of Gori's Società Colombaria in 1747 and exchanged letters with him between 1741 and 1755; see, esp., Fino 2004, 88–95, for Paciaudi's letters during his periods at Naples. For a catalog of Paciaudi's correspondence, and his correspondents, archived in the Biblioteca Palatina di Parma, see L. Farninelli, *Paolo Maria Paciaudi e i suoi correspondenti* (Parma 1985). For a review of Paciaudi's work in Parma, see M. C. Burani, "Un direttore illuminato: Padre Paolo Maria Paciaudi (1763–1785)," in S. Pennestri, ed., *Complesso monumentale della Pilotta. Il Medagliere I. Storia e documentazione* (Rome 2018) 79–94.

60. Paciaudi 1761, I.49. The table of contents labeled this section of the text "Horologium φορηματικον Herculanese nunc primo vulgatum describitur" (portable sundial from

FIGURE 1.9 Paolo Maria Paciaudi (1710–1785). Oil painting on canvas by Giuseppe Lucatelli (1751–1828), dated to ca. 1786–1787. Museo Glauco Lombardi, Parma. Public domain—CC-BY 4.0.https://catalogo.beniculturali.it/detail/HistoricOrArtisticProperty/0800405989.

what grounds Paciaudi based his claim. Two book dedications in the *Monumenta* bear the date April 1761, providing a *terminus post quem* for the volume's publication.[61] This was some four years after Paderni's letter to the Royal Society and d'Alembert's entry in the *Encyclopédie*. While he may have been unaware of those publications, the parallels between the phrasing of Paciaudi's subsequent remarks and the letter in the *Memorie* are striking and raise the intriguing possibility that, if it had not been Gori, Paciaudi may have played a crucial role in transmitting the letter to the *Memorie*. As a Venetian working on the Nani collection, he likely maintained ties with

Herculaneum now for the first time described publicly), while the entry in the index (Paciaudi 1761, II.270) reads "Horologii Herculanensis schema nunc primo editum" (form of the Herculaneum sundial now edited for the first time).

61. Caylus et al. 1877, 258, 260: Barthélemy expressed enthusiasm for receiving a copy of the *Monumenta* in a letter to Paciaudi dated July 18, 1761, and complained in a later letter dated October 16, 1761, that everyone else already had a copy. Tanucci, the prime minister, had obtained his copy in July; see Chapter 1.e.

34

THE PROSCIUTTO SUNDIAL

scholars like Zanetti, Serimàn, and Calogerà, the Venetian editors of the *Memorie.*

> I am setting forth before the eyes and sight of men an example, delineated according to the credibility of the autograph, of a pensile and portable sundial which I received from a learned and noble man in these very days in which, near the feet of Vesuvius, from the huge and most valuable ruins of Herculaneum, this notable monument of old was removed and carried into the nearby royal palace.

Even if the "learned and noble man" is not identified specifically as being English, the temporal phrase "in these very days in which" (*per eos ipsos dies*) the sundial had been recovered and taken to the palace indicates that the discovery of the sundial and the writing of the exegesis that follows were chronologically closer than the six years that had actually elapsed. The resulting erroneous details about the sundial might then be ascribed to a hastily formulated, word-of-mouth report circulated soon after its discovery or a summation of the key points from a dinner conversation. The almost ceremonial image of the instrument being carried into the royal palace suggests a narrative originating in Paderni, one he had recounted in his own letter to the Royal Society and perhaps was in the habit of retailing in the museum, that had been transmitted to Paciaudi through his source, which, given the circles in which they moved, may well have been Sir James Gray.

The most significant difference in Paciaudi's presentation, and what validates his claim of novelty, was that he illustrated his discussion with a large, detailed engraving. This makes vividly obvious to "the eyes and sight of men" just how flawed was the information on which Paciaudi had relied (Figure 1.10). That virtually nothing is correct makes it difficult to believe that the illustrator had seen the original object, yet the engraving faithfully reproduces the same errors in the verbal narrative that appear in the *Memorie* and the *Encyclopédie*, suggesting that a prototype of this illustration had served as the basis of those descriptions, if it was not simply a rendering by Paciaudi based on his own understanding: as he says, "delineated according to the credibility of the autograph." The caption states it came "from the ruins of Herculaneum in 1754" (*Ex Herculanei reliquiis an. 1754*), although it had in fact been found in 1755. The shape is too round and symmetrical and, clearly, does not even come close to representing a prosciutto. The suspension ring at the top is too small and oriented in the wrong direction. While the gnomon is on the correct side, its

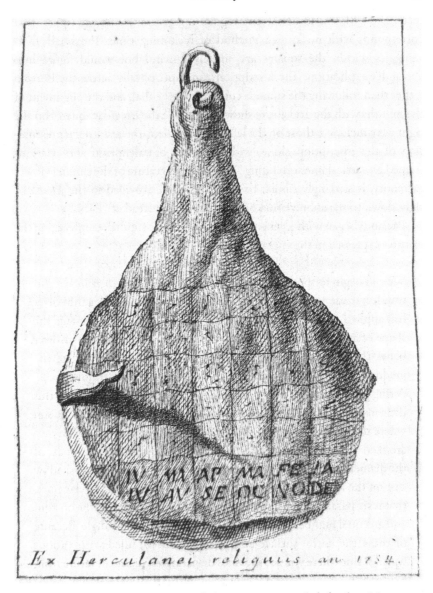

FIGURE 1.10 Paciaudi's engraving of the prosciutto sundial for his *Monumenta Peloponnesia* (Rome 1761). "From the remains of Herculaneum, year 1754." The earliest published representation of the sundial.

shape and size are incorrect. The dial most clearly captures the earlier inaccuracies concerning its design. Its six horizontal lines extend the full width of the instrument and are intersected perpendicularly by six vertical lines creating squares that are too few in number, too rectilinear, and arrayed too symmetrically on the surface. This yields an odd number—twenty-five equilateral

36 THE PROSCIUTTO SUNDIAL

rectangles with an additional ten irregular squares on the left and right—that corresponds with no known method of reckoning time. The small, faint squiggles within the squares are entirely fanciful but would figure into Paciaudi's explication. The inscription runs horizontally across the bottom rather than following the sinuous curvature of the dial, and the alignment of the months with the grid above shows that while the irregular squares on the right were included, those on the left were excluded. The resulting representation of the inscription alone, with the pairs of calendrical abbreviations framed by vertical lines extending as far as the curvature of the sundial's lower extremity, is strikingly similar to the illustration provided in the *Memorie*, right down to the abbreviation of Iunius as "IV" instead of "IVN."

Paciaudi began with a general description of the sundial, emphasizing the features expressed in the engraving:

> We will begin to speak about this shadow catcher (*skiatherium*) in this way. It is made of thin bronze, and the work is improved by a thin silver leaf applied to the front. Its height is about one palm. It presents the shape of a flask or, so to speak, a small cask, because it has a narrower belly. The circumference of the dial, leaving aside the lower sphere, gradually tapers in an extended curve and ends in a thinner neck. A ring passed through a hole at the top offers the convenience that this little instrument can be suspended and turned to the region of the sun where the solar beam might meet the gnomon appropriately. A stylus, crooked in the manner of a flame and about equal to a fourth part of the diameter, is attached to the left side at a right angle, and the shadow cast on the lines indicates the hours. Altogether there are twelve lines, that is six parallel lines drawn from the top of the sphere to the bottom and an equal number transversely intersecting at a right angle. Because of these the entire surface appears crisscrossed, filled with various squares. But where you see the lines deviating by slight curvature from a straight line, know that our engraver did this considerately and deliberately so that he might express the unevenness of the surface and set more openly and vividly before our eyes the various pummelings this temporal zodiac reader has suffered unjustly.

Paciaudi confirmed what Paderni and Bayardi had stated earlier: that the bronze core was covered in silver. Apart from Paderni, he was the first to offer any dimensions, but it is not clear what he meant by his statement that "its height is almost a palm" (*altitudo prope palmaris est*). As a standard unit of measurement in his day, the Neapolitan *palm* was 0.264 m while the actual height of the sundial, without its ring, measures only 0.118 m.[62] The instrument does fit in the palm of a hand, and Alcubierre had described it as "about the size of a hand," but then the term would be more applicable to its overall size than its height alone.

The parallels between Paciaudi and the *Memorie* continue in the comparable phrasing used in their physical descriptions of the dial. For example, in the *Memorie* the gnomon is said to be "not straight but in the manner of a flame," while Paciaudi says it is "crooked in the manner of a flame." Both state that the gnomon's length is one-quarter the sundial's diameter, even though it appears shorter in Paciaudi's engraving and the stub of the gnomon surviving today is significantly less than that. Both state that the dial is composed of six by six crisscrossing lines, perhaps in an attempt to equate them with a twelve-hour day and a twelve-month year, the number in reality being seven by seven. Paciaudi differs in adding that the lines' waviness was an attempt by the engraver to capture the irregularity of the damaged ancient surface, but this might also be seen as representing the artist's vague memory of the angular manner in which the horizontal lines traverse the dial's face to intersect the vertical ones at the hour marks.

Paciaudi then turned his attention to how the sundial's design related to its function:

Nor ought it seem remarkable to anyone that numerical indications of the hours are absent from our shadow catcher. For they are absent from practically all sundials which we have recorded thus far; because, on the basis of these very few lines, it is easy to deduce the measurement of time and the hours, to which Persius alludes: "...when the shadow touches the fifth line".[63]

The craftsman's knowledge and accuracy shine even more in the notation of the twelve months which he showed by means of the first two initials of each and which he represented in *boustrophedon*. But he associated the months in such a way that these might be in the same squares of the calendar which accord with the length of the hours and the measurement or quantity of the shadow. For January is united

---

62. One Neapolitan *palm* = 0.26455 m; the *palm* (plural *palmi*) was in turn divided into twelve units called an *oncia* (plural *oncie*) = c. 0.02 m.

63. A footnote provides the reference to the text of Persius *Satires* 3.4: *quinta dum linea tangitur umbra*.

clearly with December, February with November, March with October, April with September, May with August, June with July, based on the observation of astronomers. This disposition of the months, however, was derived from the teaching of Palladius Rutilius who scrupulously compared one with the other and ascribed one by one how long the shadow is in each one.

It is in this passage where Paciaudi began to cast a glaring light on just how completely he failed to comprehend how the gnomon's shadow interacted with the dial to mark time over the course of the day and the year. While he is correct that the hours generally are not marked numerically on Greek and Roman sundials, his literary reference to Persius is useless because he does not identify which line on the dial is the "fifth," much less define with which hours it corresponded. Like d'Alembert, though without d'Alembert's specific reference to established astronomical facts, he appears to understand how the *boustrophedon* pairing of the months related to the manner in which the shadow was cast across the face of the dial: longer in the months of December and January and shorter in June and July. He based this knowledge on the work of the fourth-century agricultural writer Palladius Rutilius, who had paired the months in the same fashion seen on the prosciutto and whose calculations of the length of shadows cast by the sun across the hours of the day similarly illustrated that these were longer in the winter and shorter in the summer.[64] His reference to it as a "temporal zodiac reader" shows that, in contrast to d'Alembert, he knew its design was based on the zodiacal, not the lunar, calendar. What appears to have befuddled Paciaudi most was precisely how the gnomon indicated the individual days and hours. Rather than pursuing this topic at this point, he first sought to establish the instrument's ancient pedigree and his own perceived role in publicizing it:

> Here I ought to congratulate myself particularly because I first am presenting to the eyes of learned men a monument not so much marvelous to use, and yet unknown, as splendid in the excellence of its remarkable old age. For, not to dispute anything about the shape of some of the letters inscribed on it in which the full nature of antiquity is detected, without a doubt the form of this portable sundial seems captured in the verses of Bato in Athenaeus. For this comic poet, nearly equal to Plato, in order to ridicule a certain avaricious man and make a

---

64. For more on Palladius Rutilius and his work, see Chapter 2.d.

Discovery 39

fool of him in biting words, said: that man is so covetous of wealth, so desirous of frugality, that he cautiously carries around a phial of oil and examines it with prying eyes, as if wielding a sundial:

> Then in the early morning you carry around an oil flask (*lekythos*)
> Diligently observing how much oil there is; and those
> Who see you think you are carrying a sundial (*horologion*) not a flask.

But who does not notice that the poet in these verses is imitating our sundial neatly and fittingly? For what other kind of sundial can it be which can be compared truly and appropriately with an oil flask, whether you see its sketch or shape or consider the manner by which it must be handled?

Paciaudi again followed closely the outline of points made in the *Memorie*'s letter. He mentioned the potential for dating the letter forms through paleographic evidence only in passing and instead turned to an obscure literary reference to establish that the sundial's shape itself proved its antiquity. The premise of the passage from Athenaeus from which Paciaudi drew his quote concerns a Cynic philosopher who spurns the good life but nevertheless lives off invitations to dinners where he can expound his philosophical discourses, his portable sundial serving to remind him of the next mealtime. The character in question uses his *lekythos* as a *clepsydra*, technically a vessel with markings that indicated the passage of time depending on how much of its liquid contents had drained out.[65] While Paciaudi may have been led astray lexically by the use here of the Greek term *horologion* to mean a *clepsydra* rather than an actual sundial, he clearly believed that the shape alone of *lekythoi*, with their wide bodies and narrow necks, was sufficient for the ancient observers of this comedic character to conclude he was using his *lekythos* as an actual sundial.[66] For Paciaudi, this was proof that sundials could bear this shape.

---

65. Athenaeus, *Deipnosophistae* 4.163b–c. Athenaeus took this quote from *The Murderer*, a play by Bato, a third-century BCE author of New Comedy. The preceding sentence in the play argues that sober and avaricious ways deprive both gods and men of their livelihoods, that not buying and consuming wine, for example, cheats both farmer and merchant.

66. The visual association would have to have been with the "squat" form of a *lekythos* whose height averaged about 0.20 m. This may, in turn, explain Paciaudi's mistaken belief that the Herculaneum sundial was a *palm* (0.264 m) high.

40   THE PROSCIUTTO SUNDIAL

That the character carried his *lekythos* around with him, scrutinizing its contents in the same manner as would be required to read the prosciutto sundial, was sufficient for Paciaudi to conclude this passage established the existence of portable sundials already in the third century BCE, while other literary references confirmed their continued use through Roman times. In this he disputed the conclusion of Isaac Casaubon, the sixteenth-century commentator on Athenaeus, who doubted this was a reference to the kind of portable sundials referenced by Vitruvius (Vitr. 9.8.1: *viatoria pensilia*) arguing that such flasks (*ampullae*) for oil were traditionally made of leather, ceramic, or bronze and therefore were not transparent.[67]

> Moreover [continued Paciaudi], so that the hours on this kind of sundial might most securely be determined it was necessary to hold it suspended vertically, to level it with precision, to turn it exactly, to observe it accurately, or, in a word, with the same care as the greedy man handles his oil flask, to turn it by hand. From these matters it is possible to understand that in the time of Bato portable sundials already were carried, and to recognize on the basis of ours of what sort these were, and accordingly to shed light mutually upon the songs of the poets and the Herculaneum sundial, with the result that, after the sundial has been observed, the comic's wit becomes evident and, with the verses read again, the antiquity of the sundial is divulged.
>
> How is it surprising, however, that this was dug up among the ruins of Herculaneum, where the extravagance of Roman men imported all Greek elegance? Thereafter the luxury of portable sundials grew to such a degree that in the time of the emperor Hadrian they were commonly reckoned among the household utensils of a noble family. Indeed, the old jurists, in order to avoid or break off litigation, laid down in law how this kind of wealth should be valued, just as is read in the *Digest*: "A bronze sundial, which is not fixed, Papinianus believed constituted a domestic instrument, following Ulpian."

---

67. The sixteenth-century commentator on Athenaeus, Isaac Casaubon (1559–1614), in his *Animadversiones in Athenaei Deipnosophistas libri* XV (1600) 185, had been the first to associate Bato's object with Vitruvius' *viatoria pensilia*. He further observed, "Today the most elegant sundials are made in many forms which even proletarian women hang around themselves not for the sake of utility but for ornament: to such a point is the discipline of public morals corrupt or rather non-existent."

Paciaudi next turned to his explanation of how the sundial functioned. In this he departed from the author of the *Memorie*'s letter, who had cautioned against relying solely on an illustration in the decipherment of an object. It is here where the errors of his engraving most clearly confounded Paciaudi's attempt to understand how time was marked on the sundial, for while he appears to have understood some general principles, he could not apply them to the evidence he had before him.

But enough about the shape and antiquity of the sundial: now it seems best to inquire about its use. Admittedly the roles of expounder and philologist that we have assumed for the moment by no means demand that we, who are following solely the ancients in this book, allow our discussion to digress to gnomonic principles. If anything must be ventured in a matter foreign to our studies and in the scholarship we abandoned twenty years ago, this perhaps was the value of the Herculaneum sundial. Six drawn squares occur on it, whose size, separated off between lines, will most suitably correspond to indicating the six hours before noon and likewise after noon, according to the varying position of the sundial. If therefore the gnomon as the tracker of the shadow is turned to face the expanse of the world where the star of the sun emerges above the horizon, after it has scarcely risen, its shadow will spread itself all the way to the last square IA. DE. where the first hour of the day is considered to be. When, however, as the altitude of the sun increases, the length of the shadow diminishes an equal proportion, that will actually, in touching obliquely in its retreat the squares placed above the diagram of the months, designate the second, third, fifth hours, until it hits the sixth and the sun reaches midday, when it makes its shortest shadow. With this first part of the day complete, the sundial should be turned toward the opposite, setting region; and so the dial itself faces the setting sun, from here it will be possible to take the six remaining hours. For because the shadow whose measurement decreases before noon increases afterwards the afternoon hours can be noted in the same squares but in the opposite order, so that where the sixth morning hour ends, there begins the first afternoon hour, and thus with the shadow projecting longer across the squares until the sun snatches itself from the eyes of mortals, that line which had welcomed the first light will show the end of the day.

42 THE PROSCIUTTO SUNDIAL

As a young man Paciaudi had studied under Francesco Maria Zanotti (1692–1777) at the Università di Bologna, where he appears to have been exposed to astronomy, if not by Zanotti himself then through his brother Eustachio Zanotti (1709–1782), a leading Bolognese astronomer and director of the observatory there.[68] This must be what Paciaudi is referring to in his remark about scholarship he had abandoned twenty years prior. It is clear he was familiar with the Roman method of reckoning time, which divided the day into six hours before noon and six hours after regardless of the season. He knew that the gnomon had to follow the sun in order to cast its shadow and likewise understood the principle that the sun casts a longer shadow in the morning and the evening when it is lowest on the horizon while it casts a shorter shadow midday. Because of the incorrect placement of the gnomon and the flawed representation of the dial's grid, Paciaudi believed that the grid constituted these six hours with the first and last hours represented by the squares on the right, corresponding with the squares above Ianuarius and December, and midday represented by those on the left. The problem is that these principles are valid for traditional horizontal sundials or meridians, with their gnomons oriented vertically.

That Paciaudi perceived that something was amiss is evident in his subsequent comments, where he noted that the days and hours were shorter in the winter than in the summer months by referencing their corresponding zodiacal signs but could not square this fact with the grid of equilateral rectangles in front of him:

> Finally, it has been observed and shown that the day lengthens as the sun ascends from Capricorn to Cancer, and the night shortens; on the other hand, however, the nights become more drawn out and the days shorter when the sun descends from Cancer to Capricorn; wherefore in order to retain the number of twelve hours, by which the ancients consistently were dividing all the light of the day and the whole night, in whatever time of year, they made by necessity the wintry hours shorter and the summer hours rather long. For this reason, the distance traversed by the shadow in the winter days ought to be rather contracted. And so, in the sundial of Herculaneum, although the lines are parallel the distance throughout is equal. Therefore, it does not seem adjusted for all the hours of the year.

---

68. *Dizionario Biografico degli Italiani* 80 (2015), s.v. *Paciaudi, Paolo Maria*; *Zanotti, Francesco Maria*; and *Zanotti, Eustachio*. Both Zanottis were members of the Royal Society of London as well.

Discovery 43

His example of the inversion of the length of days in the zodiacal reckoning of time is curious since it indicates he was aware of the difference between the lunar and zodiacal calendars. Even so, because the information before him was so limited, he, like d'Alembert, could not have applied the necessary recalibrations of the dial that might have led him to understand the sundial's operation. Instead, he speculated that small markings visible within the squares and represented in the engraving—but for which there is currently no evidence on the actual instrument—somehow helped resolve this evident defect in the design, yet he remained unable to explain precisely how these might have worked:

> But there is nothing we cannot explain more easily. In the individual squares certain worn signs and corroded engravings still shine faintly which I suspect were added not so much for ornament as for the perfection of the sundial. And since larger gaps enclosed by the squares are marked out in the smaller parts, a suitable division of the twelve shorter hours could be adopted very easily from a given point in the penultimate square toward sunset, and therefore establish a "universal" sundial.

In the end, Paciaudi theorized that some of these problems could be resolved by suspending the sundial vertically, though this appeared to have been what he had been thinking all along in comparing the prosciutto sundial to the *lekythos* in Bato's comedy. He recognized, however, that this method introduced the problem that the gnomon reconstructed in his illustration, because of its upright design and placement at the midpoint of the grid, would only cast its shadow on the lower portion of the dial, to the exclusion of the squares in the upper half. Paciaudi concluded this conundrum, captured in his nonsensical characterization of it as a "universal sundial," was better left to others in the same way that the author of the *Memorie*'s letter had placed his hopes for a detailed explication of the sundial in Mazzocchi:

> Nonetheless I see that there are many things which might undermine all this speculation and I confess that the explanation is entangled still by the gravest difficulties. For if the sundial were vertical, as indeed the structure seems not only to recommend but absolutely prove correct, and the lines drawn below the stylus served to indicate the hours, would there be any use for assigning the other lines incised above the gnomon and never cast in shadow by the gnomon? These and others of their kind which might be cited we must freely permit to be discussed and resolved by the mathematicians, by whom we are prepared in the

44 THE PROSCIUTTO SUNDIAL

end to be refuted without resentment, and for us it will be most pleasing to divulge an old mathematical footnote which for the more learned copiously furnishes material for exercising the intellect.

Paciaudi's claim to be the "first" to publish the prosciutto would come back to haunt him. After living and working in Naples from 1743 to 1750, under the patronage of Cardinal Giuseppe Spinelli, he visited there again only in late 1761. He documented this trip in letters to the French antiquarian the Comte de Caylus (1692–1765), with whom he had corresponded for years and for whom he would serve as an agent for obtaining antiquities in Italy for some ten years beginning in 1757.[69] In particular he wrote to Caylus of the eight glorious days he spent in early September 1761, in the ruins of Herculaneum and the museum at Portici:

> I spent eight days in the caverns, and examined, contemplated, and studied the King's museum with the greatest attention. I have never passed more beautiful days. Oh, what treasures! What rarities! What marvels! I dare say I have learned more in these eight days, by looking through the eyes of this museum, than by studying ten years, and that he who has not seen the monuments of Herculaneum lacks an infinity of knowledge.[70]

That he had gained access to the ruins and the museum in the first place is remarkable, however, given that the Accademia by now had known for months of his unauthorized publication of the prosciutto, while the clear parallels in language between the descriptions in the *Memorie* and d'Alembert's entry in the *Encyclopédie* likely enabled them to connect Paciaudi to those publications as well, even if only indirectly.[71] This is clear from a letter Tanucci, the prime minister, wrote to King Charles III in Spain (Figure 1.11). The letter's date—July 28, 1761—proves that the *Monumenta Peloponnesia* was already in circulation by then.

> Another bout of antiquarian rage (*un' altra collera antiquaria*) hit me this week. Two volumes on Greek antiquities in *quarto* by the Jesuit

---

69. Ridley 1992, 362–75.

70. Paciaudi et al. 1802, 260: letter from Naples dated September 12, 1761.

71. Paciaudi may have benefitted from a years-long relationship with Paderni: the archives of Paciaudi's correspondence includes a letter from Paderni dating to May 5, 1757; see L. Farninelli, ed., *Paolo Maria Paciudi e i suoi corrispondenti* (Parma 1985) 151.

FIGURE 1.11 Bernardo Tanucci (1698–1783), prime minister of the Bourbon court of Naples. Portrait from 1787 by Nicola Matraini in the Palazzo alla Giornata, Pisa. https://commons.wikimedia.org/wiki/File:Nicola_matraini,_ritratto_di_bernardo_ tanucci,_1787_(pisa,_palazzo_alla_giornata).jpg under a Creative Commons Attribution 3.0 unprotected (CC BY 3.0).

Paciaudi, friend of Cardinal Spinelli and a preacher, a native of the Ecclesiastical State, who should not be totally unknown to Your Majesty. Leafing through it I came upon some engravings of objects from Herculaneum, among them the very rare sundial which forms the Preface of the third volume [of *Le Antichità*, as yet unpublished].

THE PROSCIUTTO SUNDIAL

He was not well served by the drawing and he made some blunders in his explanation. I tried to get Pasquale [Carcani, secretary of the Accademia Ercolanese] angry too, and we will get our revenge, but it will have to be done by adding a note to the Preface and therefore one folio [of the planned volume] will have to be withdrawn.[72]

Tanucci's anger with Paciaudi simmered for months, with attacks on him recurring in letters to Ferdinando Galiani, the ambassador to France from the Kingdom of Naples, well into 1762. Paciaudi's *Monumenta* had forced the Accademia to alter significantly the contents of the third volume of *Le Antichità di Ercolano Esposte*, the formal publication of the antiquities from the excavations, which was already well along in production. This, in turn, delayed the volume's publication, the timely appearance of which Tanucci considered one of his official obligations to Charles III. Most galling was Paciaudi's appointment as chief antiquarian to Philip of Bourbon, duke of Parma and Charles III's younger brother, soon after the appearance of the *Monumenta*.[73] Tanucci complained that he could not fathom how this Jesuit could be considered an antiquarian when his volumes had demonstrated he was ignorant and overpaid, his grammar was poor, and he was a brazen plagiarist who did not credit his sources, like Martial's Fidentius (Martial *Ep.* 1.29).[74]

But in the midst of his tirades Tanucci had extracted some revenge in rejecting Paciaudi's request for another visit to the museum to see the most recent discoveries. Paciaudi explained the circumstances in a letter to Caylus, though he appears oblivious to the true cause:

---

72. B. Tanucci, *Epistolario*, ed. M. G. Maiorini (Rome 1985) 9.871–72 #706 (July 28, 1761). Tanucci mentions Paciaudi's relationship to Cardinal Spinelli, archbishop of Naples (1734–1754), because King Charles had forced him out of the diocese in 1754 after Spinelli's attempt to introduce the Inquisition to Naples had led to violent uprisings.

73. Philip had initiated excavations in 1760 at the Roman city of Veleia, his own "Herculaneum," and Paciaudi was put in charge of the investigations, in addition to serving as court librarian; on this work see C. Nisard, *Correspondance inédite du Comte du Caylus avec le P. Paciaudi, Théatin* (Paris 1877) lxxxiii–lxxxv.

74. B. Tanucci, *Epistolario*, ed. M. G. Maiorini (Rome 1988) 10.35 #27 (August 22, 1761); 10.87 #63 (September 5, 1761); 10.178 #140 (October 3, 1761); 10.437 #394 (January 3, 1762), all addressed to Galiani in Paris. Other aspects of Paciaudi's *Monumenta* annoyed Tanucci as well. For example, Paciaudi had referred (236 n. 1) to Caylus' "most wise" discussion of the famous Hercules in Arcadia painting from Herculaneum (MANN 9008), published in the first volume of *Le Antichità* (*PdE* 1.27–30, pl. 6), and then offered an interpretation of the scene that differed from the Accademia's. Martial had accused Fidentius of reciting Martial's poems as his own.

Discovery 47

The Marquis Tanucci has forbidden me to be given entrance to the Royal Museum. And for what reason? Listen. Several remarkable things have been found in Stabia (Castellamare) and Pompeii. The jealous Academicians told the Marquis that I examine everything with accuracy, that I take note of everything, and that, having literary correspondence and having to travel, I publicized these antiquities before they were clarified and published by the Accademia. I could not prevent myself from doing so and replied that the motive for such a refusal was very flattering to me, for they confessed that I had more knowledge of this kind than they themselves. Don't say a word of this to abbé [Ferdinando] Galiani: he's Tanucci's spy.[75]

Even if Paciaudi did not have a hand in the first unauthorized description of the sundial published in the *Memorie*, the limited facts, the common errors and misconceptions, and the similar images, both mental and physical, about the prosciutto's design circulating via the published sources highlight how effective the Bourbons' constraints on scholars had been in limiting the accuracy of the information that was transmitted. It was precisely because of Paciaudi's associations with the network of scholars known by the Accademia Ercolanese to be trafficking in both information and antiquities that made it possible for them to suspect Paciaudi as, at the very least, d'Alembert's indirect source.

Yet it is odd that the connections among French, British, and Italian antiquarians in this period did not result in more accurate details about the prosciutto sundial being disseminated more quickly to resolve the glaring errors in these published descriptions. Paderni had been elected a member of the Royal Society of London in 1755, shortly before his letter on the prosciutto was read into the record. His election was in recognition of "the frequent accounts sent by him, and printed in the *Philosophical Transactions*, relating to the late Discoveries of Antiquities at Herculaneum, & by his Readiness to oblige our Countrymen, who visit the Kingdom of Naples."[76] Paderni's

---

75. Paciaudi et al. 1802, 269–70: letter from Naples dated November 3, 1761.

76. Royal Society of London, *Archives EC/1755/03*: Recommendation for election of Paderni to the Society by members Anthony Askew, Robert More, and John Ward. Sir James Gray was not a member. Paderni's election came in the same year as that of Bayardi, the only other figure closely attached to the Bourbon excavations to be elected to the Royal Society in the eighteenth century; see M. B. Hall, "The Royal Society and Italy, 1667–1795," *Notes and Records of the Royal Society of London* 37.1 (August 1982) 63–81. The president of the Royal Society in 1755 was George Parker, Second Earl of Macclesfield (1696–1764), an astronomer who had built an observatory at Shirburn Castle, his home, together with the astronomer James Bradley,

THE PROSCIUTTO SUNDIAL

earliest letters on the excavations at Herculaneum had been to his friend Alan Ramsay, the Scottish painter, whom he met through their teacher, the Roman painter Francesco Fernando d'Imperiali.[77] Paderni had profited from d'Imperiali's ties to the circle of English artists and scholars living in, and passing through, Rome and Naples long before he had any official role in the Neapolitan court, and he had continued to nurture those ties thereafter.[78] If an "English gentleman" really had been the source behind the *Memorie*'s and Paciaudi's contributions, this would confirm Paderni's reputation for accommodating scholars of that nationality.[79] D'Alembert, a member of the French Académie des Sciences, was also a member of the Royal Society, since 1748, but there is no evidence he was aware of Paderni's letter to the Society or had communicated in any way with an Englishman. Nor was d'Alembert a known correspondent of Paciaudi, whose principal French contacts, the Comte de Caylus and the abbé Barthélemy, were instead members of the Académie des Inscriptions et Belles-Lettres, of which Paciaudi would become a member only in 1769. In this case, the academies did not foster the exchange of information. Moreover, while the personal correspondence between Caylus and Paciaudi, Barthélemy and Caylus, and Barthélemy and Paciaudi was extensive, there is no evidence they had passed any information about the

and had been instrumental in introducing to the British Isles the change to the Gregorian calendar in 1752.

77. Zanni 1993, 65–77.

78. Among Paderni's early works were drawings for George Turnbull's *A Treatise on Ancient Painting* (London 1740), and drawings that were engraved for Joseph Spence's *Polymetis, or, An Enquiry Concerning the Agreement between the Works of the Roman Poets, and the Remains of the Antient Artists, Being an Attempt to Illustrate Them Mutually from One Another: In Ten Books* (London 1747). Another of d'Imperiali's students was James Russel, the English painter who lived in Rome from 1740 to 1763. Russel was responsible for discovering that the prohibition against taking notes in the museum "was occasioned by Camillo Paderni, who was thereupon immediately sent packing" (De Vos 1993, 104–105). Joseph Spence, a history professor at Oxford, received reports from Paderni of new discoveries, including one dated December 7, 1753, that "workmen were then just entering on some nobleman's house, as appeared by the rich mosaic pavements, etc. and that they were in hopes it would prove a very good new mine," a reference to the start of excavations in the Villa dei Papiri (J. Spence, "Extract of a Letter of the Reverend Mr. Joseph Spence, Professor of Modern History in the University of Oxford, to Dr. Mead, F. R. S. (December 7, 1753)" *PhilTrans* 48.48 (1754) 486). On Paderni's work for Turnbull, see D. Burlot, *Fabriquer l'antique: Les contrefaçons de peinture murale antique au XVIIIe siècle* (Naples 2012a) 53–56; and for Paderni's work as a restorer in general, see D. Burlot, "Le Faux: Une tentation d' expert? Le cas de Camillo Paderni (1715–1781), artiste, restaurateur et antiquaire napolitain," in N. Étienne, ed., *L'Histoire à l'Atelier: Restaurer les oeuvres d'art, XVIIIe-XXIe siècles* (Lyon 2012b) 41–62.

79. Similarly, Winckelmann 1762, #34 (Mattusch 2011, 87), to whom Paderni had given wide access to the finds in the museum at Portici, disputed the claims of certain English gentlemen that they had been allowed to see the marble statue of Pan copulating with a she-goat (MANN 27709), famously locked away from public view.

sundial to Paciaudi, who doubtless would have been surprised to learn that they were in possession of the kind of basic information about the prosciutto that would have spared him from humiliating errors.

1.f) Les tablettes vides et la mémoire pleine: *The Accounts of Barthélemy and La Condamine*

The abbé Jean-Jacques Barthélemy had become keeper of the Cabinet des Médailles in Paris in 1753 and, in August 1755, had traveled to Italy at the invitation of the French ambassador to Rome, Étienne-François de Choiseul, marquis de Stainville, and he remained there until 1757 (Figure 1.12). Throughout this period, he sent letters to Caylus documenting his experiences and, in June 1756, reported meeting Paciaudi in Rome for the first time.[80] As a numismatist Barthélemy's primary interest was to acquire additional coins, and as part of his inquiries he made several excursions to Naples and Portici between December 1755 and April 1756. In January, he wrote to Caylus from Naples concerning his visits to the antiquities at Portici:

> All I can assure you in general is that I have been to Portici four or five times, that I have always spent several hours in a row, and that I would need to go back many more times to check what I saw. What I can add still is that this collection has surpassed my expectations, and that I have testified so to the King who has done me the honor to ask me my opinion.[81]

In his next letter, written on his return to Rome, he underscored the difficulty of providing a detailed catalog of what he had seen:

> Recall that Portici, where these antiquities are preserved, is four miles from Naples; that in this respectable sanctuary it is only permissible to satisfy one's sight, and that one returns to Naples with the notebooks blank and the memory full (*les tablettes vides et la mémoire pleine*)... I have been to Portici several times to check what I have seen, to examine what I have not seen.[82]

---

80. These were later edited and published as the *Voyage en Italie de M. l'abbé Barthélemy... ecrites au Comte du Caylus,* A. Serieys, ed. (Paris 1801). He continued to correspond with both Caylus and Paciaudi after his return to France.

81. Barthélemy 1801, 64: Letter 11 from Naples, January 1, 1756.

82. Barthélemy 1801, 76–77: Letter 13 from Rome, February 2, 1756.

FIGURE 1.12 The French numismatist and epigrapher Jean-Jacques Barthélemy (1716–1795) in a lithograph by Francois Seraphin Delpech (1778–1825) in the Wien Museum Inv.-Nr. W 448, Creative Commons CC0 https://sammlung.wienmuseum.at/en/object/246548/.

He then summarized for Caylus the variety of antiquities housed at Portici, beginning with the charred papyrus scrolls and moving on to paintings, statues in bronze and marble both large and small, and the artifacts of daily life like carbonized bread, paint pigments, surgical instruments, cooking utensils, bronze tripods, and jewelry, offering details on only a few select items. He concluded his paragraph that cataloged the small finds in silver with a description of an object that had clearly caught his eye:

Discovery 51

A small ham-shaped sundial, whose tail serves as a style, at the bottom of the divisions are plotted the names of the months, in this order
IV. MA. AP. MA. FE. IAN.
IV. AV. SE. OC. NO. DEC.

Barthélemy had studied astronomy as a young man, so he must have had some idea of the function of the "divisions" of the dial even if he did not provide details about its design or how it functioned.[83] As an epigrapher offering a first-hand transcription of the inscription, Barthélemy too fell victim to the Bourbon restrictions and erroneously abbreviated Ianuarius and December with three rather than two letters and Iunius as only two letters and not three.[84] Evidently in this case he had not employed effectively the creative solution he had adopted to circumvent the prohibition against note taking when he sought to produce an accurate transcription of twenty-eight lines of text from one of the papyrus scrolls that Paderni had allowed him to examine. He later described the process in his *Memoires*:

> I read them five or six times, and, under the pretext of nature calling, I went down into the courtyard, and traced them on a piece of paper, preserving as best as I could the arrangement and form of the letters. I mentally compared my copy with the original and found a way to rectify two or three small errors that had escaped me.[85]

---

83. Barthélemy 1801, 84–85: Letter 13 from Rome, February 2, 1756.

84. Barthélemy's early publications included an epigraphical treatise on the script and texts of Palmyra (*Reflexions sur l'Alphabet et sur la Langue dont on se servoit autrefois a Palmyra*, Paris 1754), and his writings on the material remains from Herculaneum and Pompeii often dwelled on the aspects of Greek and Latin script these illustrated.

85. J. J. Barthélemy, *Memoires sur la vie et sur quelques-uns des ouvrage de JJ Barthélemy* (Paris 1830) 55. La Condamine 1755–1756, 623, attested to the effectiveness of Barthélemy's time-consuming approach in a letter to the Royal Society dated March 11, 1756: "The Abbe Barthélemy has read well a page [of one of the papyri], except for a few words, which he had not time to study." On the other hand, Goethe, who visited the Portici museum on March 18, 1787, remarked on the positive aspect of being liberated from the pressures of taking notes or drawings in the museum: "We entered the museum well recommended, and were well received: nevertheless we were not allowed to take any drawings. Perhaps on this account we paid the more attention to what we saw, and the more vividly transported ourselves into those long-passed times, when all these things surrounded their living owners, and ministered to the use and enjoyment of life... In the hope of being able to pay a second visit, we followed the usher from room to room, and snatched all the delight and instruction that was possible from a cursory view" (J. W. von Goethe, *Travels in Italy, France, and Switzerland*, trans. R. D. Boylan, Boston 1883, 258–59). Goethe does not mention the prosciutto sundial.

He correctly recalled, however, that the prosciutto's inscription was not framed by the dial's grid in any way. Nevertheless, as a consequence of Barthélemy's errors, an accurate transcription of what was inscribed on the sundial was still not available. Moreover, only Caylus would have been privy to this limited information about the sundial, and communicating it to others would have fallen to him: Barthélemy's letters to Caylus during his time in Italy remained private until they were edited and published collectively in 1801 as the *Voyage en Italie de M. l' abbé Barthélemy... ecrites au Comte du Caylus.*[86]

Barthélemy was among the antiquarians supplying Caylus with material for his *Recueil d'Antiquités Égyptiennes, Étrusques, Grecques, Romaines et Galouises*, a seven-volume work published between 1752 and 1767 that sought to place the study of ancient art and artifacts on a more scientific footing. As a member of both the Académie Royale de Peinture et de Sculpture and the Académie des Inscriptions et Belles Lettres, Caylus' scholarship merged the aesthetics of the former with the historical and literary interests of the latter in the analysis of archaeological material. Caylus was interested in particular in the techniques of manufacture, in deciphering the underlying intellectual processes of the artist, and in illustrating how artifacts could be employed to illustrate aspects of ancient culture and society. In his first volume of the *Recueil* in 1751, Caylus had published two painting fragments in his possession that he believed came from Herculaneum. His ongoing efforts to build his collection of antiquities also drew him into a controversy over ancient paintings said to have come from Herculaneum and Rome that were actually counterfeits produced by an artist in Rome, Giuseppe Guerra.[87] Barthélemy had served as one of Caylus' agents in Italy, obtaining for him at least one of Guerra's paintings, while Paciaudi also sent him paintings and antiquities he had acquired in and around Rome throughout the late 1750s and early 1760s.[88]

---

86. Barthélemy 1801, 84–85: Letter 13 from Rome, February 2, 1756.

87. Caylus 1752–1767, I.149–51, pl. 55.1: a fragment of a heroic warrior with a portion of another, though Caylus does not describe the scene, focusing instead on the color scheme and comparing it to the Aldobrandini Wedding frieze in the Vatican museum (Cat. 79631); and Caylus 1752–1767, I.152, pl. 56.1: a winged Amor armed with a spear ("amours à la chaise"), a fragment Caylus claims was given to him by the French architect Jacques-Germain Soufflot, who had visited the ruins with Abel-François Poisson de Vandières, marquis de Marigny. Paciaudi et al. 1802, 254 (August 1, 1761): Paciaudi told Caylus that the Neapolitan court was angry that Caylus was in possession of antiquities from Herculaneum, believing, or wanting to believe, that theft was impossible, and advised Caylus to assuage Tanucci's anger by praising him in his next volume of the *Recueil*. See also E. Pagano, "Caylus e le pitture ercolanesi," *Anabases* 6 (2007) 113–34; Parente 2007, 17–23.

88. Barthélemy 1801, 92–97: Letter of Barthélemy to Caylus, from Rome (February 10, 1756), discusses the paintings. See also Burlot 2012a, 75–80; and Griggs 2011, 471–503.

FIGURE 1.13 The French astronomer Charles-Marie de La Condamine (1701–1774). Pastel on paper by Maurice La Tour (1753). Frick Art Museum, Pittsburgh PA (1970.40). Image courtesy of Frick Art and Historical Center, Pittsburgh.

Another member of this circle of academicians was Charles Marie de La Condamine. La Condamine (1701–1774) is most famous for his participation in the 1735 expedition to Peru to measure the length of a meridian arc at the equator (Figure 1.13). A skilled astronomer and scientist, he was a member of both the Académie Royale des Sciences (1730) and, since 1748, the Royal Society of London. He was also a contributor to Diderot and d'Alembert's *Encyclopédie*. In late 1754, La Condamine traveled to Italy, eventually crossing paths with Barthélemy in Rome and, like Barthélemy, corresponding with Caylus throughout his time there.[89] On April 20, 1757, he submitted his "Extrait d'un Journal de Voyage en Italie"

---

89. La Condamine 1755–1756, 622–23: Letter of La Condamine to Dr. Maty, from Rome (March 11, 1756), "The Abbé Barthelemi [*sic*], who is here, has been at Naples." Barthélemy 1801, 98–107: Letter of La Condamine to Caylus from Rome (February 17, 1756), discusses Guerra's paintings. Barthélemy likely learned of the Guerra incident from Piaggio. For La Condamine, see *Biographical Encyclopedia of Astronomers* (New York 2014) 538–39, s.v. *de la Condamine, Charles-Marie* (A. Ten). La Condamine had corresponded with Barthélemy about acquiring paintings in Rome for the Comte de Caylus; see Barthélemy 1801, 98–107: Letter of La Condamine to Caylus, from Rome (February 17, 1756); and, in general, Burlot 2012a, 65–70.

54 THE PROSCIUTTO SUNDIAL

to the Académie Royale des Sciences. For various reasons, publication of that year's volume of the *Histoire de L'Académie Royale des Sciences*, in which his report was to appear, was delayed until 1762, although knowledge of the report's contents must have been circulating in French scientific circles already in 1757.[90]

During the course of his travels, La Condamine had arrived in Naples in the spring of 1755 and had gone immediately to Herculaneum. He reported on the charred papyrus scrolls and furniture, the glass vessels and window glazing, and the engraved gemstones and the rings in the museum at Portici before turning to the striking prevalence of obscenity in domestic works of art evidenced by the finds from the Vesuvian cities. As a particularly flagrant example, he stated that,

> I was present just as they brought into the cabinet of antiquities at Portici a bronze tripod which had just been discovered and surpassed all those that had been found previously: it was as remarkable for the beauty of its craftsmanship as for the impudence of the three figures of satyrs that supported the brazier.[91]

The reference is to the famous bronze tripod supported by three ithyphallic satyrs recovered intact in the closet-sized shrine (*sacrarium*) painted with images of Egyptian and Roman deities from the *Praedia Iuliae Felicis* in Pompeii (Regio II.iv.1–12) (Figure 1.14). As was the case with the prosciutto, the discovery of this magnificent object was thoroughly documented in the Bourbon-era records. Paderni inventoried the tripod's arrival at the museum on June 13, 1755, two days after he had accessioned the sundial.[92] It was entered into the weekly inventory of finds from the royal excavations prepared by the

---

90. La Condamine 1762, 336–410. D'Alembert had two contributions in this same volume of the *Memoires*, both relating to an article by the astronomer abbé Nicolas-Louis de la Caille entitled "Mémoire sur la Théorie du Soleil"; see Chapter 2.a. The volume also contained contributions by another French astronomer, Joseph Jérôme Lefrançais de Lalande, who would later publish his *Voyage d'un François en Italie, fait dans les années 1765 & 1766* (Paris 1769); see Chapter 2.a. All three, along with Barthélemy, were also members of the Royal Society of London: *Archives EC/1748/13* (d'Alembert, 1748); *EC/1755/09* (Barthélemy, 1755); *EC/1759/12* (de la Caille, 1759); *EC/1763/12* (Lalande, 1763). An English edition of La Condamine's "*Extrait*…" was published in London and Dublin in 1763 as *Journal of a Tour to Italy*, 96–107 (for Herculaneum).

91. La Condamine 1762, 369–70.

92. Paderni *Diario*, June 13, 1755; for the context and history, see Parslow 2013, 47–72. This is the same tripod Bayardi had referenced in his *Catalogo* at entry 709.

FIGURE 1.14 Bronze tripod adorned with ithyphallic satyrs. Found on June 13, 1755, in the *Praedia Iuliae Felicis* (Regio II.iv.1–12), Pompeii. 0.88 m high. MANN 27874. Su concessione del Ministero della Cultura—Museo Archeologico Nazionale di Napoli.

excavators for the prime minister for the week ending June 15, 1755.[93] Finally, Paderni described the tripod in the same letter of June 28, 1755, to the Royal Society of London in which he discussed the prosciutto sundial and specifically states that the tripod had been discovered on June 13. La Condamine therefore must have been in the museum on June 13, or very soon thereafter,

---

93. *PAH*, 20–23 (June 15, 1755).

56 THE PROSCIUTTO SUNDIAL

when the tripod had just been extracted from the pumice and ashes covering Pompeii and transported straight to Paderni in the museum at Portici. This would be roughly eight months before Barthélemy's visit.

This, in turn, provides the context for what he described subsequently, for at the time he witnessed the tripod being handed over to Paderni, La Condamine was examining the prosciutto sundial, itself recovered from Herculaneum and cleaned by Paderni only two days earlier. The brief observations in his travel journal therefore constitute chronologically the second-oldest written description after Paderni's diary entry, though they would not be published until seven years later.

> I was at the very moment occupied in considering a monument of another kind: a small silver ham, weighing two or three ounces, on which was traced a sundial or the horary lines, the numbers marking the hours, and the initial letters of the names of the months distinctly engraved; the animal's tail, since the ham represented a thigh, served as the style for the dial. I had neither the liberty nor the ability to examine for which latitude the dial had been made, which would have been difficult to determine exactly, because the small size of the ray would not permit great precision in the angles. We can better judge when all the monuments discovered at Herculaneum, which are to be engraved and described, will be published; the drawings were well advanced in 1755.[94]

It is a remarkable coincidence that one of the great French astronomers of the day arrived in Portici so soon after the sundial's discovery and was among the first foreign scholars to examine it and offer a description of it. Earlier in his travels, La Condamine had inspected the derelict meridian of 1510 in the cathedral of Santa Maria del Fiore in Florence together with the local astronomer Leonardo Ximenes and offered advice that had led to its restoration soon thereafter.[95] In Portici, while his reputation no doubt facilitated his access to the sundial, La Condamine's comments make it clear that this access

---

94. La Condamine 1755–1756, 370.

95. La Condamine 1755–1756, 350–52. La Condamine had compared the Florentine meridian, designed by Paolo dal Pozzo Toscanelli in 1468, to that in the Basilica of St. Petronio in Bologna, designed in 1655 by Giovanni Cassini, in order to emphasize that the Florentine example was larger. Ximenes, who oversaw the renovation of the meridian, was a Jesuit priest, mathematician, geographer, and astronomer for whom the Osservatorio Ximeniano in Florence was named.

had been as limited as Barthélemy's and any other scholar's: the Bourbon restrictions effectively thwarted even a professional astronomer. Nevertheless, the description he rendered from memory was more detailed and accurate than any by an outsider of the Bourbon court from this period, even if a few inaccuracies crept in. His overall description was spot on: the instrument bears the shape of a ham, which he later specifies as the thigh, with the tail functioning as the gnomon, although he did not note that the gnomon was broken, which is good evidence it was missing already when it was found. He states that the names of the months are abbreviated but does not offer a transcription and does not mention the *boustrophedon* arrangement. Evidently the silver plating was sufficiently striking for him to believe it was constructed entirely of silver, not bronze. His estimate of its weight at two or three French *onces* is inaccurate since this would amount to only between 60 and 90 grams and it is in fact closer to 285 grams.[96] While he correctly identifies the "horary lines," he was incorrect in recalling that the hours were numbered on the dial. But La Condamine was particularly curious to determine the latitude for which the sundial had been designed, a detail crucial to understanding how the sundial functioned that he was the first to consider. This would have required the execution of mathematical computations for which he lamented not having had the time or the *liberté*, which here, in the context of the Bourbon court, must mean "permission," but he hypothesized that the small scale of the dial would diminish the instrument's accuracy. Ultimately La Condamine had consoled himself with the expectation that his questions would be resolved by the Accademia's formal publication of the finds, a process he noted was underway at the time of his visit.

1.g) Prosciutto Mania: Mazzocchi, Martorelli, and Piaggio's L'Orologio Solare

La Condamine likely was unaware of the growing tension within the museum caused by the discovery of the sundial. Its arrival had exacerbated long-festering hostilities between members of the Bourbon court that may only have been surpassed by the discord ignited earlier by the recovery and handling of the carbonized scrolls from the Villa dei Papiri. In both cases, the chronicler of the resulting disputes had been Antonio Piaggio (1713–1796), the Jesuit priest brought from Rome in 1753 to tackle the task of unrolling and copying the text on the scrolls (Figure 1.15). A skilled calligrapher and artist,

---

96. The French *once* before the Revolution in 1789 was the equivalent of about 30.59 grams.

FIGURE 1.15   Antonio Piaggio (1713–1796). Copy of portrait currently in the Officina dei Papiri, Biblioteca Nazionale di Napoli. https://commons.wikimedia.org/wiki/File:Padre_Antonio_Piaggio.jpg

Piaggio had come on the recommendation of Giuseppe Assemani, the librarian of the Biblioteca Vaticana, over the strenuous objections of Paderni.[97] This had set the tone for their relationship for the next forty years, despite some early, superficial appearances of cooperation.[98] From his table in the royal

---

97. A. Travaglione, "Padre Antonio Piaggio: Frammenti biografici," in M. Capasso, ed., *Bicentennario della Morte di Antionio Piaggio* (Naples 1997) 15–48. Piaggio related that Paderni had claimed the papyri could not be unrolled; see D. Bassi, "Il P. Antonio Piaggio e i primi tentativi per lo svolgimento dei papiri ercolanesi," *Archivio Storico per le Province Napoletane* 32 (1907) 666. Piaggio had inked a magnificently ornate letter of congratulations from Charles III to King Othman III on his ascension to the Turkish throne on October 23, 1755; see M. Capasso, "Un omaggio dei Borboni al Padre Piaggio," in M. Gigante, ed., *Contributi alla Storia della Officina dei Papiri Ercolanesi*, vol. 1 (Rome 1980) 63–69.

98. Travaglione 1997, 26: Paderni had drawn and Piaggio had engraved the vignettes adorning the frontispiece, the title, and the end page of the Preface, and the final page of Bayardi's *Catalogo* in 1754, a project Travaglione characterized as "not as a moment of relaxation but as an excuse that hid a different motivation." Piaggio is listed together with Paderni in G. G. Gandellini's *Notizie Istoriche degli Intagliatori* vol. 3 (Siena 1808) 41, s.v. *Piaggio*, "intaglio

Discovery 59

museum at Portici and through his interactions with foreign visitors, Piaggio observed, participated in, and at times even baited the partisan politics, shifting jealousies, and jostling for preeminence among Paderni, Mazzocchi, Martorelli, and even Weber. His eyewitness accounts on a number of the key discoveries and long-standing disputes in this period survive in a collection of detailed diaries and personal correspondence with acquaintances, most written down in his final years and all unpublished during his lifetime, though it seems doubtful he would have been able to repress retailing these anecdotes to anyone with the patience to listen. In the case of the papyri, he leveled his scorn principally at Paderni, both for claiming to have been the first to identify the scrolls when Piaggio knew it had been Weber down in the tunnels, and then for destroying a number of them in his ignorance.[99] He certainly had bent the ear of Winckelmann, to whom he had given room and board during his visits in 1758 and 1762. Winckelmann, who considered Piaggio the "world's greatest gentleman," had been taken in by his knowledge and unique perspective on both the papyri and the museum's internal affairs and, as a result, some of Piaggio's critical observations had found their way into Winckelmann's own published correspondence.[100]

Piaggio recorded his account of the sundial's discovery and early days in the museum in a manuscript entitled *L'Orologio Solare del Museo Ercolan(e)se* ("The Sundial of the Herculaneum Museum"), a 179-page diatribe in the form of a letter, though its style is more that of a descriptive narrative characterized by long digressions. Its intended recipient is not named, but the manuscript was bound together with, and cross-references specific matters discussed in, four other narratives, the first of which alone was addressed to Giuseppe

---

ad acqua forte nel 1755 dall' invenzione di Camillo Paderni Romano un rametto apposto nel frontispizio del libro intitolato: [Bayardi] *Catalogo . . .*"

99. Piaggio addressed his *Memorie del Padre Ant(oni)o Piaggi* [sic] *impiegato nel R(eale) Museo di Portici relative alle antichità, e Papiri p(er) anno (sic) 1769. 1771*, written between 1769 and 1771, to Count Guglielmo Maurizio Ludolf (1747–1789). The manuscript is in the library of the Società Napoletana di Storia Patria, Ms. XXXI.C.21. Substantial portions were published in Bassi 1907, 636–90, and Bassi 1908, 277–332. On Piaggio, see also A. Travaglione, "Testimonianze su Padre Piaggio," in *Epicuro e l'Epicureismo nei Papiri Ercolanesi* (Naples 1993) 53–80; F. Longo Auricchio, "La figura del P. Antonio Piaggio nel Carteggio Martorelli-Vargas," in M. Gigante, ed., *Contributi alla Storia della Officina dei Papiri Ercolanesi*, vol. 2 (Rome 1986) 17–23; and F. Longo Auricchio and M. Capasso, "Nuove accessioni al dossier Piaggio," in *Contributi alla storia della Officina dei Papiri Ercolanesi* (Naples 1980) 17–59.

100. For Piaggio and Winckelmann, see K. Justi, *Winckelmann und seine Zeitgenossen*, 3 vols. (Leipzig 1923) 2.202–16. On the other hand, the prime minister, Tanucci, called him "essentially the slowest servant of the monarchy" (B. Tanucci, *Epistolario*, ed. M. G. Maiorini, Rome 1985, 9.356 #257, February 3, 1761).

THE PROSCIUTTO SUNDIAL

Melchiorre Vairo, a professor of chemistry at the Accademia Medico-Chirurgica in Naples.[101] It is also the only one lacking a date, but the three bound before it bear dates between July 1790 and October 1791, and the one following is from November 1793, so it likely was written sometime between 1791 and 1793.[102] Piaggio was in his late seventies at the time and died in 1796, meaning the events he recounted in this manuscript had taken place roughly thirty-seven years earlier. Some lapses of memory doubtless crept in, he clearly erred entirely on certain aspects, and he doubtless had conflated his own rec-ollections with his familiarity with the results of later scientific analysis of the prosciutto. Yet the vividness of his recollections, both here and in the other manuscripts in this collection, ranging from small details to direct quotations from his principal characters, lend credence to their overall accuracy. Nevertheless, his writings as a whole leave little doubt that he would have seized any opportunity to elevate himself and his contributions above those around him, especially Paderni. In the case of the prosciutto sundial, Piaggio felt humiliated and robbed of his due recognition for its initial identification and appalled by the ignorance of the key players in the drama that had unfolded before him. That he had taken liberties with and exaggerated some of the details seems certain.

Piaggio's narrative is especially noteworthy for the way in which it lays bare the interactions among the individual personalities within the court, their idiosyncrasies, and the intellectual tools they wielded in their responses to the discovery and explication of this singular object. It raises questions

---

101. Vairo was a member of Ferdinand IV's Reale Accademia Napoletana, founded in 1780. His interest in geological phenomena would have made him a natural colleague of Piaggio, who at this time was keeping a detailed journal of observations he made of volcanic activity on Vesuvius for William Hamilton; see C. Knight, "Un inedito di Padre Piaggio: Il Diario Vesuviano (1779–1795)," *Rendiconti dell'Accademia di Archeologia Lettere e Belle Arti di Napoli* (1989–1990) 59–131. For Vairo, see R. Di Castiglione, *La Massoneria nelle Due Sicilie e i "Fratelli" Meridionale del '700* (Naples 2008) 386–87, and E. Choisi, "'Humanitates' e Scienze: La Reale Accademia napoletana di Ferdinando IV, Storia di un progetto," *Studi Storici* 30.2 (1989) 441.

102. Biblioteca Nazionale di Napoli, Ms. IX.F.51. The five documents are entitled: (1) *I Piccioni*, addressed to Giuseppe Vairo, July 16, 1790; (2) *L'Iscrizione della Porta della Città di Pompeja*, October 16, 1791; (3) *Le Tavole di Eraclea, seque dall' Inscrizione della Porta della Città di Pompeja*, Resina, October 19, 1771; (4) *L'Orologio solare del Museo Ercolanese*; (5) *Candelabri, Caccia di Persano, Funerale*, Resina, November 21, 1793. M. G. Mansi, "Di alcune fonti sette-centesche per l'archeologia e l'antiquaria in epoca borbonica nei fondi manoscritti della Biblioteca Nazionale di Napoli," in P. Giulierini, A. Coralini, and E. Calandra, eds., *Miniere della Memoria: Scavi in archivi, depositi, e biblioteche* (Sesto Fiorentino 2020) 238–39, notes that the manuscript came into the possession of the Biblioteca in 1855 from the collection of Francesco Maria Avellino (1788–1850), a member of the Accademia Ercolanese who oversaw the excavations at Pompeii from 1839 to his death.

Discovery

about many of the details that can be divined from the published sources such as who knew what about the sundial, when they knew it, and how they came to know it. If the latter part of his manuscript devolves into rambling digressions, the initial portion, presented largely in the form of dialogue that evolved over an unclear period of time, merits detailed scrutiny for the unique perspective and information it provides.

After forty days of home convalescence due to a bout of rheumatism, Piaggio had been prodded to return to the museum when his former teacher, Padre Andrea Sauli, arrived in Portici from Genova and asked to be shown "something special in the museum." Sauli, like Piaggio, was a Jesuit, considered "one of the most learned men in all the sciences."[103] Consequently, Piaggio felt duty bound to oblige him, but doing so would require him to seek this favor from Paderni, who was the only one authorized to present the museum's objects.

> A more odious duty he could not have asked of me, but he had been my teacher, my director, my superior, in sum he was one of the most distinguished characters in my religion. I had to gather myself together and go ask Don Camillo [Paderni], the custodian, that he kindly attend to him in person. He received me in good manner after so long a time that we had not seen one another, but he began to detain me concerning the tripod of fauns found at Gragnano, raising many aspects concerning it, but I begged him to keep this for another time, urging him to quickly serve the Padre.

Although Piaggio identified its provenance as modern Gragnano, and would later ascribe it to ancient Stabiae, Paderni clearly was talking about the satyr tripod from Pompeii, the very same tripod La Condamine had witnessed being brought into the museum (see Figure 1.14). This reference to the tripod helps situate the discussion in question at least two days after the prosciutto sundial was found, and close enough to the date of the tripod's discovery that it was still an object of curiosity and fascination.

Putting aside the tripod, Paderni turned Piaggio's attention to another object:

---

103. D. Gasparini and M. Peloso, *Le istituzioni scolastiche a Genova nel Settecento* (Genova 1995) 207. Little else is known about him. Piaggio spelled his name "Saoli" throughout the manuscript, but the identification is certain.

"And I want to show you something else and hear your thoughts, and this for me too must be quick—before you leave—because it is something that weighs much on me." Saying this he led me toward a window and showed me the sundial made on a prosciutto which is designed on a quadrant...[104] After having examined it briefly, "This," I said to him, "is a sundial." He began to laugh and said, "You're wrong because sundials are fixed to walls." "And those," I replied, "are called 'vertical sundials' because they are turned toward and located according to certain hours of sunlight; horizontal ones, on the other hand, stand on a level plane; but this one you have shown me is called a 'portable vertical sundial' and they are made in many ways." "But how do you know this?" he asked, and I replied, "If I didn't know it, I wouldn't talk about it. I know it because I have studied it and there are some in my house since there are those who have studied this before me, and to prove it I'll get you either the copies or the originals." "This I'd like to see," he returned to say, "but look, these gentlemen don't see it this way." "And who are these gentlemen?" I inquired. "Bayardi," he answered, "the Canon [Mazzocchi], and [Nicola] Ignarra, and everybody who has seen it, and from all of them I've gotten nothing but endless conjectures. One says it's a votive tablet, another says it's an abraxas, another that they are magical characters, another that it was made for sacrifices to Venus in memory of the boar that killed Adonis, another for Mars, another in memory of the gluttony of the ancient Romans who did not hold dinner parties if there was no boar."

"And I tell you," I concluded then, "that you tell all these gentlemen that I say this is a sundial and that the design it has [of a prosciutto] is simple caprice because [the dial] has the shape of a quadrant, as I said. Now, to finish, let's do this. This Padre whom I commend to you is most skilled in all mathematical matters. I won't return home, we'll send for him, and I won't speak further with him. Let him see this just as suddenly as you did with me and see what he says." So we stayed put, and after Padre Sauli arrived, as soon as he saw it and was asked his opinion he said what I had said, and pointing to the tail of the prosciutto he said, "This was the gnomon, it is crooked and missing. Concerning its crookedness, this was done

---

104. The addressee of Piaggio's manuscript clearly was familiar with the sundial, so Piaggio accordingly described it only briefly in this way.

according to nature and ingeniously: naturally, because if it was straight as are other gnomons, or styles, it would be deformed, since only the tip is useful. Concerning it being missing, it is the tip that must be found by referencing the elevation of the pole, or other fundamental rules."

Don Camillo, finally appearing to be persuaded and showing himself completely content said, "Since it is so, I promise you that I wish to honor you before the King who for such a long time has not known whom to believe and will be pleased that this discovery was made by you and by the chance arrival of this Padre, and this evening in fact I want to take it to him again. Leave it to D[on] Camillo."

Padre Sauli was quite pleased not only in having participated in this discovery but in the honor for both of us before the King in this encounter which both of us considered a stroke of particular good luck.

Piaggio's version of events raises several questions. That he immediately recognized the object as a sundial and that the quadrant's shape lay behind the decision to inscribe it onto a prosciutto is consistent with his, and Sauli's, Jesuit learning, which included the kind of mathematical and astronomical training required to do so.[105] So, too, was his ability to distinguish it from the more familiar horizontal and vertical types. Even his application of the Vitruvian label of "portable vertical sundial" (Vitr. 9.8.1) shows that he recognized its similarities to the so-called shepherd's or altitude sundials, popular since at least the medieval period and still in use in his day, of which he evidently owned a version along with other types of sundials.[106]

Yet Piaggio's claim that he was the first to identify it as a sundial to Paderni is suspect, even if Paderni's misconception that sundials were found only on walls evidently had prevented him from immediately recognizing it as such on his own. La Condamine, for example, knew it was a sundial, though it is not known whether he recognized this on his own or Paderni had sought to impress this famous astronomer by sharing it with him soon after learning what it was. While it is not clear on what date he had returned to the museum, Piaggio's charge that Paderni remained ignorant of its true function, and even had questioned Piaggio's identification initially, together with Paderni's

---

105. A. Udias, *Jesuit Contribution to Science: A History* (Springer 2015) 23–53, "Mathematics, Astronomy and Physics in Colleges and Observatories."

106. For altitude dials and their similarities with the prosciutto sundial, see Chapter 2.a.

remark that the King had not known what it was "for such a long time," contradicts Paderni's own writings. In his diary entry cataloging its arrival in the museum, Paderni had written, "this prosciutto served as a sundial." Either he had in fact recognized it as such, contrary to what Piaggio now claimed years later, or he had emended his diary after Piaggio had illuminated him.[107] His letter to the Royal Society, dated seventeen days after its discovery, does not help clarify the chronology, since by then Piaggio appears to have enlightened him. In Piaggio's version Paderni told him he planned to show it to the King "again" now that he knew what it really was, and this implies that his previous presentation to the royal couple had been limited to amusing them with its odd shape.

Paderni's catalog of bizarre alternative interpretations, moreover, suggests that at the time he confronted Piaggio he was still clueless after his usual consultants had sought to conceal their own bafflement in convoluted explications. Piaggio had singled out Bayardi and Mazzocchi together with his assistant Nicola Ignarra as the main culprits in having led Paderni astray. Alcubierre must have contributed to the mix as well, given his initial description of the prosciutto as having "some stripes and signs." Bayardi had written the entry for his *Catalogo* soon after the discovery of the satyr tripod, so it is unclear if he had arrived at his identification of it as a sundial independent of Piaggio. He rarely visited the museum, so his first-hand knowledge of the prosciutto may have been limited. Alessio Simmaco Mazzocchi (1684–1771) was a scholar of Greek and Latin with a specialty in Latin inscriptions (Figure 1.16). Now in his seventies, he had just completed his latest book, a two-volume commentary on the *Tabulae Heracleenses* (MANN 2481). These two bronze tablets, found in the 1730s in Heraclea Lucania, bore on one side a long inscription in Latin spelling out the municipal laws of this Roman *municipium* and, on the other, an earlier inscription in Doric Greek relating to the sanctuaries of Dionysus and Athena Polias.[108] It was a testament to his skills as

---

107. The version of Paderni'a *Diario* in the ASN is a later copy so any emendations he may have made to the entry are no longer discernible.

108. A. S. Mazzocchi, *Commentarii in Regii Herculanensis Musei aeneas tabulas Heracleenses*, 2 vols. (Naples 1754–1755). An earlier work had dealt with the ruins and inscriptions at Roman Capua, especially those related to its amphitheater: *In mutilum Campani amphitheatri titulum aliasque nonnullas Campanas inscriptiones commentarius* (Naples 1727). Such epigraphical works are what qualified Mazzocchi to be considered an archaeologist in that period. For Mazzocchi, see also G. Ceserani, "The Antiquary Alessio Simmaco Mazzocchi: Oriental Origins and the Rediscovery of Magna Graecia in Eighteenth-Century Naples," *Journal of the History of Collections* 19.11 (2007) 249–59. Heraclea Lucania is the modern Montalbano Jonico/Policoro (Matera); see C. Carpaci and F. Zevi, ed., *Museo Archaeologico Nazionale di Napoli: La Collezione Epigrafica* (Verona 2017) 138–39. Mazzocchi knew of the extant fragment

FIGURE 1.16 Alessio Simmacho Mazzocchi (1684–1771). Engraving by B. Cimarelli. Frontispiece from Mazzocchi's *In vetus marmoreum Sanctae Neapolitanae Ecclesiae kalendarium commentarius* (Naples 1744).

a linguist and epigrapher. Later in 1755 Charles of Bourbon would appoint him one of the founding members of the Accademia Ercolanese, while in the following year he was elected to the French Académie Royale des Inscriptions et Belles-Lettres. Ignarra was Mazzocchi's disciple, biographer, and successor in his ecclesiastical and scholarly appointments; multilingual; also an epigrapher appointed to the Accademia Ercolanese; and, in 1782, director of the

---

of the tablets in England but did not include its text in his publication. The fragment had returned to Italy only in 1759 thanks to the negotiations of Sir James Gray.

Regia Stamperia, the royal publishing house.[109] He worked alongside Mazzocchi, who had been tasked with deciphering the texts on the papyri from Herculaneum. They, unlike Bayardi, were both regular presences in the museum. Piaggio could barely endure either of them, and he filled his writings with bitter criticism of both. He attacked Mazzocchi for mishandling the fragments of papyri he had unrolled, his disorganization, and his inability to complete an edited text. He claimed their eyesight was so poor—"cow-eyed" Ignarra, in particular, needed two sets of convex glasses—they asked him to dictate the letters on the papyri despite the fact that he did not read Greek. Piaggio could not understand how both of these epigraphers could have seen magical or Coptic characters on the sundial and not the actual Latin ones; after all, he observed wryly, "this idiom was the forte of both of them, and the Canon [Mazzocchi] held a monopoly [in Latin] over everyone."

Equally puzzling were the other interpretations of the prosciutto, which Paderni ascribed to the scholars he consulted. That it functioned as an amulet or abraxas must be extensions of seeing magical characters in the inscription or inside the dial's grid. The other interpretations looked to its shape alone in seeing it as alluding to porcine themes. One was that it was used in some way in sacrifices to Venus and symbolized the wild boar that killed her beloved Adonis (Ovid *Met.* 10.708–39), a boar said to have been sent by Mars out of jealousy in one version of the myth.[110] More fanciful was that it somehow symbolized boar as an essential entrée of the Roman banquet, with the odd notion that the Romans "did not hold dinner parties if there was no boar." This appears to be a misconstrued reading of Martial's poem (Martial *Ep.* 1.43) about Mancinus' dinner party, to which he had invited sixty guests but placed only one small boar on the table—so small it could be carved by an "unarmed midget" (*a non armato pumilione*), though in fact Mancinus intended only that it be admired, not eaten.[111]

---

109. Mazzocchi and Ignarra (1728–1808) were both buried in the same chapel in the Church of S. Restituta in Naples. Ignarra had penned a "life" of Mazzocchi along with his epitaph. For Ignarra, his close relationship with Mazzocchi, his rivalry with Martorelli, and his contributions to the study of the papyri, see M. Mansi, "Per un profilo di Nicola Ignarra," in M. Capasso, ed., *Contributi alla Storia della Officina dei Papiri Ercolanesi*, 3 (Naples 2003) 11–85.

110. According to Ovid, Venus had sworn to Adonis as he died that annual commemorations of her mourning would conclude with a re-enactment of his death, and this may constitute the connection between the myth and the prosciutto made by Paderni's source. Piaggio refers to it as a boar (*aper*) rather than a pig throughout his manuscript.

111. Martial may be equating the situation at the dinner party with spectators of boar hunts in the arena and, in so doing, figuratively condemning Mancinus to the same fate as one Charidemus, named in the epigram's final stanza, who may have died in a reenactment in the gladiatorial arena of some myth such as the Calydonian boar hunt.

Piaggio continued his narrative by explaining the favor that Paderni had asked him in return.[112]

> "But since things are so," resumed Paderni, "while I am attending to the Padre, do me a favor and write me a brief description, but do it well, adding the reasons that you put forward, and then I will tell you what it's for." So I sat at a small table and dedicated myself to serving him with all my effort while he committed himself to serving my companion, adding a few thoughts I had in my head, concerning the Tropics, the elevation of the pole, the gnomon, etc.

Piaggio spent the following day polishing his description, assuming it was destined for the eyes of Charles of Bourbon and so was surprised to learn of Paderni's actual intentions when he next encountered him.

> "You need to know that I am sending this to England to Milord Acri. He is my special friend and old correspondent. With this gentleman I have acquired some real gems because beyond the gratitude he shows me, he sends all the Englishmen he knows to me. All the most special things which are found he wants to know about first, and in a letter which I wrote to him concerning other matters and news comes the opportunity to send him this curiosity which will be most dear to him. Add your observations and the discovery of the misconceptions of these gentlemen who don't know if they are alive and are eating the King alive together with his whole crown."

The reference Paderni is making here must be to his letter read before the Royal Society on June 28, 1755. While Piaggio identifies the recipient as "Milord Acri," Paderni in fact had addressed his letter to Thomas Hollis. Hollis, an English squire (1720–1774), was not a member of the Royal Society at the time, so Robert Watson, a doctor and a "gentleman of great merit and learning, well versed in philosophical researches," had translated the letter and submitted it to the Society.[113] Hollis had toured the continent from 1748

---

112. Piaggio alludes elsewhere to the transactional, *quid pro quo* manner in which Paderni operated in his role as custodian of the museum: visitors, like those from England, were expected to tip.

113. Royal Society of London, *Archives* EC/1750/15: Watson's (1720–1756) nomination to the Royal Society, dated February 7, 1751, was supported by Robert Mead, Anthony Askew, Thomas Birch, and five others.

68 THE PROSCIUTTO SUNDIAL

to 1749 with his friend Thomas Brand and was in Italy again from 1750 to 1753. While in Venice from December 1750 to February 1751 he had dined with Sir James Gray. During his time in Naples from April to September 1752, he likely crossed paths again with Gray and certainly met Paderni as well.[114] A series of letters addressed to him from Paderni were read before the Royal Society and subsequently published in the *Philosophical Transactions*, while Hollis' own correspondence specifically references his friendship with Paderni. In a letter of 1752 Hollis calls him "my friend Camillo Paderni of Portici" and describes him as "a sensible, ingenious, and worthy man, a great lover of the English, the King's principal designer at the Herculaneum [Museum], and in his rank a great favorite of the King" and then quotes extensively from a letter to him from Paderni.[115] Further evidence that Hollis had maintained contact with Paderni is confirmed by the language of his 1757 nomination to the Society. This describes Hollis as "well versed in several parts of polite literature, zealous in promoting all useful knowledge, and by his foreign correspondence capable of becoming a very worthy and valuable member; as may appear from the several curious and early accounts relating to literary affairs abroad, which have already been communicated by him to the Society."[116] Paderni's comment that he wanted to add Piaggio's description of the sundial to a letter he already had written is supported by the fact that this letter to Hollis begins with a detailed report on the findings from the *Praedia Iuliae Felicis* in Pompeii that describes, in particular, the satyr tripod, "one of the most beautiful antiquities in the world" (see Figure 1.14).[117]

---

114. *A Dictionary of British and Irish Travellers in Italy, 1701–1800* (New Haven 1997) 512–13, s.v. Hollis, Thomas.

115. Hollis 1843, 389–94 (Hollis to Ward, December 25, 1752). Hollis ultimately left his estate to Brand with the stipulation that he assume the name "Hollis." Brand-Hollis (1719–1804) was elected to the Royal Society before Hollis, on June 3, 1756, with the support of Ward, More, Ellicott, and two others (Royal Society of London, *Archives* EC/1756/09). For Brand-Hollis, see J. Disney, *Memoirs of Thomas Brand-Hollis* (London 1808).

116. Royal Society of London, *Archives* EC/1757/11: Recommendation for election on June 30, 1757, proposed by John Ward, Robert More, John Ellicott, Thomas Birch, and five others. Paderni wrote five letters to Hollis beginning in April 1754, and ending in February 1758, all published in the *Philosophical Transactions*. Burlot 2012b, 59–61, fig. III.4, identified as a counterfeit a painting in Hollis' collection (now in the British Museum) that he claimed to have acquired in Pompeii in 1753. Burlot suggested it was the work of Guerra and surmised that Paderni had given or sold it to Hollis, presumably during his second visit to Italy.

117. Paderni 1755–56, 490–98. In this, Paderni references a letter he had written to Hollis earlier in May, of which there is no trace, and, after describing his efforts to expand his display of the antiquities in the museum, remarks, "I hope to have the pleasure to see you again in Italy, to admire this treasure, with the sole care of which his majesty hath been pleased to honor me."

Nothing is known of Piaggio's "Milord Acri," whom he names specifically at several points in this manuscript. More likely he muddled the name of Anthony Askew (1722–1774), a London physician famed for his *Bibliotheca Askeviana*, the collection of rare books, primarily Greek and Latin classics, which filled his house.[118] Askew had traveled through Italy in 1749, the same year in which he had been elected to the French Académie Royale des Inscriptions et Belles-Lettres, so he may well have become acquainted with Paderni at that time. He had nominated the translator of Paderni's letters, Robert Watson, to the Royal Society in 1751. More germanely, Askew had supported Paderni's own election to the Royal Society in 1755.[119]

Piaggio complained later in his manuscript that Paderni had copied the description he had written into his letter to Askew: "It was I who described the sundial. To me he said he was writing about it to England with other thoughts to Milord Acri who wanted to be the first to have them. To me he said he was waiting for a gift, as in fact came..." According to Piaggio, one of the "gems" Paderni anticipated receiving in return from Askew eventually came in the form of a telescope, though the astronomical associations between this and the sundial may have been entirely coincidental.[120] Paderni had mounted it on a cabinet next to a window that faced the entrance to a room in the museum. Attracted by "that beautiful machine (perhaps not seen before in this country)" as the first thing visitors would see when entering the room, they would inquire whether it was ancient. This must have driven

---

118. Royal Society of London, *Archives* EC/1749/24: Elected February 1, 1749; nomination supported by his friend Richard Mead, also a physician and collector, as well as Thomas Birch and seven others as "a gentleman of universal knowledge, and particularly well skilled in philosophy, natural history, antiquities, and polite literature." His biography by W. Munk appears on the website of the Royal College of Physicans (https://history.rcplondon.ac.uk/inspiring-physicians/anthony-askew). Mead and Askew supported Watson's nomination to the Royal Society.

119. Royal Society of London, *Archives* EC/1755/03: The nomination was submitted on January 16, and election took place on April 24, 1755. Askew's name appears first, and joining him was Robert More, a parliamentarian and botanist who may have met Paderni during his own travels through Italy in 1750. The third member named was John Ward (c.1679–1758), professor of rhetoric at Gresham College, one of the original trustees of the British Museum, who had tutored Thomas Hollis and supported the election of both Hollises to the Royal Society.

120. Perhaps a small "reflecting telescope" of the type popularized in this period by the British astronomer and optician James Short (1710–1768), also a Fellow of the Royal Society (see Chapter 2.e). For the type, see, e.g., the two-and-a-quarter-inch Gregorian reflecting telescope of brass pillar-mounted on a rectangular wooden case signed by Short (1735) now exhibited in the National Museum of Scotland (T.1984.33; see also T.1967.93).

70 THE PROSCIUTTO SUNDIAL

Piaggio to distraction, as did the fact that Paderni had flaunted it as a symbol of his international scholarly connections.

The following day Paderni collared him excitedly with an additional task: to reconstruct the sundial and demonstrate to the King how it functioned.

> "I did not sleep all night and I thought of something that, I believe, would be better. You have said that at one time you had made these sundials, and that Padre [Sauli] said that the broken style was easy to fix. I therefore thought you could do a demonstration for the King which, if you're successful, would be a shock."
>
> He thought he would catch me off guard with this sudden task, but I was not frightened. Quickly, by attaching a pendulum with a bit of string, and by extending with a bit of wax the broken tail of the boar which served as the gnomon, or rather the style, I went looking for its tip at random, and I did so much that, rotating it and turning it to the sun, I got the idea of it and I marked the time which it was at that moment on the sundial of Don Camillo. He was most content, and he renewed his promise, made to me and Padre Sauli a bit earlier, repeating several times, "Leave it to me, leave it to me (in Roman slang), you'll see who Don Camillo Paderni is."

Piaggio's explanation of how he reconstructed the sundial reveals that he did not really understand how it functioned, despite his professed familiarity with gnomonic principles. It also helps explain the origin of the most confusing detail in Paderni's letter to the Royal Society, for which it is clear Piaggio's demonstration and written description must be credited. Piaggio had led Paderni astray by asserting that this ancient vertical sundial functioned like a contemporary quadrant rather than that only the shapes of their dials were similar. Sundials were commonly called quadrants in this period, but in this case Piaggio evidently conflated these with the quadrants used in astronomy and navigation to measure altitudes. Their canonical shape—two radii joined by the arc of a circle—recalls the shape of the prosciutto's dial, but otherwise they operated in entirely different manners. Since Edmund Gunter's (1581–1626) modifications to the basic quadrant, these instruments also could be used to read the hour of the day according to the day of the year.[121] A plumb bob attached to the

---

121. E. Gunter, *The Works of Edmund Gunter* (London 1673) 97–128: "An appendix concerning the description and use of a small portable quadrant for the more easie finding of the hour and

Discovery 71

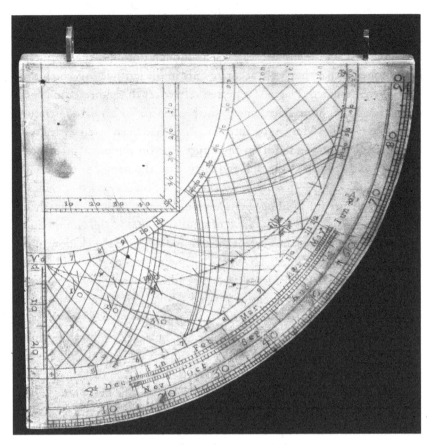

FIGURE 1.17 Horary quadrant designed on projection by Edward Gunter (1581–1626). The month abbreviations appear in the second-lowest arc from the bottom, reversed from those on the prosciutto. The hour lines are in the central arc. The string of the plumb bob was attached through the hole in the upper left-hand corner, and the sights are arranged along the top of the instrument. Seventeenth century, Great Britain, designed for 52° latitude. Boxwood. 12.7 cm × 13.1 cm × 1.0 cm. Courtesy of the Division of Medicine and Science, National Museum of American History, Smithsonian Institution.

upper left-hand corner of the instrument was allowed to fall along the appropriate point on a calendrical scale marked on the arc's edge and held in place there. Sights fixed along the top of the instrument were aligned to the position of the sun in the sky while a bead attached to the plumb bob's string would fall along a secondary scale indicating the hours of the day (Figure 1.17). Unlike a horizontal sundial, Gunter's quadrants, and horary

azimuth, and other astronomical and geometrical conclusions," with a figure of the quadrant on page 98.

quadrants designed on the same principles, depended on the tilt of the instrument and the force of gravity on the plumb bob and did not rely on a shadow cast on the horary scale itself. Though portable, their principal limitation was that they were accurate only in the latitude for which their scales were designed. Piaggio's repeated references to the need for a "pendulum," and that he attached one to the prosciutto for his demonstration, show that he believed the prosciutto originally had been fitted with a similar plumb line, suspended from the upper ring on the prosciutto's hock, which would fall across the dial and align with the names of the months inscribed below. Paderni's letter to the Royal Society stated specifically that to this ring "for holding the sundial in balance" had been attached a "plumb line which in a similar kind of sundial"—that is, the Gunter quadrant—"one lets fall on the current month to determine the shadow of the gnomon on the hour lines." Given the information provided, and lacking even a basic illustration, Paderni's audience at the Royal Society also would have accepted that the prosciutto worked similarly to the quadrants with which they were familiar.

Piaggio's reconstruction does not account, however, for how the plumb line would operate in tandem with the shadow cast onto the dial by the gnomon-tail to yield the time of day. Yet his observation that the position of the tip of the tail was an essential detail shows that he knew it needed to cast a shadow, not merely be oriented to the sun like the sights on a quadrant. Rotating the entire prosciutto so that gravity would cause the plumb line to align with the desired month, as Piaggio's hybrid device would require, distorts the manner in which the gnomon's shadow falls on the dial's grid, yielding an inaccurate reading of the hour of the day. Piaggio may have shown Paderni how it could "mark the time" by reconstructing the tail with a piece of wax, but it likely was not "the time which it was at that moment." While Piaggio had acknowledged the ingenuity of the prosciutto's design, he had failed to recognize its simplicity; that it was a wholly integrated instrument requiring only a string attached to the upper ring to suspend it vertically through the force of gravity: the prosciutto itself was the plumb bob.

How many days later the King finally came to the museum is not clear from Piaggio's narrative; perhaps only two days after Piaggio first saw it. Mazzocchi and Ignarra had arrived much earlier than usual, and, pacing up and down the rooms of the museum, they had conversed in secret, which made Piaggio nervous they were going to ask him some favor, as they had in the past. Once the King had arrived, they gathered around a small table near

Discovery

a cabinet displaying ancient instruments such as weights and measures, evidently the fourth room of the museum.[122]

The Canon [Mazzocchi] was fidgety, seemingly having something to say. The King turned to him, "And so," he said, "what have we got that's new, Signore Canon?" Don Nicola [Ignarra] was pulling at his coat when [Mazzocchi] finally spoke in this way:

"We have, Your Royal Highness, after much effort, found that there is the idea of a clock written on that piece of metal that represents a prosciutto [and also] we have encountered the fortune of the papyri, that is to say, of having a treasure which serves no purpose. So say I, speaking of those things, but speaking of the clock, it's useless because the ancients possessed mathematics *grosso modo* [more or less], so that I can say that it was better that this monument remain in this way without adornment for the moment, because a prosciutto is agreeable to no one because of the vileness of the subject, and so much less is it suitable to one of the most sublime sciences. Now, asking this science to be treated by one of the best talents and applying it to an animal among the most ignorant and the vilest of all is crass ignorance to use in this way and, I mean, [shows ignorance of] what mathematics is. And to make fun of and debase the sciences is the same as not knowing what they are. That this sundial is useless is certain because it lacks its gnomon. This gnomon should be straight and because of this it is called the style, and from this rises another ridiculous inaccuracy because the boar's tail should be twisted and, in this form, cannot serve as a style.[123] Moreover, this is certain, I say, that these lines and characters, although executed in this manner more or less represent a sundial – it's this that we have discovered. Therefore, Your Majesty must know that the sun rotating through the zodiac makes its path between two limits which are called the Tropics, one called Cancer and the other Capricorn. From these and from the zenith the elevation of the pole is revealed and the line that describes the sun is called the parallax."

...

---

122. P. D'Alconzo, "La *Memoria dell' Osservazioni fatte sopra gl'antichi monumenti d'Ercolano l'anno 1769* di Camillo Paderni: Un istantanea del Museo Ercolanese di Portici," *CronErc* 49 (2019) 271.

123. Mazzocchi was arguing that the Greek word στῦλος, from which the term "style" derived, meant "pillar" or "column," an upright form that by nature could not be twisted.

The Canon was shy, difficult at making himself understood. He was dealing with material that one could see he had learned by heart, that it was a lesson that he had spent the entire morning rehearsing with [Ignarra]. This material wasn't in his realm, and this material was of little use to him because he had lost these talents, which was a phrase he kept repeating about himself publicly. Here, then, between the Tropics, the zenith, and the parallax he began to trip himself up and look to Don Nicola who in the meantime was pulling at his coat, as was his custom. Now, can you believe that all of this occurred in my presence and that of Don Camillo?

The King realized on the one hand [Mazzocchi's] confusion and on the other he cared little to learn something he could not understand. "Signore Canon," he said smiling, "You know these are not matters in my realm, and I do not understand them, therefore I beg you not to fatigue yourself in making me understand. It is enough that we have come to terms with this question and we understand at least what it is and what is the reason for it." And here, cutting short the discussion, he concluded smiling, "Today we have a most beautiful day, so I hope we can bag a good number of quail."

The incident is extraordinary for providing a first-hand account of the presentation to the King of an object from his excavations and his response to it; such scenes doubtless took place on a regular basis when the King was in residence at Portici. It could not have gone worse for Piaggio, however. He was livid not only that Mazzocchi had robbed him of the opportunity of revealing to the King his own discovery that the prosciutto was an ancient sundial but that Mazzocchi had done so in a way that was so imprecise and so dismissive of the ingenuity behind its design and its application of astronomical principles. Why Paderni had not allowed him to demonstrate the sundial in operation for the King was a mystery to Piaggio. Adding insult to injury, Ignarra had asked another favor of Piaggio: that he employ his skills as a calligrapher to engrave a copy of the prosciutto for them to present to the King.

Piaggio devoted much of the remainder of his manuscript to criticizing Mazzocchi's presentation of the sundial to the King. His lengthy attack can be boiled down to three main points. First, Mazzocchi's remark that the ancients knew science *grosso modo* ignored the complexity of the prosciutto's design. In particular, fashioning the gnomon from the tail, which Piaggio assumed was crooked, was not an imperfection and did not make it useless, as Mazzocchi had stated in the belief, like the English gentleman and Paciaudi, that a

Discovery 75

gnomon needed to be vertical in order for a sundial to function properly. On the contrary it was evidence of the ingenuity behind its design since correctly positioning the tip of the tail to ensure the sundial functioned properly required a firm grasp of mathematics, geometry, and astronomy. Indeed, anyone could fashion a horizontal sundial; this portable vertical one was vastly more complex. Second, Piaggio argued that the selection of a pig's thigh for the object on which to engrave the quadrant did not detract from that ingenuity. Mazzocchi had condemned the pig as a vile animal, but Piaggio objected that pigs were no more objectionable than a mouse, whose tail is straight but would be too long, as would that of even that most noble of beasts, the lion. Perhaps, mused Piaggio, the designer had in mind the Erymanthian boar and the fourth labor of Hercules, an appropriate mythological reference given that hero's associations with the Vesuvian region. Finally, Piaggio scoffed at Mazzocchi's remark that it was puerile and demeaning to both pigs and science to engrave a sundial on a prosciutto: the sundial had its required shape of a quadrant, and this corresponded to the shape of the prosciutto. Such hybrid creations combining diverse subjects were a hallmark of ancient creativity and design. Piaggio needed to look no further than the cabinets of antiquities around him for his examples: anthropomorphized steelyards bearing the heads of Minerva or, better, Mercury, the patron deity of commerce; lamps in the shape of masks with open mouths to receive the wick; a waterspout in the form of a dolphin; bronze pitchers formed from goats—another "vile" creature—whose horns served as handles. Best of all for illustrating the skill of the ancient artisans for the nobleness of its composition and artifice was the satyr tripod (see Figure 1.14). This was a *tour de force* study in design and anatomy, clearly by an accomplished sculptor and bronze caster, an exuberant fantasy by a talent of the best taste—in a word, a monument "no less capricious as beautiful." In a like manner the prosciutto sundial "assumes a knowledge of astronomy, and this assumes a knowledge of geometry and arithmetic, it assumes an ingenious metal worker, and it assumes finally an accomplished professional horologist."

Piaggio attributed Mazzocchi's poor presentation of the sundial to the King to his ignorance of astronomy, despite Ignarra having coached him.[124] Paderni, too, had memorized his presentations of objects in the museum but, in Piaggio's view, hated giving tours to those who spoke more Italian than him

---

124. This is in keeping with the impression of senility Winckelmann noted upon meeting him in 1758: "Mazzocchi has lost himself in Hebrew and Phoenician etymologies; he is an old man, almost an imbecile." Winckelmann *Briefe*, vol. 1 (Berlin 1952) 360–61.

76   THE PROSCIUTTO SUNDIAL

and "could not say more than the [museum's] gatekeeper." Many were foreign visitors brought by Martorelli, to whom they would turn when they became frustrated with Paderni's presentation, so Paderni routinely begged off, claiming he was otherwise occupied, or would simply hide when Martorelli showed up with visitors.

Giacomo Martorelli (1699–1777) was another frequent presence in the museum in this period, but his days of unfettered access were numbered. A leading Neapolitan intellectual, a scholar of Latin and geometry, and a professor of Greek at the royal university in Naples, he had been Mazzocchi's student, but they had long been at odds (Figure 1.18). His antiquarian interests had led him to include observations on some of the early finds from the excavations at Herculaneum in his correspondence with Gori in Florence. Martorelli is an unlikely candidate as Gori's source about the prosciutto, however, given the expression of optimism in the letter published in the *Memorie* that Mazzocchi's work would provide the necessary clarification.

As had been the case with Paderni, Martorelli's divulging of the royal discoveries had not been received well in the Bourbon court.[125] Nevertheless, he appears to have continued to have access to the museum, presumably with the stipulation that he publish nothing. Despite ready access to the papyri, he famously maintained that they were only documents like contracts and decrees, not literary texts, a conclusion that made Piaggio view his reputation for intellectual acumen with suspicion.[126] At the time of the sundial's discovery, he was putting the finishing touches on his *de Regia Theca Calamaria*, a two-volume study of a small octagonal container that he was convinced was an ancient inkwell and not, as other scholars argued, merely a container for an unspecified liquid. While this had not been found in the Vesuvian region, it resided in the museum at Portici.[127] In the course of his pedantic 738-page excursus, Martorelli not only surveyed the use of ink in writing from ancient Egypt to Rome, but also dealt with astrological matters in order to explain that the figures engraved on the vessel's sides were the planetary deities representing

---

125. Gori 1748, xiv–xviii, 75–80. According to Gori, Martorelli began corresponding with him in 1747 and sent him regular missives. Gori appears not to have realized the consequences of exposing Martorelli as one of his sources. Their correspondence is in the Biblioteca Marucelliana in Florence (Mss. B-VII-18 and A-LIV; http://www.maru.firenze.sbn.it/gori/a.f.gori.htm). For Martorelli, see also Dodero 2019, 161–64, with additional references.

126. Winckelmann 1762, #70; Mattusch 2011, 122, who explains that Martorelli believed the room in which the scrolls were found was actually a public archive.

127. MANN 7509: Recovered in a tomb in Terlizzi, Puglia, in 1745.

FIGURE 1.18 Giacomo Martorelli (1699–1777). Engraving by Francesco Morelli. From *L'Album, Giornale letterario e di belle arti* 16 (September 1849) 252.

the days of the week.[128] He concluded it dated from the Augustan era and had belonged to an astrologer from Neapolis. In 1754, presumably in the course of his research for this book, he had been caught in the act of transcribing the text of a papyrus fragment and evicted from the museum, evidently at the demand of Bayardi.[129] By including unpublished artifacts from the royal excavations in his book, moreover, he incurred the wrath of the "Mazzocchiani." To his everlasting chagrin, not only had the Bourbon court prohibited Martorelli from circulating his book for several years, but it subsequently kept

---

128. The book swiftly became infamous for the uselessness of its erudition; e.g., G. L. Bianconi, *Lettere sopra A. Cornelio Celso al celebre abate Girolamo Tiraboschi* (Rome 1779) 155: "I don't know if it's more celebrated for its erudition or for the confusion that reigns there."

129. J. Martorelli, *De Regia Theca Calamaria* (Naples 1756) 273: he had managed to read it and would have offered a transcription "if the custodians, whom one could rather call enemies of the republic of letters, had not most fiercely prevented me as if I were a book thief" (*nisi custodes, quos melius litterariae reipub(licae) hostes dicas, me tanquam libellorum raptorem accerrime prohibuissent*); Travaglione 1997, 30.

78 THE PROSCIUTTO SUNDIAL

him at arm's length and never made him a member of the Accademia Ercolanese.[130]

Soon after the episode between Mazzocchi and the King, as news of the discovery had spread through the Neapolitan intellectual community, Martorelli appeared in the museum trailed by a group of foreigners, all anxious to see the sundial.[131] Martorelli's interest had been piqued especially because Mazzocchi's revelation had contradicted Mazzocchi's own original interpretation of the inscribed characters as magical or Coptic. According to Piaggio, Martorelli also knew that Mazzocchi "did not know the principles of gnomonic science while in all the other sciences except this one he had been taken for a genius." Martorelli's group had headed straight to the cabinet where Paderni stored the sundial.

> "This is the sundial," said Martorelli, "that I and these gentlemen are looking for. Do me the favor of letting me consider it a bit according to my method because I no less than they have come for this more than for anything else." Paderni could not refuse but, on the other hand, not having expected this surprise he had not had a warning to remove it from where it was, in the state in which it was when he had shown it to me, that is, without a pendulum and without a gnomon, since he had removed both, having no care nor respect for me who had to see it in this condition every time I passed by after I had fixed it.
>
> When Martorelli saw the sundial in this state, [he said] "So, this was analyzed in this condition?" "So it was," said Don Camillo. But then Martorelli really flew into a rage, without looking for who it was. "My Don Camillo," he replied, "You know that I speak clearly and that I have always spoken in this way. Who led you to listen to this falsehood? If you don't know, I know, and I will make you see that this can't be and it will never be because here there is no gnomon, no pendulum, without which this proof cannot be made. Tell me, including who says they have analyzed this sundial whom

---

130. Martorelli himself accused Mazzocchi's adherents in a letter to A. Piscopo dated July 18, 1771; see F. Strazzullo, *Il carteggio Martorelli-Vargas Macciucca* (Naples 1984) 263–66. Winckelmann 1762, #5; Mattusch 2011, 66–67, believed it was because Martorelli had attacked Mazzocchi's scholarship in the book. While the frontispiece bears the publication date of 1756, the delay in distribution had allowed Martorelli to insert another hundred pages in the first volume in the form of several addenda, indices, and even a rebuttal to Winckelmann's criticism.

131. Piaggio noted that Martorelli only brought foreign, not Neapolitan, scholars to the museum, all of them on the recommendation of his correspondents.

Discovery 79

I know not only does not know the principles of that which they described but does not know the subject on which this has been executed. This is a prosciutto, is it not true? Now, know that if you ask him [Mazzocchi] what ham is he doesn't even know. The reason is that he has never eaten anything other than salted anchovies. But this fabrication, my Don Camillo, seems to me to have passed into barbarity ([*Piaggio*:] with this term he had complimented him another time) because leaving this monument so imperfect, without restoring it, when all the other things more vile and useless are restored: Must it be said that there is no one in Naples who knows how to make a sundial? It doesn't take anything – I'll come one day without anyone and within a half hour I'll fix it for you."

Paderni and Martorelli then went back and forth about whether the King's permission was needed to restore the sundial, with Martorelli arguing that Paderni did so routinely with much less interesting objects. "I know," concluded Martorelli, "that the King will take more pleasure in you than he has when he has seen the other things of so much less importance restored by you." "You can imagine the pleasure that I too will have," added D. Camillo, "…the first chance I get I want to speak to him about it and persuade him to let you fix it. I'm certain that he will listen to me, and I don't doubt that I will send for you immediately."

Poor Martorelli, entirely content, said to him, "To show you that I know what I'm talking about and that I am not burdening you in vain I will now show you how to read it and how to operate it." Here he began to explain the whole thing clearly, and, a bit in vernacular and a bit in Latin, he described all that was inscribed on it. Then he proceeded to demonstrate its use supposing that the gnomon and pendulum were there. Don Camillo, feigning to understand everything, showed himself most content both with the vernacular, of which he knew nothing, and the Latin, of which he knew even less.

All of this evidently had transpired in front of Martorelli's visitors, causing Paderni to become alarmed that one of them would publish a detailed description of the sundial. La Condamine fit the profile of the kind of scholar Martorelli might have enjoyed taking to the museum, though this would require compressing the chronology of events into fewer days than Piaggio's narrative allows. Paderni related his fears to the King, underscoring his role as

80   THE PROSCIUTTO SUNDIAL

devoted guardian of the royal findings while, according to Piaggio, acknowledging at the same time his own network of foreign correspondents and the
personal benefits he reaped from them. Paderni persuaded the King to issue a
royal edict prohibiting Martorelli from entering the museum, though how
long this endured is unclear; Martorelli remained on cordial terms with
Paderni years later.[132]

Piaggio found it especially vexing that Paderni had not reconstructed the
prosciutto in the manner he had demonstrated and Martorelli had described
but had left it on display in its incomplete, forlorn state. This was how
Barthélemy had seen it in February 1756, while the sundial referred to by
Anne-Marie Fiquet du Boccage in a letter dated October 1757 is likely the
prosciutto.[133] Paderni had instead further infuriated Piaggio by engaging one
of the royal marble carvers to fashion for him a modern horizontal sundial
designed for Portici's latitude, which he placed near a window in the museum.
Thereafter he would proclaim proudly, "At least we know what time it is
because the sun doesn't lie!"

The Prosciutto Sundial: Casting Light on an Ancient Roman Timepiece from the Villa dei Papiri in Herculaneum.
Christopher Parslow, Oxford University Press. © Oxford University Press 2024. DOI: 10.1093/9780197749418.003.0001

132. Martorelli had spent his autumn vacations in Portici from at least 1764 and died there in
1777. He visited Paderni in his home in 1766, when he was shown a bronze head of Neptune;
had lunch there in 1767; and conversed with him as late as 1772. See Strazzullo 1984, 184,
241, 283.

133. A.-M. Fiquet du Boccage, *Letters Concerning England, Holland and Italy* (London 1770)
2.87: Letter 29 from Naples (October 15, 1757). "Sundial" appears in a list of objects from daily
life accompanied by a footnote stating that, "In the year of Rome, four hundred and sixty-two,
according to Pliny, Papirius Cursor caused the first sundial to be placed near the temple of
Quirinus." This is the same reference to Pliny *HN* 7.60.212–15, quoted by the English gentleman in the first published description of the sundial.

2

Analysis

THE ACCADEMIA DID not produce their own treatment of the sundial until 1762, seven years after its discovery and the same year in which La Condamine's journal was finally published, but in doing so they broke in an extraordinary and pointed manner from their established practice. The Accademia's previous publications were two volumes of *Le Pitture Antiche d'Ercolano e Contorni*, the first in the series entitled *Le Antichità di Ercolano Esposte*. The format called for collecting the finds by genre, as the Accademia perceived was in keeping with contemporary aesthetic taste, so the first four volumes produced were devoted exclusively to the ancient paintings; the first of two volumes of small bronzes, among which the prosciutto sundial could have been counted, did not appear until 1767. Each volume was filled with engravings by the best artists working for the Bourbon court accompanied by detailed explications showcasing the deep erudition of the Accademia's members.[1]

They had used the preface to the first volume, addressed to the King and published in 1757, to set forth the rationale behind this exclusive focus on the paintings: not only were these "the envy of the most illustrious museums" and "awaited with great impatience by the curiosity of the erudite," but the paucity of other examples of ancient paintings rendered these examples particularly precious.[2] In their preface to the second volume in 1760, the Accademia argued that their work helped establish the pedigree of the paintings, preserving the value of the royal collection and insulating it from the poorly produced

---

1. *PdE*, I.preface: "The very brief explanations (*spiegazione*) [accompanying each engraving] have as their object the reawakening to reflection those readers who want to examine things themselves; the notes will alleviate the toil of those who are content with our thoughts." Their methodology was viewed as slow and laborious: cf., e.g., La Condamine 1755, 622–23: "An academy of Antiquaries is just founded at Naples, for explaining all the antiquities dug up at Herculaneum; but according to their method of discussing things in their assemblies, they will not explain two dozen antiquities in a year. They will alter their method and find that such kinds of works, and perhaps all others, are not to be done by committee."

2. *PdE*, I.preface.

forgeries circulating on the antiquarian market, specifically citing Guerra's forgeries. The Accademia had acquired three of Guerra's paintings and forced him, as part of a legal settlement, to copy the famous *Education of Achilles by Chiron* (MANN 9109), a magnificent large-scale painting from Herculaneum. These they hung side-by-side in the museum at Portici to allow the public to draw their own conclusions.[3] All of this, they claimed, had prevented them from focusing on the finds in metal, so as a taste of the marvels to come they had offered the engraving of an exquisite gold coin, recently extracted from the ruins at Pompeii, with only a brief description of its design (Figure 2.1).[4]

FIGURE 2.1 Gold coin of Augustus (obverse) and Diana (reverse). Engraving by Paolo Campana from a drawing by Camillo Paderni. Preface to volume 2, *Le Antichità di Ercolano* (Naples 1760).

---

3. Piaggio recounts the incident in his usual detailed manner. He was flabbergasted that Guerra had duped so many artists and connoisseurs; see D. Bassi, "Altre lettere inedite del P. Antonio Piaggio e spigolature dalle sue 'Memorie,'" *Archivio Storico per le Province Napoletane* 33.1 (1908) 277–332. Barthélemy 1823, 237–38, remarked on how the Guerra episode reflected the paranoia of the Bourbon court: "We must be on our guard and full of mistrust, conclude our scholars [of the Accademia], against those who claim to have paintings taken from the excavations of Herculaneum: *In tanto è ognuno nel obligo di diffidare, quando si senta vantar pitture che sieno uscite dalle scavazioni d'Ercolano* [In any case everyone is obliged to be suspicious when one boasts about paintings that have come out of the Herculaneum excavations]." The full incident is treated in detail by Burlot 2012a, 75–80; Griggs 2011, 471–503; and summarized in D'Alconzo 2021, 82–87.

4. *PdE* II.Preface. The coin (MANN 3692; see Figure 2.1) bears the image of Augustus on the obverse and Diana on the reverse; see Cantilena 2012, 459–76. The original copper matrix of the engraving, drawn by Paderni and engraved by Pietro Campana, is in the MANN 351416. Barthélemy's review of the second volume of the *PdE* (*Oeuvres complètes: avec une notice sur la vie et les ouvrages de Barthélemy*, 4 vols. [Paris 1821] 4.358) included a discussion of this coin that went further in placing it in its historical context.

As Barthélemy's time in Naples had demonstrated, ancient coins were extremely popular with antiquarians, but their relatively simple images could be readily reproduced and their legends easily transcribed and debated in scholarly circles, so in this instance too the Accademia clearly was seeking to pre-empt the transmission of this spectacular object through the clandestine network.[5]

Now, in this Preface to the third volume, the Accademia focused exclusively on the prosciutto sundial, calling it "not only among the most beautiful and most curious for its design, but also unique for its kind and in every sense of truly singular worth." They justified its appearance in this volume of paintings as fully in keeping with their promise to provide to the public "some piece of ancient metalwork from the museum of the King in the prefaces, which, whether new or rare, could be looked forward to with greater eagerness and less suffering."

2.a) Una Cosa Inedita: *The Prosciutto vs. the Egyptianizing Base in* Le Antichità d'Ercolano

In featuring the sundial here, the Accademia was clearly responding to its previous unauthorized publication and seeking to reassert their own authority over the royal collection of antiquities by preempting any further attempts to assess it as an ancient scientific instrument. This is clear from the fact that the prosciutto had not been the first choice for the third volume's preface. That distinction had initially been reserved for a small rectangular base of bronze, its squat form with sides angling in from its square plinth before curving out in a high cornice at the top evidently meant to be evocative of Egyptian architecture (Figure 2.2).[6] This had been recovered from outside the so-called Palestra in Herculaneum on September 13, 1760, but, in a curious echo of his discovery of the prosciutto, Paderni claimed he had not recognized its significance until some days later when he removed the thick volcanic material encrusted on it.[7] The exotic quality imparted by the figural images of an

---

5. As if in confirmation of their anxiety, in this same year Caylus published his *Numismata Aurea Imperatorum Romanorum e Cimelio Regis Christianissimi* (1760), a collection of sixty-eight plates of coins from the Roman Empire, without any descriptive text. References to the coin of Augustus appear in most descriptions of the museum from this period.

6. MANN 1107/76384: 0.273 m L x 0.193 W x 0.082 H, thickness of walls = 0.01 m. *StErc* 318 (September 13, 1760); Ruesch 1908, #383; Tran Tam Tinh 1971, 52–55; De Caro 2006, 126 #II.83.

7. Paderni, *Diario* September 13, 1760: "Moreover I received [from the excavations in Herculaneum] a bronze base 12 ½ [*oncie*] long and 9 [*oncie*] minus one wide and 4 [*oncie*]

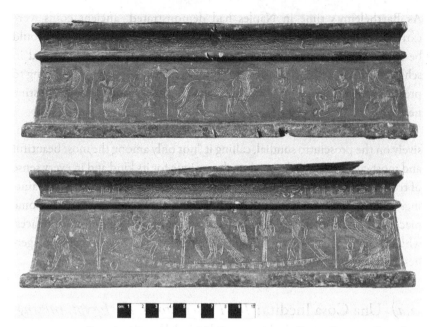

FIGURE 2.2 Egyptianizing base with Egyptianizing décor. Bronze. Found in Herculaneum on September 13, 1760. MANN 1107/76384. Su concessione del Ministero della Cultura—Museo Archeologico Nazionale di Napoli.

Egyptianizing style cluttering its four sides must have rendered it an irresistible object for featuring in the *Le Antichità*, for already by November Tanucci had alerted Charles III in Spain that it would serve as the preface to the third volume. This decision, evidently taken at the September meeting of the members of the Accademia, appears to have been motivated in part by the recent publication of Barthélemy's *Explication de la mosaïque de Palestrine* (Paris 1760), the first detailed study of the famous mosaic depicting the flora and fauna of the Nile delta discovered in the early 1500s at Palestrina, Italy. In the course of his discussion Barthélemy had referred to several engravings of small paintings published by the Accademia in the first volume of *Le Pitture* that

minus one high. To this there is another base of a column which was soldered on top of the larger base, and the base is laminated. After several days, taking this base in hand, which was covered in a finger's thickness of tartar, and desiring to observe in which state the bronze was and if it had suffered from calcification, I took tools and removed the outer layer of patina and I found not only that the bronze was in an optimal state of preservation but moreover I saw that it had been engraved on all four sides and that each of these were full of hieroglyphs, figures, animals and Egyptian characters." Tran Tam Tinh 1971, 52, equates the form with a miniature Egyptianizing *naos*, or shrine, but there are no exact architectural parallels. Most scholars refer to it as a base, either for a statue or, as Tran Tam Tinh, for some small cult object.

Analysis 85

contained Egyptianizing features.[8] The Accademia had offered little more than descriptive commentary on these paintings, while Barthélemy's approach had been somewhat more analytical, more in the manner of Caylus, arguing in particular that the Graeco-Roman architectural elements featured in these paintings did not preclude the possibility they depicted Egyptian landscapes simply because they did not include pyramids. Yet Tanucci had interpreted Barthélemy's remarks as critical: "On page 31 is said something about the Egyptian paintings of the first volume of the Paintings of Herculaneum, which we also said were Egyptian, but not with the certainty that Monsieur l'Abbé would like, with his French frankness."[9]

Evidently the Accademia had viewed the elucidation of this bronze base as a means not only of reasserting their credentials on matters Egyptian but also of showcasing the multicultural span of the finds entering the royal collection from the Vesuvian excavations. The fact that Caylus' publications dealt in depth with Egyptian antiquities must have figured as well in the Accademia's calculations to fast-track its publication. Moreover, like the gold coin featured in the preface to the second volume, the base could be trumpeted as among the most recent discoveries, while rapid publication would thwart its transmission on the clandestine network: the fear that Caylus might get his hands on it must have figured in its selection for the Preface. Paderni had been tasked with drawing it, from which the necessary engraving would be made, a process Tanucci had lamented was bound to drag out through January 1761, given the dilatory nature of the engravers. Indeed, as January drew to a close Tanucci was still awaiting a copy of the engraving to send to Charles III.[10]

By March, however, the Egyptianizing base was out and the prosciutto sundial was in. This was not entirely due to the dilatoriness of the artists as the extant evidence confirms that the Egyptianizing base would progress

---

8. Barthélemy 1760, 31–32. Barthélemy discussed four small landscapes whose details—crocodiles, papyrus plants, hippopotami, Egyptianizing figures—he argued were intended to represent Egyptian life and, especially, architecture: *PdE* 1.87, plate 16; 1.133, plate 26; 1.253, plate 48; 1.257, plate 49; 1.263, plate 50.

9. B. Tanucci, *Epistolario*, ed. M. G. Maiorini (Rome 1985) 9.123 #86 (November 18, 1760). B. Tanucci, *Epistolario*, ed. M. G. Maiorini (Rome 1988) 10.291 #247 (November 14, 1761): Tanucci was still fuming over Barthélemy a year later in a letter to Galiani in Paris: "Barthélemy, with his Egyptian system, with his elevating Paciaudi above Mazzocchi, and with some remarks made about the museum that Paderni recounted to me, has lost his reputation."

10. B. Tanucci, *Epistolario*, ed. M. G. Maiorini (Rome 1985) 9.330 #231 (January 27, 1761): "Campana [the engraver] still has not given me the engraving of the Egyptian monument which is to be explained in the Preface."

86 THE PROSCIUTTO SUNDIAL

considerably along in the process toward publication. Paderni had completed scaled drawings of all four sides of it. Some years later he incorporated these into his *Monumenti antichi rinvenuti ne Reali Scavi di Ercolano e Pompei*, a manuscript he prepared for Charles III highlighting finds from the excavations over the period 1759 to 1768 (Figure 2.3). In the meantime, he had reimagined his drawing, with the base, shown in perspective at the top of the image, used as a weight to hold down an unraveled scroll on which the other three sides were depicted. This had been used by the engraver Ferdinando Campana for the copper plate. How long this had taken and when Tanucci finally had it in his hands are unknown, but in the end it joined the stack of other so-called *Rami Inediti* from this period that never appeared in any volume of *Le Antichità* (Figure 2.4).[11]

Tanucci explained to Charles III that the change had come about because such Egyptian objects "are found in other museums." Instead, continued Tanucci, the goal was to feature "something unpublished" (*una cosa inedita*): a sundial of silver in the shape of a prosciutto (*un orologio d' argento di figura di prosciutto*).[12] Charles III already had met the prosciutto, of course, thanks to Paderni's nocturnal presentation of it to him in the palace garden soon after its discovery, followed by Mazzocchi's confused explanation witnessed by Piaggio. The change in object suggests the Accademia ultimately realized that presentation of the base would not distinguish their publication sufficiently from the mélange of Egyptian, Greek, and Roman antiquities featured in Caylus' *Recueil*. They would have known this when they initially selected the base, however, so there must have been a more compelling reason for altering the topic of the Preface after production had already begun. The recognition of the near-impossibility of deciphering the images on the base and rendering a meaningful explication, one that would not open them up to criticism, may have played a significant role.[13] No member of the Accademia

---

11. The sole copy of Paderni's *Monumenti antichi rinvenuti ne Reali Scavi di Ercolano e Pompei* manuscript he prepared for Charles III is in the library of the École Française de Rome (Manuscript 26). For a reproduction of the full manuscript, see U. Pannuti, *Monumenti antichi rinvenuti ne reali scavi di Ercolano e Pompej & c[ontorni] delineati e spiegati da d. Camillo Paderni romano* (Naples 2000), and see also C. Parslow, "Camillo Paderni's *Monumenti antichi* and Archaeology in Pompeii in the 1760s," *Rivista di Studi Pompeiani* 18 (2007) 7–22. The Egyptianizing base is plate 9.A&B (long sides) and plate 10.A&B (short sides). The copper plate is in the MANN (Catalogo Generale 1500352347).

12. B. Tanucci, *Epistolario*, ed. M. G. Maiorini (Rome 1985) 9.439 #331 (March 3, 1761).

13. Paderni's caption to his drawing (*Monumenti antichi*, pls. 9 & 10) illustrates the potential pitfalls the Accademia confronted in deciphering the images on the base: "I will leave to the more qualified the explication of these Egyptian characters and hieroglyphs, admitting my

Analysis

FIGURE 2.3 The Egyptianizing base (MANN1107/76384) as drawn by Camillo Paderni for his *Monumenti antichi rinvenuti ne Reali Scavi di Ercolano e Pompei* (1768), Plates 9 and 10. Courtesy of the Bibliothèque of the École Française de Rome (Ms. 26).

FIGURE 2.4 Original copper plate engraving of Egyptianizing base (MANN1107/76384) found in Herculaneum on September 13, 1760. Engraving by Ferdinando Campana from a drawing by Camillo Paderni. MANN 352347. Su concessione del Ministero della Cultura—Museo Archeologico Nazionale di Napoli.

ignorance of them. I can only say that in the month of September was found in Herculaneum a bronze base which was consigned to me the same day it was found, about which one did not take much attention for the reason that it came covered in quite a thick coating of the hardest patina. After a few days I returned to inspect this base and with tools I removed a piece of that coating and, in that instant, I discovered there were stamped engravings, and with further work

Analysis . 89

was versed in matters Egyptian, an area of study still in its early stages of development, and analysis of Egyptian imagery remained rooted largely in literary sources.[14] The prosciutto, on the other hand, allowed the Accademia not only to redress d'Alembert's ham-handed treatment of it in the *Encyclopédie*, a wound that had been festering for some three years, but to ground their explication in the context of the latest astronomical discoveries.

Tanucci had already penned his dedication of the third volume of *Le Antichità* to Charles III in late March, doubtless in anticipation of its forthcoming publication, but the engraving of the prosciutto for the Preface remained unfinished in the first week of April, and Tanucci anxiously awaited other engravings for the volume through July.[15] In the meantime Caylus had

on it all four sides were revealed which are all [shown in the drawings] at the scale of the original. **A:** Front of the base, which can be called Isiac because in this first part one sees a kind of small boat, desiring thereby to indicate the ship which brought Isis to Egypt, and since Jupiter gave Mercury the task, he led her into Egypt and therefore one sees in this ship Anubis in the act of trying to stop this ship, intending under the name of Anubis to mean Mercury. **B:** the back side [of the base]. Plate 10. The sides A.B." According to V. Tran Tam Tinh, *Le culte des divinités orientales à Herculanum* (Leiden 1971) 53–55, the central figure on this side of the base is Horus-Râ in the form of a falcon, and the papyrus ship is piloted by a monkey, an animal sacred to the sun. More importantly, the images on the base are not, in fact, hieroglyphs nor even properly Egyptian, but Ptolemaic-era symbols similar to those on the so-called *Mensa Isiaca* (*aka* the Bembo Tablet) now in Turin, which dates to the first century; see E. Leospo, *La Mensa Isiaca di Torino* (Leiden 1978). As Tran Tam Tinh 1971, 55, concluded, "The 'hieroglyphs' are instead fanciful signs, imitating sacred writing of the [Nile] Delta, selected not to express ideas but to fill out the space and to give, it seems, the monument both a mysterious and a sacral cachet."

14. The Accademia did not completely abandon their Egyptianizing base nor shun Egyptian imagery. Perhaps in response to Barthélemy's perceived criticism of their earlier analyses, they included in the third volume three paintings of architectural perspectives (*PdE* 3.301–13, pls. 57–59) and three small vignettes (*PdE* 3.333–35, pls. 26–28), all with Egyptianizing elements. Plate 58, in particular, depicts an aedicula with a sphinx perched on the roof. The Accademia describes the sphinx (*PdE* 3.306, n. 7) as having the body of a lion and the head of a woman with locks of hair falling upon her breasts. They compare this to a similar sphinx on the *Mensa Isiaca* in Turin that at that time had been identified as a representation of Isis, the sphinx bearing a male head representing Osiris. "Similarly," they noted, "in the most beautiful *Mensa Isiaca* of the Museo Reale, recovered recently in our excavations, one sees in one part Osiris with a mortarboard on his head and the body of a winged lion, and then on the other Isis (if it is not the same Osiris) with the body also of a winged lion and a human head. And from this one deduces that in Egyptian monuments when the sphinx was represented without wings it depicted the animal itself which, as we have seen, was believed to be real and extant; when wings were added, the deity was being denoted." These appear to be the figures depicted on one of the long sides of the base that Tran Tam Tinh 1971, 54–55, identified as a pterophoros divinity and the lion-god Miysis, both of whom serve protective roles.

15. B. Tanucci, *Epistolario*, ed. M. G. Maiorini (Rome 1985) 9.520–21 #402 (no date): dedication; 9.729 #576 (June 9, 1761), to Charles III: "The third volume is composed for the most part of paintings found in 1759 and a few from 1760. So, mounting them in frames, drawing them, engraving them, explicating them is a work of more than two years"; 9.807 #648 (July 7, 1761):

90 THE PROSCIUTTO SUNDIAL

published his third volume of the *Recueil*, and Tanucci was on the brink of war with him for having declared Herculaneum "a little provincial city" and dismissing *Le Antichità* as featuring only mediocre works of art because Caylus himself had not published them.[16] The appearance in July 1761 of Paciaudi's detailed treatment of the " unpublished" sundial accompanied by his egregiously wrong illustration, with the Accademia's Preface and the third volume itself well along in preparation, had sparked Tanucci's apoplectic fit.[17] In the end, production of the third volume dragged on for another year: the formal publication of the prosciutto sundial had occurred only in late August 1762.[18]

While the text of the Preface with its copious footnotes replete with references to the literary sources was clearly penned by the Accademia's members, they had been forced to outsource the sundial's astronomical analysis. The lone physical scientist in the Accademia was Giovanni Maria Della Torre, who in 1754 had published a history and description of the volcanic activities of Vesuvius.[19] But Della Torre was not an astronomer. Indeed, in this period the study of astronomy was still in its infancy at Naples and the international community of astronomers relatively small and engaged in independent and collaborative research and active debates. When Charles of Bourbon had named Pietro di Martino as the first Chair of astronomy at the Università di Napoli in 1735, Naples still lacked a proper observatory. The same conditions prevailed when Felice Sabatelli became Chair upon di Martino's death in 1746. In the meantime, Nicola Maria Carcani (1716–1764) had become professor of astronomy and rhetoric at the Collegio Reale delle Scuole Pie at San Carlo alle Mortelle: like Piaggio, he was a Jesuit. Here he installed an *osservatorio astronomico*, the first of its kind in Naples, eventually equipping it with a

---

"The engravers for the Herculaneum volumes are tormenting me...I finally lost my patience with Pozzi. I have arranged that, after threatening him repeatedly, he be paid only half a month's salary and the remaining half at the end of the year when he will have handed over three or four engravings, which Your Majesty sees will each cost 200 ducats." The slow pace of the artists producing the engravings for the third volume was a recurring topic in Tanucci's letters to Charles III.

16. B. Tanucci, *Epistolario*, ed. M. G. Maiorini (Rome 1985) 9.835–36 #678 (July 18, 1761), to Galiani in Paris.

17. B. Tanucci, *Epistolario*, ed. M. G. Maiorini (Rome 1985) 9.871–72 #706 (July 28, 1761).

18. B. Tanucci, *Epistolario*, ed. S. Lollini (Rome 1990) 11.346 #334 (August 24, 1762).

19. G. M. Della Torre, *Storia e Fenomini del Vesuvio* (Naples 1755). He was the director of the Reale Biblioteca, the Reale Tipografia, and the Museo Farnesiano at the Villa Reale di Capodimonte.

Analysis   91

wall quadrant, two small telescopes, and a pendulum clock.[20] Working along with Sabatelli in 1751 he had contributed to the French astronomer Nicolaus-Louis La Caille's (1713–1762) calculations to determine the solar parallax.[21] Sabatelli and Carcani had also recalculated the latitude of Naples at the Collegio Reale as 40° 50′ 15″.[22] In 1752 he produced a pamphlet of astronomical tables of readings taken at the Collegio for establishing the elevation of the sun and moon at any hour of the day for 1753 that would allow for determining the time of day in Italy and beyond, but his calculations had been criticized for corresponding with those of neither Cassini nor Zanotti.[23] He had published his observations on a solar eclipse in 1753 and took part in an international effort to track the transit of Venus across the sun in 1761 in order to establish the astronomical unit that could be used to measure distances in the solar system.[24] This latter project was coordinated by the French astronomer Joseph-Nicolas Delisle, who was instrumental, together with La Caille, in

---

20. F. Amodeo, *Vita matematica napoletana: Studio storico, biografico, bibliografico* (Naples 1905) I.96–97; L. Santoro, "Two Neapolitan scientists in the XVIII century: Felice Sabatelli and Nicola Maria Carcani," *Società Italiana degli Storici della Fisica e dell'Astronomia. Atti del XXXVIII Convegno annuale/Proceedings of the 38th Annual Conference* (Pavia 2018) 17–22. Lalande 1769, 6.143–144, saw the quadrant and meridian during his visit in 1765 and noted that Carcani, who had died shortly before his arrival, had acquired fame among astronomers.

21. N.L. La Caille, "Sur la parallaxe de soleil…faite en l'annee 1751…," and "Memoire sur la parallaxe de soleil," *Histoire de L'Académie Royale des Sciences avec les Mémoires de Mathématique et de Physique* (1761) 108–10 (*Histoire*), and 90–92 (*Mémoires*); M. Gargano, "The Status of Astronomy in Naples before the Foundation of the Capodimonte Observatory," in S. Esposito, ed., *Società Italiana degli Storici della Fisica e dell' Astronomia, Atti del XXXVI Convegno Annuale* (Naples 2016) 205–14; M. Gargano, "The Development of Astronomy in Naples: The Tale of Two Large Telescopes made by William Herchel," *Journal of Astronomical History and Heritage* 15.1 (2012) 30–41; M. Capaccioli, G. Longo, and E. O. Cirella, *L'astronomia a Napoli dal Settecento ai nostri giorni: Storia di un' altra occasione perduta* (Naples 2009) 99–102. For La Caille, see *Biographical Encyclopedia of Astronomers* (New York 2014) 536–37, s.v. *de La Caille, Nicolas-Louis* (M. Murara); his name is currently spelled "Lacaille."

22. The current latitude is 40° 50′ 19″ (40.83°).

23. F. A. Zaccaria, *Storia Letteraria d'Italia*, vol. 7 (September 1752–June 1753) 137–38, with a table highlighting the differences: review of Carcani's *Tavole astronomiche calcolate al meridiano del Collegio delle Scuole Pie, e loro uso, cosi per trovare i luoghi o sieno i gradi del Sole e della Luna, in qualsivoglia ora di ciascun giorno dell' anno 1753, come per ben regolare l'orologio tanto all' Italiana quanto all' oltramontana* (Naples 1752).

24. N. M. Carcani, "*Solaris defectus*, observatio habita Napoli die Octobris 1753," *Memorie per Servire all'istoria Letteraria* 2.4 (1753) 77–78; letter dated November 24, 1753, containing Carcani's observations provided to an "amico stimatissimo" because it was too cloudy for him to have observed it himself.

92 THE PROSCIUTTO SUNDIAL

sponsoring Carcani's election as a corresponding member of the French Académie des Sciences in 1762.[25]

Carcani was the brother of Pasquale Carcani (1721–1783), the Accademia's secretary and Tanucci's right-hand man, so the task of analyzing the prosciutto from an astronomical perspective had fallen to him.[26] Carcani therefore was the first true astronomer given the *liberté* denied to La Condamine to examine the sundial firsthand and perform the empirical testing required to understand how it functioned, a task for which he had adopted the same creative approach Piaggio claimed to have employed but with more accurate results. He also had the necessary credentials to highlight the flaws in d'Alembert's discussion of the sundial, though d'Alembert's reputation for lack of clarity and precision in his own work made him a relatively easy target and one that, as a leading figure in the French Enlightenment, the Accademia would have relished besting.[27] Moreover, d'Alembert was embroiled at that time in a very public dispute concerning solar observations with La Caille, Carcani's advocate in the Académie des Sciences.[28]

The brothers Carcani had collaborated on the Preface, Pasquale ensuring that the affronts of the earlier treatments were properly addressed and Nicola handling the astronomical analysis, while Tanucci had softened the tone of their text on the prosciutto when he found it a bit crude (*un poco crudo*) and the Accademia as a whole had placed their stamp of approval on it. "The

---

25. N. M. Carcani, "Passaggio di Venere sotto il Sole, osservato in Napoli nel Real Collegio delle Scuole Pie, la mattina de' 6 giugno 1761," *Novelle Letterarie pubblicate in Firenze* (1761) col. 696–702 (= #44 of October 30, 1761), 714–20 (= #45 of November 6, 1761). D. Martuscelli, *Biografia degli uomini illustri del regno di Napoli*, vol. 1 (Naples 1813), s.v. *Pasquale Carcani*, note "a," believed his work on the prosciutto had also contributed to his election to the Académie des Sciences, but this had occurred on January 30, 1762, prior to the publication of the third volume of *Le Antichità*. For more on the roles of Delisle and Carcani in documenting the transit of Venus in 1761, see Chapter 3.a.

26. B. Tanucci, *Epistolario*, ed. M. G. Maiorini (Rome 1985) 9.124 #86 (November 18, 1760) wrote to the King that Pasquale Carcani was responsible for nine-tenths of the work on *Le Antichità*. On the Carcani brothers in general, see *Biografia Universale Antica e Moderna, Supplemento*, vol. 4 (Venice 1839) 384–88.

27. T. Hankins, *Jean D'Alembert: Science and the Enlightenment* (Oxford 1970) 63–65: "d'Alembert was apparently incapable of writing a careful, methodical piece of work."

28. N. L. La Caille, "Mémoire sur la théorie du soleil," *Mémoires* 108–44, and J. D'Alembert, "Résponse a un article du Mémoire de M. L'Abbé de la Caille, sur la théorie du soleil," *Mémoires* 145–79, in *Histoire de L'Académie Royale des Sciences avec les Mémoires de Mathématique et de Physique, 1757* (Paris 1762), with a summary of the dispute provided in the *Histoire*, 111–20. These articles appear in the same volume as La Condamine's "Extraite d'un journal de voyage en Italie." This opens up the possibility that La Caille had alerted Carcani to La Condamine's discussion of the sundial, which had been read to the Académie in 1757.

Analysis 93

refutation of Paciaudi on the sundial in the Preface is complete," Tanucci observed to Charles III in early September 1761.[29] This indicates that the bulk of research and writing on the sundial had taken roughly six months, beginning in March when the sundial had become the designated topic for the Preface.

Not surprisingly, the Accademia's rancor with the previous unauthorized publications was unequivocal and their disparagement of the analyses of d'Alembert and Paciaudi withering, even if they relegated their criticism to the Preface's second footnote. They were so dismissive of these earlier treatments that they considered the sundial essentially unpublished, "the public up to this point not having seen its true design, nor its exact description." Their detailed critique helped highlight the more sophisticated level of science and inquiry underpinning their analysis of the sundial's operation while emphasizing the consequences of basing the study of archaeological finds on verbal reports and second-hand sources. Though they knew exactly who the authors of the earlier treatments were, they dispensed with names and addressed their remarks to the publications themselves, beginning with the statement, "The learned author of the article *Gnomonique* in the seventh volume of the Encyclopedia wanted to give an idea of it and explained it in this way," and followed by quoting the entry word for word.

Emphasizing that the *Encyclopédie*'s article was based on "false reports," they demonstrated that even the most basic details provided about the prosciutto's design were flawed, from its overall shape to the "toothy" form of the gnomon and the arrangement of the grid:

The false reports which he followed have tricked him, as oftentimes has happened to others who with more enthusiasm than wisdom and travail have hastened to speak about the antiquities of Herculaneum and made them write that which is not and give a most false report on this bronze. Since in the first place the two surfaces of our sundial are neither convex, as he supposes, nor concave but irregular precisely like those of a prosciutto which at one point rise and in another lower themselves and in another part are flat. "The toothed gnomon," which he recalls, and which according to him forms "the fourth part of the diameter" of the instrument, is in truth a piece of the truncated tail of the prosciutto, which does not have teeth of any sort, nor does he say

---

29. B. Tanucci, *Epistolario*, ed. M. G. Maiorini (Rome 1988) 10.102 #74 (September 8, 1761) to Charles III.

of which diameter might be this "fourth part." It is wrong, too, that the "upper surface" is covered in silver, since not only this but the entire piece shows that at one time it had been silver plated through the clear traces visible throughout, and especially on the "lower surface" and between the ripples of the skin near the fat of the prosciutto. It is also false that the upper surface is divided by twelve parallel lines that form as many small squares because the lines, as everyone sees, are not twelve but fourteen, of which seven are straight and parallel and the other seven are not entirely straight nor parallel but composed of small straight lines variously inclined one to the other, and therefore it is clear from the combination of the first with the second that the surface cannot be divided in squares. It is also false that the squares are a bit hollowed out, while the nature of the portions of the surface comprised by the aforementioned lines is the same as the nature of the whole surface, that is, in part convex, in part concave, in part flat. It is moreover false that the six last squares are terminated by the circumference of the circle, of which, in our bronze, no trace is found.

The Accademia had faithfully reproduced the *Encyclopédie*'s rendering of the inscription in their footnote, and this added visual impact to the target of their next assault, the calendrical inscription: "Nor are the initial characters of the months contained in squares and disposed in the manner represented in the figure put forth in the Encyclopedia, their disposition on the bronze being diverse and the characters not enclosed nor divided by any line." The underlying science necessitated the arrangement of the abbreviations of the months, not some archaic mode of inscribing or the whim of the artist: "There is nothing mysterious or extraordinary, finally, in the disposition of the months that is pointed out so often and characterized by the name *boustrophedon*."

The Accademia also took aim at the *Encyclopédie*'s attempt to explain the science behind the sundial, in which the author not only had provided it with an upright gnomon, as if it were a horizontal sundial, but also had failed to recognize that its dial functioned according to the zodiacal calendar rather than the lunar one. They spelled out the principles on which the sundial was designed, emphasizing first such basic concepts as that it was vertical not horizontal and had its dial oriented according to the manner in which the gnomon cast its shadow across the surface throughout the zodiacal cycle.

Our sundial, which is a vertical one, must necessarily be described with the shadows reversed, their length in the entrance of the sun in each sign of the zodiac represented, according to gnomonic rules, by the seven parallel and vertical lines. Now, the author of the instrument, wanting to make the point of the tail of the prosciutto serve as the gnomon and having located this on the left, by necessity had to locate to the right, in the last place, the shortest shadow of the solstice of Capricorn, which is the first of the ascending signs, and to the left, in the first place, the longest of the solstice of Cancer, which is the first of the descending signs, and between these successively the other five, each of these corresponding to the start of two signs, the one ascending and the other descending, which for being equally distant from the first two have the same declination and the same shadow. Therefore in the fourth place, which is that in the middle, is located the equinoctial shadow of Aries and Libra, which are distant ninety degrees from one solstice point to the other; in the second, that of the Twins and Leo, which are distant from Cancer by thirty degrees; in the third, the other of Taurus and Virgo, which are sixty degrees distant from it; in the fifth, the shadow of the sun at the start of the two corresponding signs of Pisces and Scorpio are sixty degrees distant from the solstice of Capricorn; and finally in the sixth, that of Aquarius and Sagittarius which are distant thirty degrees.

Here, and throughout their analysis, the Accademia associated the vertical lines with the traditional zodiacal signs that, because of the effects of lunisolar precession, or the precession of the equinoxes, differed a full zodiacal sign from the current values. So, for example, the vernal equinox occurred in Aries rather than Pisces and the summer solstice in Cancer rather than Gemini (Figure 2.5).

In explaining the relationship between the calendrical notations and the dial's vertical lines as the sundial's artist originally had intended, the Accademia obliquely called attention to the *Encyclopédie*'s blunder in describing the sun's behavior according to the manner in which the months had been paired and boxed together in their representation of the inscription. Instead, the names of the months on the top row of the inscription were associated with the vertical lines to their left and those on the bottom with those to their right; it was this astronomical reality that had resulted in what the earlier commentators, so blind to the instrument's technological achievements,

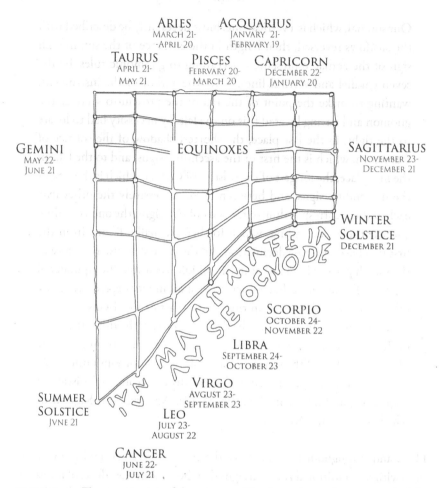

FIGURE 2.5 The organization of the current zodiacal months on the prosciutto's dial; these are one full sign behind antiquity (ancient Cancer = modern Gemini). Spring and fall equinoxes are marked by the central vertical, the solstices are the outer verticals, and readings progress from right to left and left to right over the course of the year. For the hours, see Figure 27. Author's image.

had so enthusiastically identified as the *boustrophedon* organization of the inscription.

> Beyond this, because the author of the sundial knew that the sun travels the ascending signs in the first six months of the year and the descending in the last six, in order to express the times of the successive advancement of the sun from one sign to another (that which, as will be said, matters a great deal for the use of the sundial), he could not

help but indicate the month of January between the lines of Capricorn and Aquarius, the month of February between Aquarius and Pisces, and so forth one by one for all the first six months up to June between Gemini and Cancer, and then turning back for the descending lines, putting the month of July between the lines of Cancer and Leo, August between this and Virgo, and September between Virgo and Libra, and in the same manner between the corresponding signs, October, November, and finally December between Sagittarius and Capricorn.

It was precisely the zodiacal principles underlying the design that addressed the *Encyclopédie*'s wrongheaded concern that this pairing of the months would result in inaccurate readings in the second half of these months.

> It remains also to be explained the answer to the doubt, raised at the end of said article [in the *Encyclopédie*], about the accuracy of the sundial, which is admirable precisely for the great diligence used by the author even in situating the names of the months not beneath the vertical lines but one between the other. Since, to follow the example adopted by said article, while the sun rises in April from Aries to Taurus, its shadow successively grows longer, and so, on the contrary, descending in September correspondingly from Virgo to Libra, the shadow proportionally gets shorter. But the correspondence of the shadow in these two months repeats itself not according to the corresponding days but the corresponding degrees in which the sun, finding itself equally distant from the cardinal points, has the same height and the same length of shadow.

As for Paciaudi's treatment of the sundial in his *Monumenta Peloponnesia*, the Accademia could scarcely be bothered to give it the time of day except to point out that it was egregiously wrong on virtually every detail. They evidently believed that Paciaudi had based his description on that of the *Encyclopédie*, rather than the more likely scenario that it was the other way around or that both had relied on the description provided by the "English gentleman" referenced in the *Memorie*. Neither would explain the source of the illustration Paciaudi had published, which the Accademia clearly relished the opportunity of ridiculing as representing not a prosciutto but "a carafe."

After all of this we do not believe it necessary to linger on what was said in the first part of a book entitled *Monumenta Peloponnesia*, in which one reads the same article as the Encyclopedia transcribed in good faith, without even omitting the *boustrophedon*. And truly it would have been desirable that the editor had used the same good faith in citing the illustrious author he was transcribing. But the pleasure of wanting to oblige the public in communicating an unpublished piece not only made him ignore the Encyclopedia which had preceded him in this news but also seduced him even to add to his own all that which rendered his own work even more lacking. It lacks the history, as it says he had the design of it in 1754 and the instrument was found on June 11, 1755. It lacks astronomy, which one needs in order to explain it, since he gives sufficient indications of having studied everything but that science. It lacks a drawing, since instead of a prosciutto, as is truly the design of this bronze, he gives us a carafe.

The Accademia ascribed these follies to Paciaudi's hubris in wanting to publish the sundial before them, condemning his efforts along with other scholars who had sought to do likewise. In the process, they confirmed that the *Monumenta Peloponnesia* had appeared after they had already begun work on this third volume of *Le Antichità* and in so doing directly connect Paciaudi's denial of access to the royal museum in November 1761 with his book's publication.

And if everything else were there, it lacks the circumspection, the caution, the temperance, the respect in wanting to anticipate a Sovereign who has taken care to publish his own museum (of which, beyond the *Catalogo* [Bayardi's], it is now the fifth year of the Accademia imprinting the Third Volume, the publication of which was started four months before the publication of the aforementioned book), and halt the lubricity, the headlong incontinence, and the thoughtlessness of the hasty, of whom the number is as large as is rare the number of the truly wise.

Finally, they reasserted their predominance as servants of the King over the collection of antiquities under their care, the superiority of their intellectual exegesis, and the authenticity of their presentation, ensnaring in their dragnet of errant scholars the circle of their chief nemesis, the Comte de Caylus (Figure 2.6).

These last know well the gratitude which is owed to the King for the security of the truth which the lovers of antiquity cannot find in editions which are produced by hands other than those provided with

FIGURE 2.6 Portrait of the Comte de Caylus (1692–1765). Oil on canvas by Alexandre Roslin (1718–1793), ca. 1752–1753. Courtesy of the National Museum in Warsaw (M.Ob.1017 MNW). https://cyfrowe.mnw.art.pl/pl/katalog/504741 under a Creative Commons Attribution 3.0 International (CC BY 3.0).

attention and resources by the King. This Company, content with some superficial explanation, decides nothing and, awaiting the best enlightenment from the most enlightened, strives to provide themselves with accurate reports. Sure of their gratitude, it laughs at something useless and, impatient, with childish petulance, with womanly rage, and with empty words it complains of wanting its members to be the sole Oedipuses of the Sphinxes of Herculaneum.

2.b) *Caylus and the Oedipuses of the Sphinxes of Herculaneum*

These final damning words were not merely a reference to the riddle-solving skills of the mythical Theban king but to scholars in the mold of Athanasius Kircher and their self-proclaimed abilities to unravel the secrets of the past.

THE PROSCIUTTO SUNDIAL

Kircher, in his *Oedipus Aegypticus* (Rome 1652–1654), had cast himself in the role of Oedipus deciphering the riddle of the Egyptian hieroglyphs: the frontispiece of his volume depicts him standing before the sphinx solving her riddle through his knowledge. Kircher, too, had been dependent on an international network of scholars whose correspondence supplied him with transcripts and illustrations of the myriad objects of his research. While his scholarship had been well received during his life, already by this time in the eighteenth century, its flaws had been exposed and it had become more an object of ridicule and a curious relic.[30]

The Accademia clearly viewed Caylus' scholarship in a similar light, dependent as it was on a network of correspondents whose information was fallible, their observations unreliable if not utterly wrong, and their collaborative efforts superficial and, worst of all, purposely intended to undercut the work of true scholars engaged in serious empirical research. Caylus had been forced to rely increasingly on written descriptions and illustrations supplied by his circle of friends, the Accademia having largely stanched the flow of antiquities from the Vesuvian excavations. Despite the embarrassment of his earlier incident with Guerra's forgeries and Paciaudi's warning about a black market in counterfeit objects from Herculaneum, Caylus had forged ahead in his quest for additional antiquities from the sites buried by the eruption of Vesuvius to feature in his *Recueil*.[31] In his third volume, published in 1759 after only the first volume of *Le Pitture* had appeared, Caylus had lamented that the Accademia's slothful pace of publication would scarcely keep up with the abundance of riches flowing from the sites on a daily basis; he therefore believed he was providing a useful public service. His attempts to acquire additional finds had been thwarted, however, and so he had been reduced to relying on drawings provided to him for the group of fourteen objects he featured here. He expressed confidence in their accuracy, noting that they included such factual details as the objects' dimensions and material,

---

30. D. Stolzenberg, *Egyptian Oedipus: Athanasius Kircher and the Secrets of Antiquity* (Chicago 2013). Kircher's utterly fantastical translations of the hieroglyphs would not be fully exposed as such until Champollion's preliminary decipherment of the Rosetta Stone in the early nineteenth century.

31. Paciaudi et al. 1802, 244, #59 (June 18, 1761). For details on Caylus' attempts to acquire and publish antiquities from Herculaneum, see H. Eristov, "Les Antiquités d'Herculanum dans la correspondance de Caylus," in G. Cafasso et al., eds., *Il Vesuvio e le Città Vesuviane, 1730–1860: In ricordo di Georges Vallet* (Naples 1998) 145–61. Ironically, Charles III himself would commission Paderni to acquire antiquities in Rome for shipment to him in Spain in 1763–1764; see M. C. Alonso Rodriguez, "La colección de antigüedades comprada por Camillo Paderni en Roma para el rey Carlos III," in J. Beltran Fortes et al., *Illuminismo e Ilustración: Le antichità e i loro protagonisti in Spagna e in Italia nel XVIII secolo* (Rome 2003) 29–45.

Analysis   101

and he had supplemented his own discussions of several of the objects with information culled directly from Bayardi's catalog. Several of the objects were tripods, so to provide some context he observed that Greek and Roman houses had been heated by means of tripods, like those used for sacrifices in temples, and that they came in all shapes and sizes, but were cast preferably in bronze. One of his principal examples employed sphinxes as supports while the other was the tripod with ithyphallic satyrs from Pompeii whose arrival in the museum La Condamine had witnessed (see Figure 1.14). As if to rub salt in the wound, Caylus remarked that both "have the reputation of being the most significant," and indeed the satyr tripod now adorned the first room of the museum at the royal palace in Portici. He had been careful not to reveal his source: "The fear of attracting disgust to the clever artist who communicated to me these designs, obliges me to hide his name, because antiquity in Naples is a matter of State and it is with a great anxiety that I communicate the monuments of that country."[32] Their appearance here must have infuriated the members of the Accademia, as it clearly had Tanucci, who wrote to Charles III blaming Paderni for the leak.[33] But Caylus forged ahead, openly acknowledging in his fifth volume that Paciaudi was his source for an antiquity from Naples.[34] The drawings of both tripods are remarkably poor—the sphinxes are the wrong scale, the satyrs' ithyphallic nature partially censored— bearing all the flaws and lacunae of objects reproduced from memory, in

---

32. Caylus 1752–1767, 3.142–47, plates 37–38. Bayardi 1755, 293–94 #4 (sphinx), 420–23 #709 (satyr). The Accademia had produced an engraving of the satyr tripod but never published it. For this engraving, among the so-called *Rami Inediti* (unpublished plates), see Pagano and Prisciandaro 2006, 2.333, fig. 23; and in general, Esposito 2013, 162–67.

33. B. Tanucci, *Epistolario*, ed. M. G. Maiorini (Rome 1988) 10.51 #35 (August 22, 1761). Paderni, in turn, appears to have fingered Piaggio. Tanucci had written to the King, "I spoke to Paderni about the Marchese de Caylus' book, cautioning him strongly about it, and he told me after a long time that he is not responsible because Father Antonio [Piaggio] passes through those rooms [where the finds were stored], and his assistant Merlo, about whom he spoke not at all positively. I think those two should move elsewhere so that Paderni would be responsible."

34. Caylus 1752–1767, 5.170–71, plate 61. Caylus' fifth volume appeared in 1762, the same year as the Accademia's third volume. Paciaudi et al. 1802, 100, #23 (December 19, 1759): "A scoundrel of a worker has stolen three small bronze statues representing Hercules, one of which is mediocre and has a broken leg, the other two are not worth the devil. Since they cost me little money, I am sending them to you; you can discard them or give them to whomever you please. The broken pieces are wrapped in paper. But be careful not to say where it comes from, because it could cause me a scene with the minister of Naples [Tanucci] and have the worker driven out"; 300–3, #72: In a letter from 1763, Paciaudi described a small mosaic panel (*emblemma*) depicting street musicians signed by the artist Dioskurides of Samos (MANN 9985), one of a pair of similarly exceptional mosaics discovered in Pompeii on April 30, 1763 (see Chapter 2.h), and concluded "May God bless them all: they do not deserve to be so happy."

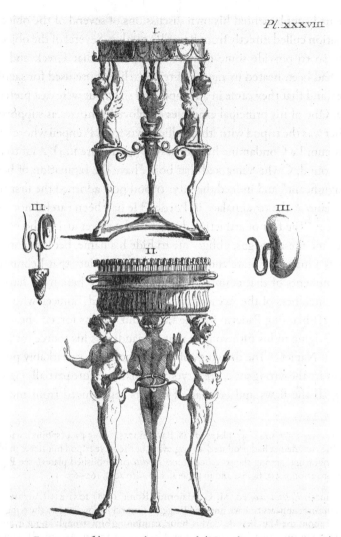

FIGURE 2.7 Engraving of bronze sphinx tripod (Herculaneum or Pompeii) and the ithyphallic satyr tripod (*Praedia Iuliae Felicis*, Pompeii; MANN 27874), with two earrings. From volume 3 of Caylus's *Recueil d'Antiquités Égyptiennes, Étrusques, Grecques, Romaines et Galouises* (Paris 1759) plate 38.

keeping with most of the unauthorized publications of the antiquities in the museum that appeared in this period, Paciaudi's "carafe" being one of the more glaring examples (Figure 2.7). By 1763, Caylus and Paciaudi were close to executing their most audacious plot: to purloin one of the papyrus fragments recovered from the Villa dei Papiri for Caylus to study and publish. "If this enterprise succeeds," observed Caylus to Paciaudi, "*this* is what will send good old Tanucci and all the Neapolitan Academicians up in smoke once and for all!"[35]

As far as the Accademia was concerned, Paciaudi was as guilty as Caylus: guilty of unauthorized publication of the Vesuvian finds and guilty of collaborating with Caylus. As late as 1761, during his time in Naples, Paciaudi had written to Caylus confiding in him that he could acquire antiquities and that he would do so even though the Neapolitan court now required an export permit.[36] Only days later, Paciaudi's own access to the antiquities in the museum had been cut off. His risible treatment of the prosciutto served as the perfect object lesson for the Accademia on the pitfalls that awaited similar Oedipuses whose scholarship was based on clandestine correspondence rather than visual inspection of the objects themselves. Caylus, for his part, remained unrepentant: the frontispiece of his sixth volume of the *Recueil* in 1764 featured a winged representation of Time, charging forth while balanced on a wheel and wielding a scythe, the surrounding landscape scattered with sphinxes decapitated, buried, or whole, and the caption *Tot Sphinges Multos Quaerunt Aedipodas*: "So many sphinxes seeking many Oedipuses"[37] (Figure 2.8).

---

35. Caylus et al. 1877, 332 (July 11, 1763). Caylus had first raised this idea in 1759; see Eristov 1998, 151.

36. Paciaudi et al. 1802, 257–59 #63 (August 29, 1761). In the same letter, he warned Caylus that he had heard the court planned to register a complaint with the French court about Caylus' unauthorized acquisitions and publications and advised him to write a retraction, observing, "Is there anything stranger? Has it ever happened that sovereigns make war on philosophers and antiquarians?" See also Caylus et al. 1877, 261–62 (October 12, 1761), in which Caylus sought to assuage Paciaudi's concerns by observing that he planned to publish only a surgical instrument and a cistern head from the Vesuvian sites in his forthcoming fourth volume of the *Recueil* (173–75, plate 58.1, cistern head; 325, plate 100.3, instrument). The Bourbon court had required a permit to export antiquities beginning already in 1755; for a summary, see D'Alconzo 2001, 507–37.

37. Caylus 1752–1767, 6. Frontispiece. Caylus described the image as representing "Time in the countryside: It is surrounded by sphinxes, some are discovered in full, others half hidden in the ground and almost destroyed, in general all their parts are scattered. Time cannot be fixed or stopped, it is standing not on a ball like Fortune but on a wheel: the movement of the latter being regular and continuous and that of the former variable and inconsequential." Caylus was signaling his intention to march on in his scholarly endeavors. Indeed, this volume featured two plates of objects from the Vesuvian sites (6.206–15, plates 66–67) about which he remarked, "I must admit that, not having the pieces before my eyes, I cannot answer for their precise proportion or their form and their material. The reader must at least be grateful to me for the pains which I have given myself to present to him some pieces of good taste, drawn from this celebrated place, and which differ from the monuments of painting." He did not name the drawings' artist. For a sampling of Caylus' acquisitions ascribed to the Vesuvian sites, though of dubious provenance and date, see the exhibition catalog *Caylus, Mécène du Roi: Collectionner les antiquités au XVIIIᵉ siècle* (Paris 2002), e.g., 144–45 #60–63.

TOT SPHINGES MULTOS QUÆRUNT ÆDIPODAS

FIGURE 2.8 "So many sphinxes seeking many Oedipuses." Representation of "Time." Frontispiece from volume 6 of Caylus's *Recueil d'Antiquités Égyptiennes, Étrusques, Grecques, Romaines et Galouises* (Paris 1764).

Analysis 105

In this case, however, the target of the Accademia's wrath was specifically Paciaudi, Caylus' acolyte in Italy, who evidently had fancied himself an Oedipus during his last visit to Naples in 1761. Word of their intention to castigate him in print had spread through scholarly circles well in advance of the volume's publication. From Rome, Giuseppe Spinelli, bishop of Ostia and the one-time cardinal of Naples who had sponsored Paciaudi's studies during their years there, had alerted Winckelmann, also a friend of Paciaudi, who in turn had passed the news along in a letter to his Swiss friend Leonhard Usteri in December 1762:

> I await the third volume of the Paintings of Herculaneum in which the proud Father Paciaudi is put down in a manner cruel and unprecedented for our civilized times. After having made general complaints in the text of the Preface against those who furtively pass around drawings of the monuments with partial explications, the Jesuit father is cited because in his publication with its sensational and pompous title intended to muscle its way through all the antiquarians [written in Italian: *col titolo strepitoso, ampulloso e da farsi largo fra tutti gli antiquari*] of *Monumentis Peloponnesiacis* [sic] included and explained quite well with the advice and aid of another [written in Latin: *consilio et ope alterius*] the bronze sundial in the shape of a small ham in the museum. After which, below the text, follows an annotation of sixty-one lines concerning this Jesuit. Cardinal Spinelli read to me from a letter which roughly says "This young man (who is nevertheless a man of some 50 years) appeared some time ago in Naples giving himself the air of Oedipus: with decisions made headlong, and trying to assert his authority with his book written in feminine terms inlaid with Greek, *when everyone knows he knows nothing about it*" [written in Italian: *Questo giovane comparve tempo fa a Napoli dandosi l'aria d'Edipo: con decisioni fatte a piombo, e procurando d'imporre col libro suo scritto con termini femminili, intarsiato di Greco quando ognuno sa che non ne sa niente*]. This young man…" it continued, but I forget the words and I remember only the substance of the matter. I then felt my heart more beset than ever by the truth, because this Eminence [Spinelli] had asked me several times my feelings about this friend [Paciaudi]. The Prince of Francavilla did everything possible to avert this terrible blow, but Tanucci has been relentless. War is declared between him and me, and I will no longer visit him. I'm speaking of Tanucci because despite

THE PROSCIUTTO SUNDIAL

our correspondence he received me in such a way that nothing could commit me to return to his home.[38]

2.c) *The Accademia Strikes Back: The Preface to Volume III of* Le Antichità d'Ercolano

Having devoted the better part of the first three pages of their Preface to this criticism of d'Alembert's and Paciaudi's unauthorized publications, the Accademia now turned to highlighting the sundial's unique qualities, especially those which previous scholars had misrepresented or bungled entirely.[39] They underscored that it was distinctive not only because it was a sundial but because it was portable and precise.

> An ancient sundial is a piece which for its rarity would do honor to any more select and rich museum, but beyond this there is none from those few about which we have news that corresponds with ours because the others are either horizontal, or fixed, or concave and of such a difficult and intricate construction that not everyone can be sure they are sundials. Ours, on the other hand, is portable and vertical, and is most unusual for its simplicity and quick and easy use, and for the detailed information it provides of the entire motion of the sun through the ecliptic in all the months of the year, identified by their names.[40]

This initial statement had required three explanatory footnotes to place it into its proper historical, archaeological, and scholarly context. Of particular relevance to their discussion was Giovanni Luca Zuzzeri's *D'una antica villa scoperta sul dosso del Tuscolo, e d'un antico orologio a sole tra le rovine della medissima trovato: Dissertazioni due* (Venice 1746), a detailed treatment of a

---

38. H. Mener and J. Schulze, eds., *Winckelmanns Werke*, 11 vols. (Berlin 1824.) 10.150–153: Letter 219, from Rome to Usteri (December 14, 1762). The prince of Francavilla was Michele Imperiali Simeana, a wealthy nobleman of Genovese decent who housed his collection of antiquities and hosted lavish gatherings, often for visiting foreigners, in the Palazzo Cellamare in Naples and a villa in Portici; a friend of Casanova, he later served as majordomo to King Ferdinand IV. Tanucci had not given Winckelmann a warm reception during his second visit to Naples in early 1762.

39. La Condamine escaped the Accademia's wrath because his description was published in the same year as this volume of *Le Antichità* and was therefore unknown to them.

40. *PdE* 3.v–viii.

horizontal or "hemicyclium" sundial of marble found intact, except for its gnomon, during excavations at Tusculum at the site of the Villa Rusinella, southeast of Rome. Zuzzeri had argued the villa may have belonged to Cicero. Cicero had written to his freedman Tiro, in residence at the Tusculum villa in 44 BCE, that he was sending him a sundial so this had the potential of elevating the significance of the find substantially.[41] Moreover, the villa's excavator, R. J. Boscovich (Ruđer Bošković), a renowned Jesuit astronomer in his own right, had concluded that the design of the sundial was based on that of Berosus the Chaldean, one of the designers of sundials cataloged by Vitruvius (Vitr. 9.8.1).[42] The sundial had ended up in the Museum Kircherianum, Kircher's celebrated collection of antiquities, natural-history specimens, and technological and scientific curiosities housed in the Jesuit Collegio Romano in Rome. As the earliest Roman sundial of its kind recovered archaeologically, on display in a museum, from a known site, perhaps having belonged once to a renowned Roman, and illustrating an example described in ancient literature, it served as the perfect foil to the prosciutto sundial.[43] The Accademia also contrasted their sundial with the circular bronze disk identified as a portable sundial by Gianfrancesco Baldini in 1741 by noting that "while this is portable, like ours, it is different not only because it is horizontal and does not have the lines representing the extension of the shadow and the course of the sun through the twelve signs of the Zodiac, but more because (if in fact it is a sundial rather than another instrument) its design is so composed and intriguing that even after the erudite illustration by the most clear Father Baldini its

---

41. Zuzzeri 1746, 61–86 (on the sundial). For the reference, see Cicero *Ad Fam.* 16.18: *Horologium mittam et libros, si erit sudum* (I will send the sundial and books, if it will be dry).

42. Boscovich had published his more detailed discussion of the sundial's underlying science in "D'una antica villa scoperta sul dosso del Tusculo, d'un antico orologio a sole, e di alcune altre rarità, che si sono trà le rovine della medesima ritrovate (Art. XIV)," which appeared in the *Giornale de' Letterati di Roma* (1746) 115–35. The Accademia also refer to the brief survey of the history of sundials in M. Falconet's "Dissertation sur Jacques de Dondis, auteur d'une horloge singulière; et à cette occasion sur les anciennes horloges," *Memoires de Litterature, tires des registres de L'Académie Royale des Inscriptions et Belles-Lettres*, 20 (1753) 440–58, which includes many of the same ancient literary references to sundials that had become standard fare in discussions of ancient sundials. Given the brevity of this latter source, and the fact that more scholarly ones were available, its selection by the Accademia may have been intended to demonstrate their familiarity with French sources, as a pointed comment to the authors of the *Encyclopédie*.

43. The Tusculum sundial routinely appears as the first example in modern histories of sundials; e.g., A. J. Turner, "Sundials: History and Classification," *History of Science* (1989) 308; Severino 2011, 44–45; Gibbs 1976, #1035G 151 (tentatively identified with Zuzzeri's); Gatty 1900, 31.

operation remains uncertain and doubtful."[44] Finally, in a lengthy footnote, the Accademia sought to differentiate their sundial from those mentioned in the literary and juridical sources, first establishing that Vitruvius' reference to *viatoria pensilia* (Vitr. 9.8.1) was the appropriate term for such portable sundials. Their exegesis began with Egyptian priests wielding sundials in their sacred rites and referenced the fifth-century work of Horapollo and his treatise on hieroglyphs in which he explains (1.16) that the representation of a sitting *cynocephalus* signified the two equinoxes because these dog-headed apes urinated once an hour twelve times a day; *cynocephali* were therefore depicted on *hydrologia*. But they dwelled in particular on Bato's philosopher referenced in Athenaeus scrupulously checking his *clepsydra* and argued it must have been made of glass in order for him to discern its contents.[45] They ended by noting Papinianus' legal determination that a portable bronze sundial was not mere furniture but a domestic instrument because, concluded the Accademia, a sundial was not ornamentation but a necessity.[46] Here, and throughout their excursus, they sought to demonstrate their deep erudition by employing many of the same sources as previous commentators but coming to different conclusions.

The sundial "represents (just as is shown in the engraving) the shape of a prosciutto," continued the text of the preface, "suspended at the foot by a moveable ring, and on the back, that is, from the part of the rind, is drawn the sundial for which the tail of this same prosciutto serves as the gnomon. And this is worked with such skill and mastery that all the parts are seen to be expressed with utmost precision and vividness." These details prompted inquiries into the literary evidence for prosciutto—the food—in antiquity,

---

44. Baldini 1741, 185–94. Baldini was an antiquarian and numismatist. Despite the Accademia's skepticism, the disk is commonly recognized as a type of sundial, though certainly of later date; see Talbert 2017, 24–28 (date uncertain; now lost).

45. In determining that Bato's *clepsydra* was glass they rejected the interpretation of the sixteenth-century commentator on Athenaeus, Isaac Casaubon (1559–1614), who had argued it was made of leather, terracotta, or bronze: *Animadversiones in Athenaei Deipnosophistas libri XV* (1612) 368.

46. *Corpus Iuris Civilis*, ed. P. Krueger and T. Mommsen (Berlin 1893) 1.470: Papinianus *Digest* 33.7.23: *Papinianus quoque libro septimo responsum ait: sigilla et statuae adfixae instrumento domus non continentur, sed domus portio sunt; quae vero non sunt adfixa, instrumento non continentur, inquit: suppellectili enim adnumerantur, excepto horologio aereo, quod non est adfixum: nam et hoc instrumento domus putat contineri* ("Papinianus even responded in his seventh book that fixed figurines and statues are not reckoned among the equipment of the house but are a part of the house; those things that are not fixed are not reckoned equipment: utensils are counted, with the exception of a bronze sundial, which is not fixed: for even this he reckoned among the household equipment").

Analysis 109

summarized in the accompanying footnotes. The most esteemed prosciutto came from Gaul (Varro *Rust.* 2.4.10) and Spain (Strabo 3.162: the hams of the Carretanians rivaled those of Cantabria; Martial 13.54). The leg was commonly referred to as *perna* (Apicius 7.9), the term *petaso* interpreted as encompassing the entire ham, though a *perna* alone could provide up to four meals (Martial 10.48). The *petaso* was served fresh, according to Apicius (7.9.3) and Martial (*Ep.* 13.55), while the *perna* was first salted and then smoked for two days, in an abbreviated form of the more detailed recipe provided by Cato (*Agr.Orig.* 162).[47] Despite maintaining that "everything that can be said about the sundials of the ancients is well-known and trite," the Academicians proceeded to summarize the main literary sources for the history of the two principal types, those using water and those dependent on the sun, devoting the bulk of their attention to Greek examples and ending with the terse observation that "regarding the Romans, the use of sundials was relatively late."

The meaning of this "artist's joke" of inscribing a sundial onto a prosciutto stimulated a circuitous inquiry that began by seeking to link the design to the name of the sundial's artist or owner. The *nomen* (family name) "Suillius," from the adjective *suillus* or swine, was attested on an inscription from Rome cataloged by Gruterus that provided the full name of P. Suillius Celer.[48] Or it may have been inspired by the *cognomen* (nickname) "Perna," or "Scrofa," sow, as came to be used of Tremellius in a story retailed by Macrobius (*Sat.* 1.6).[49] Or perhaps it was an allusion to the manners of the parasites, accustomed to measure the length of shadows to determine whether it was time to eat.[50] This reference led to another, from the *Noctes Atticae* (*Attic Nights*) of Aulus Gellius (*NA* 3.3), quoting a passage from a lost play of Plautus in which the parasite curses the inventor of the sundial for restricting his dining to the hours fixed by the sun rather than the natural pangs of his belly.

---

47. Apicius recipe was for braised *perna* with figs slow-baked in dough (7.9). A graffito (*CIL* IV.1896) scrawled on a wall of the Basilica at Pompeii observes, "When there is cooked ham, if it is served to a diner, he does not taste the ham, he licks the bowl and the cooking pot" (*ubi perna cocta est si convivae apponitur non gustat pernam linguit ollam aut caccabum*).

48. J. Gruterus, *Inscriptiones antiquae totius orbis Romani* (Amsterdam 1707) 226 (CIV.6.8) = *CIL* 6.199.

49. When his neighbor came looking for his sow killed by Tremellius' slaves, he hid it in bed alongside his sleeping wife and then swore to the neighbor that the only sow in the house was in his bed.

50. Plutarch, *Quomodo adulator ab amico internoscatur*, 4: this *Moralia* entitled "How to distinguish a flatterer from a friend" advised avoiding the man who is "never caught noting the shadow on the sundial to see if it is getting towards dinner-time." The Accademia explored this relationship between time and the length of shadows in great detail; see Chapter 2.d.

THE PROSCIUTTO SUNDIAL

Then the hungry parasite says, "Let the gods destroy him who first devised hours and who first set up this sundial here and broke up the day into tiny bits for miserable me. For when I was a boy, my stomach was the best and truest sundial of all: when this advised you, you ate, unless there was nothing; now, even what there is is not eaten, unless the sun says so. The city is so full of sundials that the majority of people creep around withered by hunger." [51]

The power of sundials to affect life's routines was further illustrated by a passage from a lost play of Menander, *Orge* ("Anger"), cited by Athenaeus (*Deip.* 6.243a), in which a man is likened to the parasite Chaerephon, who was said to have been invited to a "twelve-foot dinner" (εἰς ἑστίασιν δωδεκάποδος), understood to mean when the sun cast a shadow that could be paced off with twelve feet: "rising at dawn, examined the shadow cast by the moonlight and ran as though he were late, arriving at daybreak." [52] Sources documenting the role of slaves in announcing meal times were then explored, including the sundial coupled with a trumpeter employed by Trimalchio in his triclinium to remind him of how rapidly his life was wasting away. [53] Trimalchio's was likely a *clepsydra*, the Academicians hastened to clarify, the ability of the sun's rays to enter a dining room having been called into question, in the same way that

---

51. Aulus Gellius, *NA* 3.3: the topic of discussion at this point in the narrative is not about sundials but whether the language of the passage, said to be from Plautus' lost play *Boeotia*, is genuinely Plautine: *Parasitus ibi esuriens haec dicit:/ut illum di perdant, primus qui horas repperit,/quique adeo primus statuit hic solarium!/qui mihi conminuit misero articulatim diem./ Nam me puero venter erat solarium/multo omnium istorum optimum et verissimum:/ubi is te monebat, esses, nisi cum nihil erat./Nunc etiam quod est, non estur, nisi soli libet;/itaque adeo iam oppletum oppidum est solariis,/maior pars populi aridi reptant fame.* A. S. Gratwick, "Sundials, parasites, and girls from Boeotia," *Classical Quarterly* 29.2 (1979) 308–23, suggests Plautus modeled his parasite's cursing of sundials on a play by Menander, perhaps similarly entitled *Boeotia*, by noting the parallels in a letter (3.3) from Alciphon's "Letters of Parasites" in which the starving writer plots to throw down a sundial or bend its gnomon because it is not yet marking the sixth hour (Ὁ γνώμων οὔπω σκιάζει τὴν ἕκτην, ἐγὼ δὲ ἀποσκλῆναι κινδυνεύω τῷ λιμῷ κεντούμενος).

52. Menander, *Orge* K. 364: διαφέρει Χαιρεφῶντος οὐδὲ γρῦ ἄνθρωπος ὅστις ἐστίν, ὃς κληθείς ποτε εἰς ἑστίασιν δωδεκάποδος ὄρθριος πρὸς τὴν σελήνην ἔτρεχε τὴν σκιὰν ἰδὼν ὡς ὑστερίζων καὶ παρῆν ἅμ᾽ ἡμέρᾳ (Whoever the man is he differs not one whit from Chaerephon who, invited once to a banquet when the sun's shadow marked twelve feet, rising at dawn, examined the shadow cast by the moonlight and ran as though he were late, arriving at daybreak). Athenaeus' catalog of references to the character of Chaerephon shows that he had become the archetypal clever parasite used by other comic poets. Athenaeus (*Deip.* 1.8b–c) also retails a similar version of this story, quoting the comic poet Eubulus, but ascribing the incident to one Philocrates who had been invited to a "twenty foot dinner"; see Chapter 2.d.

53. Petronius 26: *Trimalchio, lautissimus homo, horologium in triclinio et bucinatorem habet subornatum ut subinde sciat quantum de vita perdiderit.*

Analysis   111

a properly maintained *clepsydra* had summoned Sidonius to dine in the fifth hour at his host's house in Gaul (Sidonius *Ep.* 2.9). All of this was sufficient evidence for the Accademia to connect the prosciutto sundial with dining, while Vitruvius' (Vitr. 9.8.1) examples of sundials deriving their common names from their shapes—Patrocles' ax-shaped "Pelecinum," Dionysodorus' "Cone" (*conus*), Apollonius' "Quiver" (*pharetra*)—demonstrated that the names of sundials could be derived from their shapes: "Therefore," to bring the argument full circle, "in the same manner it could be said that our sundial had been called *perna*, the prosciutto."[54]

The main discussion then moved from the general to the more specific, focusing first on the physical details rendered on the surface and then plotting them onto their temporal equivalents to explain how they functioned together. Such observed, factual details required significantly less scholarly explication.

> On the rind, then, are observed seven vertical lines, beneath which the twelve months of the year are read in two rows. Starting from the last line, which is the shortest, and moving backward all the way to the first, which is the longest, IAnuarius and beneath it DEcember, FEbruarius and beneath it NOvember, MArtius and beneath it OCtober, APrilis and beneath it SEptember, MAjus and beneath it AVgustus, IVNius and beneath it IVlius.

This listing of the months merited a reference to Censorinus (*de Die Natali* 22), who provided the proper names of all twelve months, distinguishing the "natural" months, which comprised both the solar months corresponding with the signs of the zodiac as well as the lunar months dependent on the intervals between moons, from the "civil" months, defined as being peculiar to, and determined independently by, each nation-state according to its needs. Censorinus also surveyed the evidence for whether the months had been named by the Romans themselves or they adopted them from the Latins, and he noted how Quintilis had become Iulius in the fifth consulship of Julius Caesar and Sextilis had become Augustus in the twentieth year of that emperor's reign. While subsequent emperors had sought to change the names of the months during their reigns, all had reacquired their traditional names

---

54. Gibbs 1976, 59–65, for a discussion of Vitruvius' sundial shapes; P. Pattenden, "Sundials in Cetius Faventinus," *Classical Quarterly* 29.1 (1979) 203–12, discusses this fourth-century CE architect's description of the design of the *pelecinum* and *hemicyclium* sundials.

THE PROSCIUTTO SUNDIAL

following those emperor's deaths. Having already dismissed the *boustrophedon* arrangement of the calendrical notation as "nothing mysterious or extraordinary," the Accademia entirely ignored it here.

They then explicitly correlated the horizontal lines on the dial and the calendrical notations with the zodiacal calendar and the seasonal solstices and equinoxes:

> In addition to these seven vertical lines are shown seven other transversal lines, some of which are parallel to the horizon, and some variously inclined to the horizon. The use of the first as well as the second is most clear. The seven vertical and parallel lines show with their length the extension of the shadow which the gnomon throws as the sun enters in each sign of the Zodiac, and with their position they indicate the parallel and successive passage of the sun from one to the other of the twelve celestial signs, thereby they come to represent all together the motion of the sun through the entire ecliptic. The first line, which is the longest of all, shows the length of the shadow of the gnomon in the entrance of the sun in the sign of Cancer, or rather in the summer solstice, which falls in the month of June. The month is noted expressly next to this same line. The last line, which is the shortest of all, indicates the shadow of the gnomon in the entrance of the sun in the sign of Capricorn, or rather in the winter solstice, which occurs in the month of December, therefore therein written beneath. The middle line, which is the fourth, represents the quantity of the shadow in the entrance of the sun in the equinoctial sign of Aries and of Libra in the months of March and September, which are written below. In the same manner the second line indicates the parallel of the sun and the shadow at the start of the signs of Leo and Gemini, which the sun enters in the months indicated therein of July and May. The third line points out the start of the signs of Virgo and Taurus in the months which are read there of August and April. The fifth line expresses the shadow in the entrance of the sun in the signs of Scorpio and Pisces in the months notated there of October and February. And, lastly, the sixth line corresponds to the beginning of the signs of Sagittarius and Aquarius in the months of November and January.

As a reference for their knowledge of the organization of the Roman calendar and its relationship to the zodiacal signs, the Accademia credited Lambecius' 1671 publication of the codex-calendar of 354 CE, which not only included a section on the signs of the zodiac but also offered illustrations representing

Analysis   113

the months, accompanied by Latin verses recording pagan festivals and traditions.[55] Ausonius, too, had equated the zodiacal signs with the months in his *Eclogarum liber*, a fourth-century collection of astronomical and astrological versifications in epic and elegiac meter.[56]

The vertical, "horary" lines then needed to be correlated with the division of the Roman day into twelve equal hours, six hours before and six hours after midday:

> The seven transverse lines on our sundial give the twelve hours of the day, that is, both the six hours before midday and the six hours after. In this way, the shadow of the gnomon descending step by step through each of these, in touching the second line (counting from the top to the bottom) indicates the first hour of the rising sun, the third line the second hour, the fourth line the third hour, the fifth line the fourth hour, the sixth line the fifth hour, and the seventh line the sixth hour or rather midday. After these, the shadow returning up, the sixth line signals the seventh hour (or rather the first after midday), the fifth line the eighth hour, the fourth line the ninth hour, the third line the tenth hour, the second line the eleventh hour, and the first line the twelfth hour, in which the sun sets (Figure 2.9).

The full explication of how the Romans conceived of, divided, and utilized the hours of the day required six footnotes of references, the last of these filling over a page of text in small font. The Accademia began with the statement by the Roman antiquarian Varro that the Roman day extended from one midnight to the next.[57] This was known as the "civil" day of twenty-four hours, as

---

55. *Petri Lambecii Hamburgensis… Commentariorum De Augustissima Bibliotheca Caesarea Vindobonensi Liber 4* (Vienna 1671) 289–340. Lambecius (Peter Lambeck, 1628–1680) was the keeper of the imperial library of Emperor Leopold and created a catalog of his library, completing eight of a projected twenty-four volumes before his death. The Accademia referenced as well the work of J. G. Graevius, *Thesaurus antiquitatum Romanarum: in quo continentur lectissimi quique scriptores, qui superiori aut nostro seculo Romanae reipublicae rationem, disciplinam, leges, instituta, sacra, artesque togatas ac sagatas explicarunt & illustrarunt*, 8 vols. (Leiden 1698), which cataloged the texts of other Roman calendars, such as the *Fasti Capitolini*. For the codex-calendar, see M. Salzman, *On Roman Time: The Codex-Calendar of 354 and the Rhythms of Urban Life in Late Antiquity* (Berkeley 1990).

56. Ausonius 13.16: e.g., section entitled *In quo mense quod signum sit ad cursum solis*, "In which month which sign relates to the course of the sun."

57. The reference to this definition in Varro's *Antiquitates rerum humanarum et divinarum (Antiquities of human and divine matters)* was made by Aulus Gellius in his *Noctes Atticae* (3.2). Varro's work, originally in forty-one books, is known today only in fragments, such as this quote.

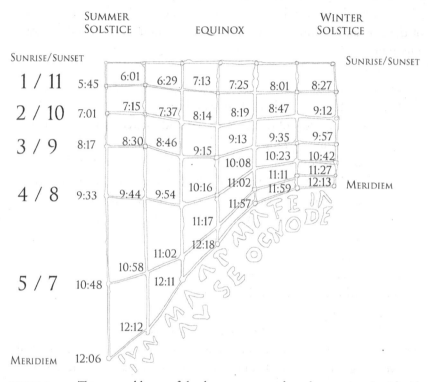

FIGURE 2.9 The seasonal hours of the day as represented on the prosciutto's grid and the times corresponding with the first six hours of the day in March, June, July, August, October, November, and January of 28 CE. Author's image.

opposed to the "natural" day, which ran from sunrise to sunset; this was how Censorinus had defined them (*de Die Natalis* 23). The natural day was divided into twelve equal parts, as was the night, also according to Censorinus. Because the hours of the day and the night are equal in length only on the equinoxes, and the longer hours of the days in the summer result in shorter hours in the night, with the opposite condition being operative in the winter, the term *hora hiberna* (winter hour; Plautus, *Pseudolus* 5.2) came to refer to a short hour while *hora aestiva* (summer hour; Martial 12.1.4) designated a long one. "All the attention, therefore, of those who made sundials was to ensure that the days, whether equinoctial, solstitial, or of another time, were divided into twelve equal parts," noted the Accademia. As to how the hours of the day

Analysis

were spent, a précis of Martial's epigram (4.8) to Euphemus was sufficient for spelling this out: the first six hours were devoted to business, the other six to care of the body and repose.[58] The first six hours of the day were considered the best, as both Vergil (*Aen.* 9.156) and juridical sources confirmed.

Next, the Accademia scoured the literary sources providing evidence for mealtimes in antiquity. After determining that the ancients ate three times a day, they explained what Homeric heroes ate for breakfast and concluded that their hardiest meal was at sunset. Similarly, the Romans originally had three meals a day, but this expanded to four and even five times a day, as established by both ancient literary sources and modern studies that discussed the Romans' eating habits. The main meal for the Romans was at sunset, after the end of work in town and country, proved by reference to the required literary sources, such as Statius' *Silvae* 4.6.3, where he had been invited to dinner at Nonius Vindex's house late in the day (*iam moriente die*) and Pliny's *Epistle* 3.1.9 in which he stated that dinner in winter was at the tenth hour, and at the ninth in summer. The Accademia subsequently built a case to demonstrate that the literary sources established the ninth hour or shortly thereafter as the ordinary time for *cena* (dinner) in Rome, which was distinguished from *prandium*, the midday meal. In the Republic, the *cena* taken at the ninth hour was the main meal of the day, but the Accademia hastened to note that there were some idle folk, given over to luxury, who ate handsomely twice a day, at *pranzo* and *cena*, a practice that became more common with "serious people," as indicated by Sidonius (*Ep.* 2.9.6), who had also documented the practice, referenced by other sources, of bathing after a meal.[59]

Having established these temporal minutiae, the Accademia's next footnote also was tied to their spelling out of the correspondence between the horizontal

---

58. Martial *Ep.* 4.8: *Prima salutantes atque altera conterit hora/exercet raucos tertia causidicos/in quintam varios extendit Roma labores/sexta quies lassis, septima finis erit/sufficit in nonam nitidis octava palaestris/imperat extructos frangere nona toros:/Hora libellorum decuma est, Eupheme, meorum,/temperat ambrosias cum tua cura dapes/et bonus aetherio laxatur nectare Caesar/ingentique tenet pocula parca manu./Tunc admitte iocos: gressu timet ire licenti/ad matutinum nostra Thalia Iovem* (The first and second hours of the day wear down those paying their respects; the third keeps the raucous lawyers busy; Rome spreads out various tasks to the fifth hour; the sixth brings rest to the weary; the seventh ends the day. The eighth to the ninth is enough for the oily palestra; the ninth enjoins us to throw ourselves on the bountifully furnished dining couches. The tenth is the hour for my books, Euphemus, when your skill tempers ambrosial feasts and good Caesar relaxes with celestial nectar and holds the little goblets in his huge hand. Then let loose the jokes: Thalia fears to approach Jupiter in the morning with unrestrained step).

59. Sidonius (*Ep.* 2.9.6): *prandebamus breviter copiose, senatorium ad morem . . .* (2.9.8) *balneas habebat*: "we ate copiously for a brief time, in the manner of senators . . . he had a bath." Oddly, they overlooked the reference to the after-dinner bath taken by Trimalchio and his guests (Petr. *Sat.* 73).

116   THE PROSCIUTTO SUNDIAL

lines on the sundial and the hours of the day. For 460 years, they noted, the
Romans had neither sundials nor any terms for designating time. This was
borne out by the fact that the Twelve Tables, the written laws of Rome dating
to 450 BCE, only referred to midday and the rising and setting of the sun;
this was verified by Aulus Gellius (*NA* 17.2), Censorinus (*de Die Nat.* 23), and
Pliny the Elder (*HN* 7.60), the latter of whom had clarified that the Twelve
Tables mention only sunrise and sunset.[60]

2.d) *Chaerephon's Twelve-Foot Dinner: Deciphering*
Greek and Roman Time

The explication of the correspondence between the horizontal lines on the
prosciutto and the hours of the Roman day, and specifically their reference to
the first line corresponding with the "twelfth hour, in which the sun sets," had
led the Accademia back to the matter of the "twelve-foot dinner" to which
Chaerephon had been invited in the lost play of Menander. They had earlier
concluded that this meant Chaerephon should arrive when the shadow cast
by the sun could be paced off in twelve feet. This seemingly primitive approach
to timekeeping now merited a closer examination of those literary sources in
which a time had been expressed in terms of the length of a shadow and how
such temporal expressions could be mapped onto the fixed hours of the day
captured on the surface of a sundial. This allowed the Accademia, in the final
footnote of this section of their commentary, to wade into a long-standing
debate carried on between modern chronologers concerning the evolution
of the methods employed by the ancients—including the Babylonians and
Egyptians—for reckoning time.

The Accademia limited their focus to the relevant Greek sources, all of
which dealt with dinner times. This was not because the Accademia had sam-
pled this evidence alone; it was because all the references to specific times in

---

60. Aulus Gellius was referring to table 1.6–9 of the Twelve Tables: "When the litigants settle
their case by compromise, let the magistrate announce it. If they do not compromise, let them
state each his own side of the case, in the *comitium* of the *forum* before noon. Afterwards let
them talk it out together, while both are present. After noon, in case either party has failed to
appear, let the magistrate pronounce judgment in favor of the one who is present. If both are
present the trial may last until sunset but no later." Censorinus stated, "That the name of the
hours has been known in Rome during at least three hundred years is probable, because
although in the Twelve Tables and the laws which followed them we find the hours named only
once, there are employed the words *ante meridiem* for the reason no doubt that the day was
then divided into two parts, separated by that which we call *meridies*." Pliny the Elder's full
statement is: "In the Twelve Tables, the rising and setting of the sun are the only things that are
mentioned relative to time."

Greek literature relate to dinner. In their earliest example, from 390 BCE, the character Praxagora in Aristophanes' *Ecclesiazusae* (647) set the start of a dinner engagement at *dekapon stoicheion*, literally when "ten feet lined up." By the second century CE, Pollux, in his *Onomasticon* (6.44), could gloss the word *stoicheion* alone as meaning a ten-foot-long shadow, adding that such a shadow signaled the time for dinner. The Scholiasts, too, had explained how measuring the length of the sun's shadow had been used to determine the time. The sixth-century Byzantine chronicler Hesychius of Miletus, on the other hand, associated the *stoicheion* with a shadow of twelve feet, perhaps through the influence of Menander's Chaerephon, and also observed that this indicated the time for dinner. A seven-foot shadow he defined as one that had been paced off in feet, from which the hour could be determined. These and other textual references provided the Accademia with the necessary links connecting shadows, the pacing off of those shadows in feet, and the conversion of the number of those feet into hours: a twelve-foot dinner meant one scheduled to begin at the twelfth hour.

Fundamental to this discussion was the work of Palladius Rutilius, whose fourth-century book, *Opus Agriculturae*, had been organized according to individual chapters dedicated to the agricultural activities of each month. At the end of each chapter, Palladius had appended a section entitled "concerning the hours" (*de horis*). These were simple tables that specified the length of the shadows cast during each of the eleven hours of daylight in each month, capturing numerically the lengthening and retreating of the shadows cast by the movement of the sun through the ecliptic, what came to be known as "temporary" or "seasonal" hours. Palladius conceived of these shadows as being cast by a vertical gnomon on a horizontal surface, though he, like the Greek sources before him, had not specified the gnomon's height and his calculations did not take into account the day-by-day variations in the lengths of the shadows that occurred during the course of each month. According to his tabulations, the length of the shadow in the first hour after sunrise was consistently the longest and was equivalent in length to the shadow cast at one hour before sunset in the eleventh hour. Similarly, the second hour's shadow was the second longest and was paired with the tenth, the third was paired with the ninth, the fourth with the eighth, and the fifth with the seventh. Only the sixth hour, constituting midday, when the sun reached its highest point and cast the shortest shadows, naturally had no corresponding hour. Palladius' days lacked a twelfth hour, however, because this corresponded with sunset, just as sunrise was the "zero" hour, when the sun was too low to cast shadows.

118   THE PROSCIUTTO SUNDIAL

Palladius also had indicated the manner in which the length of the hours in the months were mirrored through the year by, for example, pairing January's table of equations with December's, when the lengths of the shadows were the longest throughout the day, and joining June's with July's, when the lengths were the shortest. This was precisely how the abbreviations of the months appeared to be combined in the prosciutto's calendrical notation, with January seemingly written above December, and June above July. The resulting pattern must have been what led d'Alembert to believe they should be paired similarly in reading the prosciutto's dial. Paciaudi, too, had assumed the prosciutto's design reflected Palladius' tables equating the length of a shadow with the hour of the day and the month of the year. Palladius, however, also had not stated the unit of measurement he had employed for determining the length of the shadows, but because he used the term *pedes* (feet) scholars tended to assume that Palladius' figures represented the pacing off of these shadows. More problematic for the Accademia than the lack of a twelfth hour was the fact that the lengths of Palladius' shadows were significantly higher than any of the hourly figures provided in the Greek sources, especially in the first and eleventh hours of January and December when the shadows reached "29 feet."[61]

Chronologers from the sixteenth and seventeenth centuries seeking to synchronize the histories of ancient civilizations had had to grapple with issues relating to the methods employed by the ancients in measuring time, and this included this question of the "twelve-foot dinner." Consequently, the Accademia had consulted first the work of two leading luminaries in the development of these chronologies, Joseph Scaliger (1540–1609) and Claudius Salmasius (Claude de Saumaise, 1588–1653), each of whom had interpreted the expression's meaning in different ways.[62] In 1579, Scaliger had published a text and commentary on the first-century didactic astronomical poem in five books by Marcus Manilius entitled the *Astronomica*, one of the earliest extant ancient works on astronomy and astrology that sought to link the signs of the zodiac to the affairs of humans.[63] Scaliger was considered—and

---

61. T. Owen, *The Fourteen Books of Palladius Rutilius Taurus Aemilianus on Agriculture* (London 1807) 347–49, provides a table and discussion of these hours.

62. The Accademia also referenced Casaubon's commentary on Athenaeus' *Deipnosophistae*, but this contributed less to their discussion. For a fourth figure, Dionysius Petavius, see below. For a general survey of the contributions of Scaliger and Petavius to the creation of chronologies, see Wilcox 1987, 195–208.

63. J. Scaliger, *M. Manilii Astronomicon libri quinque* (Paris 1579). The edition used here is the third edition published in Strasbourg in 1655. A. E. Housman (*M. Manilii Astronomica,*

Analysis

considered himself—an authority on ancient calendars and chronology, but, as a philologist, he lacked technical knowledge of astronomical matters, his primary concern having been to produce a properly emended text.[64] This had left him open to criticism, which Salmasius had supplied a half century later.[65] Salmasius had written a detailed commentary on Solinus' *Polyhistora*, also known as the *de Mirabilibus Mundi*, a third-century CE geography of the known world emphasizing select curiosities that was based largely on Pliny the Elder's *Historia Naturalis*. Despite the apparent differences in their subject matters, and in particular the fact that Solinus' text was geographical and not zodiacal, their topics had converged in their need to associate the same natural phenomenon with a specific zodiacal event: the flooding of the Nile in the sign of Cancer, when the hours of the day begin to be diminished by the lengthening hours of the night. This had compelled both to examine the literary sources that explained how the ancients had measured the hours over the course of the seasons and tied them to the zodiacal signs, and this had ultimately led them both to the tale of Chaerephon.

In his third book, Manilius (3.247–274) had spelled out in detail the means for calculating the length of the hours of the day and night throughout the course of the year in each of the zodiacal signs. He began with a standard hour whose length could be adjusted depending on the time of year to accommodate the "diurnal" hours, when the hours of the day were longer, and the "nocturnal" ones, when the hours of the night were longer. For his standard he had used the length of the hours at the equinoxes, when all the hours of the day and night are equivalent, and he observed that in the sign of Capricorn, beginning with the winter solstice, the hours of daylight expand to fourteen

---

Cantabrigae 1932) dedicated the last years of his life to this work. On Scaliger, Housman wrote, "The commentary is the one commentary on Manilius, without forerunner and without successor; today, after the passage of three hundred years, it is the only avenue to a study of the poem. He seems to have read everything, Greek and Latin, published and unpublished, which could explain or illustrate his author; and his vast learning is carried lightly and imparted simply in terse notes of moderate compass. Discursive he often is, and sometimes vagrant, but even in digressions he neither fatigues his readers like Casaubon nor bewilders them like Salmasius. His style has not the ease and grace and Latinity of Lambinus', but no commentary is brisker reading or better entertainment than these abrupt and pithy notes, with their spurts of mockery at unnamed detractors, and their frequent and significant stress upon the difference between Scaliger and a jackass."

64. Grafton 1983, 185–226, esp. 196: "He could explain the elementary astronomy that the reader needed, but he sometimes missed astronomical points of a rather technical kind."

65. Salamasius came to occupy the same professorship at Leiden that Scaliger had held until his death.

THE PROSCIUTTO SUNDIAL

and a half while those of the night diminish in number to nine and a half.[66] Scaliger's exegesis of these passages was concise and narrow in its scope. He noted Vitruvius' initial precept (9.7.2) echoing Manilius that wherever a sundial was to be erected it was first necessary to determine the length of the equinoctial hours. Scaliger then explained that the Greek term *polos* preceded Vitruvius' use of *horologium*, and then quoted Herodotus' observation (2.109) that the Greeks had adopted from the Babylonians both the *polos* and the division of the day into twelve parts.[67] The earliest timepieces in both Greece and Rome were "winter" dials (*horologia brumalia*), by which Scaliger evidently meant *clepsydra* that could work in the darker days of winter as well as over the course of the day and night.[68] Meton's fifth-century BCE *heliotropion*, the observatory he constructed on the Pnyx, had allowed him to establish the dates of the equinoxes and the solstices.[69] It was Scaliger's final remark, however, that was of the greatest relevance to the Accademia's argument. Both "nocturnal" and "diurnal" dials were available in Menander's day, by which Scaliger meant the ability to tell time both day and night by means of the combined use of *clepsydrae* and sundials. Proof of this he found in the "charming words about a certain man invited to dinner in the XII hour," and then quoted in full the relevant passage about the unfortunate Chaerephon. "For although he had been invited to be present on the twelfth shadow," Scaliger explained, "having awoken at night when the moon was full and the light brightest, he believed not only that it was day but that the twelfth hour had passed, having perceived the shadow on the twelfth line. Which certainly is accustomed to occur on a full moon."[70]

Salmasius had seized upon Scaliger's notion that the twelfth line on a sundial could be read under a full moon, calling it "wrong in many ways, and inept" (*multis modis falsum est et ineptum*). He had confronted Scaliger's reading of the Menander passage while commenting on Solinus' section (37) on the Euphrates River and how its seasonal flooding coincided more or less

---

66. Goold 1977, 180–85. For his calculations Manilius claimed to have used the latitude of Alexandria (31° 11′ 41″), but it was in fact that of Rhodes (36° 26′ 75″).

67. *Polos* generally was understood to refer to the dome of the sky, but it came to be transferred to the carved bowl of a hemispherical sundial that was viewed as its representation in reverse.

68. Vitruvius (9.7.7) used the term *brumalia* for hours whose length conformed to the shorter days leading up to and following the winter solstice. This distinguished them from *aequinoctilia* hours for the increasingly longer days leading up to and following the summer solstice.

69. Scaliger, among others, believed that Meton's *heliotropon* had only established the dates on which the equinoxes and solstices occurred, as a means of reckoning time between those events.

70. Scaliger 1655, 228–32, esp. 228–29.

Analysis   121

with that of the Nile's, beginning when the sun is in Cancer and diminishing in Leo as it approaches Virgo. Salmasius' gloss of Solinus' text expanded to a forty-two page discourse on every imaginable and tangential aspect of the gnomonic arts, from the origins of the division of the day and the designation of the hours, to the development of sundials and the history of their appearance in Rome and the advantages of the *clepsydra*.[71] His selection of sources (both ancient and modern), his readings of those sources, and the organization of his arguments in many ways established the model for subsequent surveys of the history of astronomy in Greece and Rome and clearly served as a primary font of information for the Accademia. His principal points, reiterated here and elsewhere in the Accademia's commentary, are that even after the invention of the *polos* and the *heliotropon* the individual hours had not been determined and there remained no word for "hour." A different, more primitive method for distinguishing the hours of the day was therefore necessary, and measurement by pacing off shadows in feet had been the solution.

Salmasius scoffed at the notion that the hours on a sundial could be read by moonlight. He interpreted the passage to mean that Chaerephon awoke while the moon was shining bright and, thinking it was the sun, had paced off his shadow and headed to dinner. "Scaliger thinks the twelve-foot shadow signified the twelfth hour of the day, when the hour is read on the twelfth line…," continued Salmasius, "This is the last hour of the day, when the sun has already set. Indeed, the sun barely touches the twelfth line on sundials." As proof of this he pointed to Palladius' calculations that excluded any readings for the twelfth hour. He then asked, "What do you make of the guy who was invited to a meal when the sun's shadow was twenty feet?" Here Salmasius was referring to Philocrates, another character in Athenaeus (*Deip.* 1.8b–c) attributed to the comic playwright Eubulus, who "had been asked out to dine by some friend who told him to come when the shadow on the dial measured twenty feet. So at dawn he began to measure when the sun was rising, and when the shadow was too long by more than a couple of feet he came to dine, and said that he had arrived a little late because of business engagements—though

---

71. T. Mommsen, *C. Iulii Solini Collectanea rerum memorabilium* (Berlin 1895). Mommsen's text of the relevant passage in the *de Mirabilibus Mundi*, 37.2–3, reads: *Mesopotamiam opima inundationis annuae excessibus, ad instar Aegyptii amnis terras contegens, inuecta soli fecunditate, iisdem ferme temporibus quibus Nilus exit, sole scilicet in parte cancri uigesima constituto; tenuatur cum iam leone decurso ad extima uirginis curricula facit transitum. Quod gnomonici similibus parallelis accidere contendunt, quos pares in terrarum positione aequalitas normalis facit lineae.*

he had come at daybreak!"[72] Again Salmasius rejected the idea that there was a one-to-one correlation between the number of feet paced off and the hours on the sundial, recalling that Palladius' calculations showed that the first hour was not equivalent to a one-foot shadow but was the longest shadow of the day. He surmised that an individual's own upright body served as the gnomon and the size of their foot served as the unit of measurement.[73] The concept of, and even the word for, hours had, in fact, coincided with the invention of the sundial, while this primitive means of pacing off of shadows had endured thereafter. Finally, the line from Persius' satire (3.4), "when the shadow touches the fifth line," served as Salmasius' proof of how all these elements had evolved to work together: the shadow pointed to a line which equated with an hour.[74]

The Accademia's creative approach to resolving the discrepancies in the sources was to deploy another passage from Pollux's *Onomasticon* (1.71–72) that explained that the Greek word *sameion* (σημεῖον) was synonymous for both hour and half hour, claiming it had been used as such by Menander.[75] Eubulus' "twenty-foot dinner" thus could be viewed as equivalent to a "ten-foot dinner." As for Chaerephon, he had simply confused the twelfth hour of the night for that of the day. They concluded that this standard Greek time for dinner, originally measured in feet, had been translated directly onto the hour lines of the sundial: the twelfth, or sunset, was the usual hour, but some guests showed up early—sometimes very early. In this way they established to their satisfaction that the times given in shadow lengths actually did correspond with the lines on the dial.

To check the hypothesis that the human body served as the gnomon and provided the unit of measurement for reading these shadows, the Accademia

---

72. The passage quoted is from the fourth-century BCE comic poet Eubulus (frag. 117), but the extant text of Athenaeus here is slightly corrupt: ὅν φασί ποτε κληθέντ' ἐπὶ δεῖπνον † ὡς φίλουκαὶ τῷ τινος †εἰπόντος αὐτῷ τοῦ φίλου, ὁπηνίκ' ἂν εἴκοσι ποδῶν μετροῦντι τὸ στοιχεῖον ᾖ, ἥκειν, ἕωθεν αὐτὸν εὐθὺς ἡλιουμετρεῖν ἀνέχοντος, μακροτέρας δ' οὔσης ἔτι πλεῖν ἢ δυοῖν ποδοῖν παρεῖναι τῆς σκιᾶς, † ἔπειτα φᾶναι † μικρὸν ὀψιαίτερονδι' ἀσχολίαν ἥκειν, παρόνθ' ἅμ' ἡμέρᾳ. According to Athenaeus (*Deip.* 13.567c–d), this same Eubulus wrote a play entitled *Clepsydra*: "The *Clepsydra* of Eubulus, and the woman who bore this name, had it because she used to distribute her favors by a water-clock, and to dismiss her visitors when it had run down."

73. Salmasius 1629, 455: *Corporis quisque sui umbram propriis pedibus metiebatur...*

74. Salmasius 1629, 454–57.

75. Pollux, *Onomasticon*, 1.71–72: ὥρα δε και ἡ μιώριον, ὡς Μένανδρος, σημεῖον ωνομάζετο παρά τους παλαιοίς. και από σκιάς δε εδηλούτο, οίον δεκάπους ἡ σκιά και ενδεκάπους. The LSJ defines σημεῖον as "sign," but it came to mean a "unit of time" in later Greek, though the Accademia appears to be stretching the meaning.

Analysis 123

turned to the work of Dionysius Petavius (Denis Pétau, 1583–1652). Petavius had enhanced upon the chronological work of Scalinger in his 1627 *Opus de Doctrina Temporum* through his deep knowledge of astronomy, and over the next quarter century supplemented his work with a number of "dissertations." In 1630, Petavius had challenged several of Salmasius' conclusions concerning the measurement of shadows, as is evident from its title alone: "In which it is argued against the Plinian Exercises of Salmasius" (*In quo contra Plinianus Salamasii Exercitationes disseritur*).[76] The thesis here was that the "twelve-foot dinner" must have been referencing a standard gnomon, given the consistency of the times provided in the literary sources versus the variations in human stature. As proof, Petavius had calculated the length of a shadow at the time of the original presentation of Aristophanes' *Ecclesiazusae* at the festival of Scirophoria, during the time of the summer solstice. Using the known latitude of Athens (37°), the local time at the eleventh hour of the day (19:22), the altitude of the sun at that time (12° 16′), and, most importantly, the French foot (*pied*), he concluded that an object standing upright would need to be only 0.77 m high to cast an eleven-foot shadow (3.57 m) while an object 0.84 m high would be needed to cast a twelve-foot shadow (3.89 m)—therefore, shorter than the average human.[77]

The Accademia, however, sided with Salmasius in this dispute. Using the same astronomical settings as Petavius, they input their own variables. The gnomon was six feet high, the normal human stature, but they appear to have employed the *palm* (0.26 m) rather than the *pied* (0.32 m), and thus set their gnomon's height at 1.58 m. Inputting the tenth hour in the evening at 18:24 yielded a shadow 3.21 m long, or twelve *palmi* in length, getting them to their desired "twelve-foot dinner," and a perfectly reasonable mealtime. To get a "ten-foot dinner" they needed only to take a reading in the ninth hour (17:10) and add an additional fifty-two minutes (18:02) to produce a shadow 2.69 m

---

76. The Accademia referenced the 1703 edition of Petavius' *De doctrina temporum: Auctius in hac nova editione notis & emendationibus quamplurimis* (Antwerp 1703); his original publication of this had appeared in Paris in 1627, roughly contemporary with Salmasius'. Volume three of the 1703 edition contains the *Ad auctarium operis de doctrina temporum variorum dissertationum*, in eight books, which was originally published in Paris in 1630 as a supplement to his *Uranologion sive systema variorum authorum, qui de sphaera, ac sideribus, eorumque motibus Graecae commentati sunt*. The relevant sections in the 1630 edition are pages 269–75; in the 1703 edition they are 142–47.

77. Petau et al. 1703, 142–44. The length of the day on June 21 was given as fourteen hours, forty-five minutes, making each hour seventy-three minutes long. The French *pied* before the Revolution in 1789 was equivalent to 0.3248 m and this was the unit of measurement Petavius employed; the Roman foot is generally set at 0.296 m. The latitude of Athens on the Acropolis is 37.97°; the eleventh hour is currently 19:38 with the sun's altitude at 12.17°.

long, or ten *palmi* in length. As for Palladius' shadows, which Petavius had concluded were insufficient, they attributed some of the disparities to errors in textual transmission, others to rounding errors, and still others to differing foot sizes, all of which rendered useless any attempt to apply Petavius' calculations rigorously. He had measured the length of the shadow cast at midday at the winter solstice as "nine feet" while the corresponding shadow at the summer solstice was "two feet" long. The Accademia suggested, on the basis of later textual evidence, that the "nine" (IX) might have been transposed from what had originally been the Roman numeral for "eleven" (XI). Since the length of the shadows cast by their 1.58 m high gnomon at midday on the winter solstice would be 2.90 m and 0.57 m on the summer solstice, they were satisfied that they had found the eleven and two feet they were looking for to accept Palladius' numbers.

2.e) *Carcani's Empirical Research and Analysis*

With this initial application of modern astronomical data as a means of testing the evidence from the ancient literary sources the Accademia began to shift their approach away from their characteristically rigorous philological assessment to a more science-based analysis. It is here where Carcani's contributions become most evident, in the interrogation of the prosciutto as an astronomical instrument. Carcani based his analysis on empirical tests aimed at determining how the sundial functioned in the real world coupled with a recalibration of the astronomical equations to compensate for the changes that had occurred over the 1700 years since the sundial's design. The purpose of these final sections of the Preface was threefold: first, to explain how the sundial functioned and confirm its accuracy; second, to ascertain at which latitude it had been designed to function; and third, to demonstrate how the resulting astronomical data could be used to establish the year in which the prosciutto had been made. Demonstrating that it was possible to derive reasonably accurate astronomical data from the prosciutto would set the Accademia's inquiries apart from earlier treatments and even many subsequent ones.

The first step toward reaching these goals was to clear up the misconceptions about how the sundial worked. The Accademia explained this in clear and simple language. In order to take readings, it needed to be suspended from a string attached to its upper ring and the side of the prosciutto with the gnomon oriented toward the sun's location in the sky. The vertical lines on the dial mark the progress of the sun's path through the sky, known as the ecliptic,

Analysis 125

so the tip of the shadow produced by the gnomon had to be aligned with the vertical line corresponding with the zodiacal sign through which the sun was transiting.[78] The point of intersection between the shadow cast by the gnomon and the vertical line would indicate the hour of the day depending on its relationship to the horizontal horary lines.

> Now, in order to use this sundial, it behooves one first to suspend it by its ring, so that it rests vertically equalized by its own weight, then turn toward the sun not the face of the sundial but only the side where the gnomon rises, arranging it in such a way that the gnomon's shadow comes to meet the place of the sun in the ecliptic indicated by the vertical lines. In this way the shadow itself will point out the hour which is sought on the horary lines.

Here, the delicate interface among, and the need for careful negotiation among, the design of the dial, the earth's gravity which kept the dial properly aligned, the sun's course, and the visual acuity of the human operator was clearly spelled out for the first time.

The loss of the gnomon had posed a small challenge to Carcani's experiments for it prevented him from knowing its exact position in relation to the dial. In this he followed Piaggio's creative solution in reconstructing the missing portion using a piece of wax, rolled and formed into a solid tube with a pointed tip that he attached to the gnomon's remaining stub. This initially ran parallel to the angled side of the prosciutto but then bent toward the right to point toward the upper left-hand corner of the dial formed by the intersection of the first horizontal line, representing the horizon, and the first vertical line on the left, representing the sun in Gemini. Cognizant of the slight difference in the declination of the sun between antiquity and his own day, which resulted in variations between the lengths of the hours of daylight at other times of the year, Carcani chose the period of the spring equinox in Naples for his experiments. Only then did the twelve hours of the day in his own time correspond with the division of the day in antiquity into twelve equal parts. This allowed him to "set" the tip of the shadow produced by his gnomon so that it fell on the horizontal line corresponding to the equivalent modern hour along the fourth, or central, vertical line representing the sun in Pisces.

---

78. The ecliptic is tilted at an angle with respect to the plane of the earth's equator. The equinoxes occur when the equatorial plane intersects the ecliptic.

It being noted that only in the times of the equinoxes the ancient hours converged with ours, the twentieth of March was chosen, that is, the day of the spring equinox, to make the observations, and having tried to substitute with wax the missing portion of the tail, it was extended to the plane of the first horary line and the tip of its point disposed in such a manner that scrolling its shadow on the fourth vertical line, that is, parallel to the equinox, it came to indicate exactly the first hour of the day, counting from the rising of the sun on the horizon.

Since the sun on the equinox rises a bit after 6:00, Carcani had set his gnomon when it cast its shadow along the second horizontal line representing 7:00 and then taken readings throughout the remainder of the day. These had correlated well with the modern hours with the exception of the third horizontal line, which was equivalent to the start of the second and tenth Roman hours, or 8:00 and 16:00, a deviation Carcani attributed to a defect in the sundial's surface.

And with amazement one observed that it faithfully continued to indicate with precision all the eleven other hours of the day, with the exception of the second and the tenth, which are represented by the third transverse line, with a variation that was not more than two or three minutes; this could arise from some alteration that the surface of the sundial suffered in that part.

As a result of his observations, Carcani deemed his reconstruction of the gnomon a success in that it had accurately marked the hours over the course of the day on the spring equinox. He had limited his empirical testing to this single day, however, and did not conduct further checks on his data within this same equinoctial period, much less supplement his findings with additional empirical inquiries at other times of the year. He had been constrained in part by the narrow window available for conducting his research, which stretched from March 1761, when the sundial was substituted as the topic for the Preface, to early September 1761, when Tanucci announced the completion of the text in a letter to Charles III. While this timetable would have allowed him to take readings during the next major astronomical event—the summer solstice—he would have missed the opportunity to check his earlier data afforded by the autumnal equinox.

Instead, he derived the remainder of his data from established geometrical and astronomical calculations. To do so he had to determine the proportional

Analysis 127

relationship of each of the vertical lines representing the zodiacal signs. He explained this process in a footnote:

> The measure and proportion observed in the sundial of the seven vertical lines and in the supplied gnomon are the following. Suppose the equinoctial shadow is divided into 1000 equal parts, the gnomon, or rather the distance from the extremity of the tail of the prosciutto from said shadow, has 881 parts, the shadow of the solstice of Cancer has 1686, of Capricorn 687, of the Twins and Leo 1543, of Taurus and Virgo 1244, of Scorpio and Pisces 804, and finally the line of Sagittarius and Aquarius 691.

Carcani had taken the central equinoctial line (Line 4) and converted it into a thousand units, or "parts," rather than using its actual length.[79] Mathematically, this makes no difference provided the conversions are consistent. He then converted the other six lines in the same manner: Line 1 converted to 1686 units, Line 2 to 1543, Line 3 to 1244, Line 5 to 804, Line 6 to 691, and Line 7 to 687 units. The crucial datum is what he defined as "the distance from the extremity of the tail of the prosciutto from said [equinoctial] shadow," which he gave as 881 units. This should therefore be the length of the shadow projected from the tip of the gnomon to the fourth equinoctial line on the dial along the upper horizontal line, or the length of the diagonal measured from the tip of the gnomon to the top of the fourth line. The resulting ratio—881 units to 1,000 units—meant that a one-meter-high gnomon would produce a shadow 0.881 of its length at noon on the equinox.

From this ratio Carcani had derived the latitude for which the sundial had been designed, or what he properly termed the elevation of the pole: "With this gnomon, therefore, supposed to be correct and which stands at the fourth vertical line representing the quantity of the equinoctial shadow as 881 to 1000, one passes then to calculate the elevation of the pole, which is found to be 41° 39′ 45″." Carcani observed that this latitude of 41° 39′ 45″ was closer to that of Rome than of Naples, having established the latter of these himself at his own Collegio Reale delle Scuole Pie, and so he concluded the sundial had been designed for use in Rome. "But this elevation of the pole being greater than that of Naples, which is at 40° 50′ 15″, and as a consequence also that of

---

79. Such a conversion made sense since the Roman foot (*pes*) measured 0.296 m and was divided into sixteen digits (*digiti*), each equivalent to 0.0185 m. The Neapolitan *palm* was 0.26455 m, divided into 12 *oncie* c. 0.02 m. The meter had not yet been quantified; see Chapter 3.d.

128   THE PROSCIUTTO SUNDIAL

Herculaneum, and, conversely, a little less than that of Rome, which is 41° 54', and according to Ptolemy 41° 40' (*Geography* 3.1), it is likely that our sundial was made for the pole of Rome which at that time was the most known."

Carcani had had to clarify in accompanying footnotes why he had arrived at this figure for the latitude that differed from the one he had observed on the prosciutto: 0.881 on the prosciutto, which is equivalent to a latitude of 41° 22' 48", against the latitude of 41° 39' 45", or 0.889, he specified. He explained how he had adjusted his actual reading to account for three factors: the semi-diameter of the sun at the spring equinox and La Caille's solar refraction minus the solar parallax.

> As 1000 [units] the equinoctial shadow given to 881 distance from the point of the gnomon, so the total *sine* at 88100 tangent of 41° 22' 48", the apparent distance from the vertex of the upper edge of the sun, to which is added the semi-diameter of the sun at the spring equinox of 16' 5", and furthermore the refraction according to Signor de la Caille, less the parallax, that is 52", one has in the sum of 41° 39' 45", the true distance of the Equator from the vertex, or rather the height of the pole.

He then concluded that the difference in latitudes between Rome and Naples was sufficiently negligible and the prosciutto itself so small that any variations in the readings would not have been discernible.

> It is not surprising that a sundial although designed for a latitude diverse from ours nevertheless gives us the hours corresponding so well, as if it had been made for our own pole. One has to consider, on the one hand, that the deduced latitude of 41° 39' 45", as is seen, directly from the construction of the sundial, does not differ from the latitude of Naples but for 49' 30" and, on the other, one should consider that since the instrument is so small it results in so small a difference of the shadows and, as a consequence, of the hours, the difference in the semidiurnal arcs, given the difference in latitude, being about 1½°.

With the position of the gnomon and the latitude established, Carcani could check his initial conclusions about the sundial's accuracy along the fourth, equinoctial vertical against the design of the dial as a whole through known astronomical computations for the behavior of the sun's shadows throughout the year. In so doing Carcani discovered that the first and second vertical lines had been inscribed longer than they should be. He ascribed this error to modifications made by the engraver in transferring the grid to the irregular surface of the prosciutto. In fact, the surface here does undulate slightly, curving in before bulging out, and this doubtless had contributed as well to the slight inaccuracy Carcani observed in his empirical tests of how the shadow fell along the equinoctial line. In other words, evidently the desire to accurately represent the shape of a prosciutto had resulted in inaccuracies in the dial; form had taken precedence over function.

> And to verify by calculation the correct quantity and position of the gnomon comparing with this height of the pole the tangents of the distance from the vertex of the same northern limit of the sun with each of the six other vertical lines as the shadows of the sun in the other signs of the zodiac, from all these connections and replicated calculations the extremity of the gnomon almost always resulted in the same point. The first line only of Cancer, and the second which pertains to Gemini and Leo, give to the gnomon an extension greater than what is suitable. Because the back of the prosciutto was made natural it is not a flat surface but has irregularities and a curvature with a slight rise toward the seventh line, or rather the meridian, between the middle of the two lines of Cancer and Leo. These [lines] the author of the sundial had to lengthen a bit more than necessary so that the shadow in that area could come to touch the meridian at midday, since it was actually observed in the months of May, June, and July that, as much in the rising of the sun in the sign of Gemini as in its setting in Cancer and Leo, the shadow came as well to the line of the sixth hour in midday as it did to those of the other hours in their true times.

2.f) The Transit of Venus, the Obliquity of the Ecliptic, and Dating the Prosciutto

Satisfied that this combination of analyses had answered the specific questions they had posed and had established the sundial's overall accuracy, the Accademia took their inquiry one step further to use astronomical data to determine when it had been manufactured. Previous attempts to date it had been based on two details readily visible on the prosciutto: the names attributed to the months and the orthography of the inscription's lettering. The letter in the *Memorie* had been the only source so far to attempt to use the lettering itself as a means of dating the sundial, the open form of the "P" deemed indicative of the "oldest form" of writing among the Romans and therefore presumed to attest to the prosciutto's early date. This was more a conjecture than an established fact since only later did paleographical studies demonstrate that this letter shape had continued in use through at least the Julio-Claudian period. Nevertheless, the Accademia showed little interest in the inscription and completely ignored the letter forms. Bayardi, at least, had used the inscription to note that the abbreviations "IV" and "AV" required the sundial to date after the reigns of both Julius Caesar and Augustus.

The Accademia thus far had limited themselves to reviewing the discussion of Censorinus on the renaming of the months Quintilis and Sextilis to Julius and Augustus, respectively. They must have deemed these other approaches inconsequential compared to their methodology of ascertaining the true date of the sundial's manufacture through astronomical analysis. To do so Carcani needed to establish the obliquity of the ecliptic, that is, the angle of the inclination of the Earth's axis which is decreasing at a known rate. Determining the angle in antiquity as evidenced by the prosciutto and subtracting it from the value of the angle in their own day would allow the Accademia to compute the number of years that had transpired.

> The sundial therefore finding itself, in all the tests put to it and with direct observations and calculations, designed by its author with great care, it was determined to try another much more delicate inquiry with the lines it has on it and that is to deduce the obliquity of the ecliptic for the time in which it was fabricated, deriving it directly from the proportion of the shadows observed on it, and it was found to be 23° 46′ 30″.[80]

The terseness of their presentation of this value belied the complexity of the calculations required to ascertain it. In a footnote Carcani explained that the defect detected in the first and second vertical lines of the dial had forced him to calculate this value for the obliquity of the ecliptic from the fourth and seventh lines. In so doing he provided the only other numerical figure derived from his research rather than from the instrument itself: that the distance from the tip of his reconstructed gnomon to the seventh line along the upper horizon line was 1,482 units.

---

80. Equivalent to 23.775°. At the present time about 23.4° and decreasing at a rate of 0.013° every hundred years; for discussion of the quantity of the obliquity of the ecliptic, see Chapter 4.e.

Analysis

Not being able in this most subtle research to make use of the [first] solstice line of Cancer for the reasons given [above], the [fourth] line of Aries was chosen, or rather [the fourth line] of Libra and [the seventh line] of Capricorn, the outer points of these two lines being on the same plane. And because, as [noted above], the true distance of the Equator from the vertex was found, the true distance from the same vertex of the Solstice point of Capricorn was calculated, doing so in the same manner as the 687 [unit] shadow given for the Solstice of Capricorn to 1482 the distance of the point of the gnomon from the same shadow, in this way the total *sine* of 215720 tangent of 65° 7′ 45″ the apparent distance from the vertex of the northern limb [the upper edge] of the sun at the winter Solstice, which by adding the semi-diameter of the sun of 16′ 18″, and the [solar] refraction according to said Signor de la Caille, less the [solar] parallax, that is 2′ 12″, gives in the sum of 65° 26′ 15″ the true distance from the vertex of the solar center at the point of the Solstice of Capricorn. And comparing this distance with the other found above of 41° 39′ 45″, of the Equator from the same vertical point, the difference of 23° 46′ 30″ [= 23.775°] gives the angle of the Equator's ecliptic.

The obliquity of the ecliptic at the time of the prosciutto's manufacture in antiquity having been established to his satisfaction, Carcani could proceed to explain how he had gone on to derive the degree of diminution to his own day: "This compares with the angle of 23° 28′ 18″ [= 23.4717 °], which we currently observe is formed by the ecliptic with the equator, one sees that the obliquity has diminished 18′ 12″ from the time in which the sundial was made down to the current year." In the final footnote inserted here, Carcani again underscored the remarkable accuracy with which the artist had rendered the dial on the prosciutto: "The result of this observation is all the more notable for the mutability of the ecliptic as it is simple: because it is to be believed that the author of this precious and erudite monument, which is found to be most precise in all its parts, after having attentively observed the proportions of the shadow of the gnomon, without in fact altering them in the slightest part, marked them as such on his sundial." With their case sufficiently underpinned with the necessary scientific data, the Accademia concluded that it confirmed the prosciutto dated to the early first century: "Therefore, according to the calculations and observations of Cavalier de Louville that the inclination of the ecliptic has diminished 21′ in 2000 years, the epoch of our sundial ought to fall around the year AD 28."

132 THE PROSCIUTTO SUNDIAL

In these final sections of the Preface, Carcani had endeavored to highlight the ways in which he had applied the latest discoveries in astronomical science to his analysis of the prosciutto, effectively positioning it within some of the most hotly debated issues in contemporary astronomy. Carcani had singled out the work of La Caille, who at the beginning of 1762 had sponsored his election to the Académie Royale des Sciences. In so doing Carcani acknowledged La Caille's most recent contribution to astronomy, his table of solar refractions, while also trumpeting his own role in advancing astronomical science. These debates centered on the values for solar refraction, the semi-diameter of the sun, and, most important of all, the solar parallax and the obliquity of the ecliptic.

Solar refraction accounts for the alteration of the apparent position of the sun in the sky through the bending of light rays by the earth's atmosphere. The effects of refraction cause the sun to appear higher above the horizon than it actually is while latitude plays a role in the magnitude of the refraction. Both refraction and latitude therefore need to be taken into consideration in any solar reading. The standard reference table compensating for the effect of refraction had been established by the French astronomer Giovanni Domenico Cassini and had been in use since 1684. La Caille had published a new table in 1757 based on observations he made in 1751 at the Cape of Good Hope, which took into account the crucial effects of variations in atmospheric pressure and temperature that Cassini's values had not.[81] These he had slightly revised in an article read before the Académie in 1755 and published in their *Memoires* in 1761. While astronomers would continue to debate the method for determining refraction through the end of the nineteenth century, La Caille had given a value of 2′ 20″ for solar refraction at 65° and 1′ 0″ for the solar refraction at 42°, and Carcani had used these values in his equations.[82]

To these, Carcani had added the value for the semi-diameter of the sun, the angle between the points marking the sun's lower extremity, or "southern limb" in astronomical terminology, and its center. This value varies depending on the distance of the sun from the earth, and therefore varies according to

---

81. N. L. La Caille, *Astronomiae fundamenta novissimis solis et stellarum observationibus stabilita lutetiae in collegio mazarinaeo et in Africa ad caput Bonae Spei* (Paris 1757) 214: comparative table of refractions at Cape of Good Hope and Paris.

82. C. Wilson, "Perturbations and Solar Tables from Lacaille to Delambre: The Rapprochement of Observation and Theory, Part 1," *Archive for History of Exact Sciences* 22.1/2 (1980) 58–59. N. L. La Caille, "Recherches sur les réfractions astronomiques, et sur la hauteur du pôle à Paris; avec une nouvelle table de réfractions," *Mémoires de l'Académie Royale des Sciences, Année 1755* (1761) 547–93, 571, for the table. E.g., Lalande's table (Lalande 1771, I.237) gives a value of 2′ 5.7″ at 65°.

the time of year, but the precise values remained in dispute in Carcani's day. La Caille's table of the semi-diameters, published in 1758, set the value for the winter solstice at 16′ 19.3″.[83] Carcani, however, states that he used 16′ 18″. This suggests he had relied instead on the tables provided in the *Connaissance de Mouvemens Célestes*, an annual almanac produced by the Académie in Paris summarizing astronomical data in a given year. The *Connaissance* for 1763, which was published in 1761 and therefore would have been the most recent edition available to Carcani, gives the diameter of the sun on December 19 as 32′ 34.6″, or about 16′ 18″ for the semi-diameter.[84] For his second calculation, for the semi-diameter of the sun on the spring equinox, Carcani used 16′ 5″, which also is consistent with the *Connaissance*'s figure of 32′ 8.6″.

The last calculation Carcani added to his analysis was the value of the solar parallax. This is the angle under which the earth's equatorial radius is perceived from the sun's center. This angle helps determine the mean distance between the earth and sun, and this distance, in turn, establishes the value of one astronomical unit on which, according to Kepler's Third Law of Planetary Motion of 1619, the relative distances and sizes of the planets and stars, and therefore the size of the universe, depend. As a consequence, ascertaining the value of the solar parallax had become one of the pressing astronomical issues of the eighteenth century. In 1716, the British astronomer Edmund Halley had proposed that accurately measuring the time it took Venus to pass across the face of the sun, termed its transit, from two widely spaced points of observation at different latitudes to the north and south and then subtracting the times obtained at one site from those of the other would yield the parallax.

---

83. N. L. La Caille, *Tabulae Solares* (Paris 1758) table XVI: *Semidiametrorum Solis et motuum eius horariorum*. La Caille notes on this table that "The true value of the sun's semi-diameter has not yet been established among astronomers. We employ here what seems to correspond best with the most recent observations: see the *Memoires* of the *Académie Royale des Sciences* from the year 1754." The last reference is to La Caille's article "Diverses observations faites pendant le cours de trois differentes traversées pour un voyage au Cap du Bonne-Esperance et aux Isles de France et de Bourbon," *Memoires...l'annee 1754* (Paris 1759) 104, where, e.g., he recorded a semi-diameter of 16′ 16″ and a 7″ refraction from observations near Rio de Janiero. Lalande 1771, table XVII, gives the diameter on the winter solstice as 32′ 35.3″, for a semi-diameter of 16′ 18″, and acknowledged (in notes to his table IV) that his values were around 3.5″ lower than those published by La Caille. For Lalande, see Chapter 3.a.

84. *Connaissance de Mouvemens Célestes pour L'Annee 1763* (Paris 1761) 77 (for December). This edition of the *Connaissance* had been produced by Lalande. Carcani, "Passaggio di Venere...," II, had determined through his own observations and not on the basis of published tables that the semi-diameter was 15′ 46″ on June 6, 1761, from Naples; the *Connaissance* for 1763 set the value at 31′ 31″ = 15′ 15″ from Paris. For 2023, e.g., the value of the semi-diameter of the sun on the winter solstice in Greenwich was 16.26′ (= 16′ 16″); for the spring equinox: 16.06′ (= 16′ 4″); for the summer solstice: 15.74′ (= 15′ 44″); see *Nautical Almanac of the Stars*: https://www.nauticalalmanac.it/en/.

134 THE PROSCIUTTO SUNDIAL

He was confident that accuracy could be attained to within one second. Venus transits occur only when Venus and the earth align with the sun, allowing the passage of time between when the orb of Venus passes the left solar limb to the moment it exits its right limb to be measured from earth. Such transits are quite rare, though regular: one had occurred in 1639 and the next in 1761.[85] For the latter event, the French astronomer Joseph-Nicholas Delisle, who had sponsored Carcani's election to the Académie Royale des Sciences along with La Caille, had taken a leading role in promoting an international effort to observe the transit. Delisle had simplified Halley's method by reducing the necessary observation to only the exact moment when the orb of Venus either passed onto or exited that of the sun: if one of these moments was viewed from two points at different latitudes and longitudes, their different readings could be used in calculating the parallax. By reducing each astronomer's observation to this single datum and engaging as many astronomers as possible, all widely dispersed at key points across the globe whose latitude and longitude were known, Delisle increased the accuracy of the readings while limiting the possibility that poor viewing conditions or unforeseen circumstances would diminish the volume of viable data. Delisle's approach also compensated for the fact that the passage was calculated to occur at night in Europe, when Venus would only be exiting the solar orb: the full transit would not be visible.[86] The times obtained by the 130 observation stations, the manning of several of which had required launching elaborate expeditions equipped with the necessary astronomical instruments, would then be converted to a common absolute time, such as that of Paris, to compute the solar parallax.[87]

---

85. The bibliography on Venus transits is extensive; the sources referenced here include: H. Woolf, *The Transits of Venus; A Study of Eighteenth-Century Science* (Princeton 1959); H. Woolf, "Eighteenth-Century Observations of the Transits of Venus," *Annals of Science* 9.2 (1953) 176–90; J. Ratcliff, *The Transit of Venus Enterprise in Victorian Britain* (Pittsburgh 2016) 9–19 ("The Precedent: Transit of Venus Expeditions in 1761 and 1769"); S. Dumont and M. Gros, "The Important Role of the Two French Astronomers J.-N. Delisle and J.-J. Lalande in the Choice of Observing Places during the Transits of Venus in 1761 and 1769," in C. Sterken and P. P. Aspaas, eds., *Meeting Venus* (*The Journal of Astronomical Data* 19.1, 2013) 131–44; S. Débarbat, "The French Savants, and the Earth-Sun Distance: A Résumé," in C. Sterken and P. P. Aspaas, eds., *Meeting Venus* (*The Journal of Astronomical Data* 19.1, 2013) 109–19.

86. Delisle had prepared a detailed map of the world, called his *Mappemonde*, indicating the areas where the best viewings were possible along with the timing of the event. For Delisle's 1760 *Mappemonde sur laquelle on a marqué les heures et les minutes du temps vrai de l'entrée et de la sortie du centre de Vénus sur le disque du soleil* (Bibliothèque nationale de France), see Dumont and Gros 2013, 136, fig. 3.

87. P. P. Aspaas, "Maximilianus Hell (1720–1792) and the Eighteenth-Century Transits of Venus: A Study of Jesuit Science in Nordic and Central European Contexts" (PhD dissertation, University of Tromsø, 2012) 194–203 (http://hdl.handle.net/10037/4178).

Analysis

Tanucci knew of this international effort and had expressed his willingness to fund Carcani's travel to Jerusalem to view the event on June 6, 1761.[88] Whether Carcani's potential contributions to elucidating the sundial on the basis of these new discoveries had come up in the course of discussions about the merits of substituting the sundial for the Egyptianizing base is nowhere documented, but this was precisely the period in which the change had occurred.

Carcani ultimately had participated from his observatory in Naples. In the course of the four-hour transit, he had recorded data relating to such aspects as the ingress, the egress, the angle of incline of Venus' orbit with respect to the sun's equator, and the apparent diameter of Venus. He was among the astronomers who formally published his results, in a six-page pamphlet detailing his methodology and summarizing the data he had collected. But he ultimately refrained from publishing a fixed value for the solar parallax, presumably in the knowledge that his own data played only a small role in this collaborative effort and would be factored into the results obtained from all the stations.[89]

The desired degree of accuracy was not attained, however, despite the scale of the international effort and Delisle's precise calculations and preparations. The results varied widely, from a value of 8.56″ to 10.6″. The reading at the

---

88. B. Tanucci, *Epistolario*, ed. M. G. Maiorini (Rome 1985) 9.217 #157 (December 20, 1760), in a letter to B. Galiani in Paris. Tanucci had listed Siberia, Pondicherry, and Jerusalem as optimal viewing spots, all of which are confirmed by Delisle's *Mappemonde*. He deemed participation in the enterprise "literary idleness" through which a sovereign could earn praise at little expense. In a letter posted to Galiani a week later (9.238 #172), he expressed his concern that clouds could obscure the event and that an Italian therefore might not be interested in making such a journey. This suggests Carcani had expressed reservations about traveling to view the transit.

89. N. M. Carcani, *Passaggio di Venere sotto il Sole, osservato in Napoli nel Real Collegio delle Scuole Pie, la mattina de' 6 giugno 1761* (Naples 1761); first published under the same title in the *Novelle Letterarie di Firenze* (1761) col. 696–702 (= #44 of October 30, 1761) 714–20 (= #45 of November 6, 1761): *Il metodo che ò adoprato nella presente dilicatissima osservazione, la quale pel vantaggio, che se ne spera, della verificazione tanto delle parallassi del Sole, e di Venere, quanto della Teoria di questo Pianeta, à meritata l'attenzione non solo di tutti gli Astronomi, ma di molti Sovrani dell' Europa: è l'istesso, che propose, e praticò il chiarissimo Sig. de l'Isle pel passaggio di Mercurio de' 9. Novembre 1723. nelle Memorie dell' Accademia delle Scienze del detto anno: metodo in practica il più difficile, e intrigato di molti e fastidiosi calcoli, ma in sostanza il più esatto, e più sicuro di ogni altro*: "The method which I have adopted in the present most delicate observation, which, for the benefit which one hopes for of the verification both of the parallax of the Sun and of Venus and of the Theory of this Planet, has merited the attention not only of all astronomers but of many European sovereigns, is the same which Signore Delisle proposed and applied for the passage of Mercury on November 9, 1723, in the *Memorie* of the *Académie des Sciences* of that year: a method in practice the most difficult, and complicated by many careful calculations, but in substance the most exact and secure of all." See also N. Severino, "Quando Venere passa davanti al Sole," *Astronomia: La rivista dell' Unione Astrofili Italiani* 3 (1999) 8–12, a popular summary of Carcani's observations.

Cape of Good Hope, for example, was 8.98″, while those from Paris ranged from 9.57″ to 10.25″. In Italy, Zanotti had observed the transit in Bologna, Ximenes in Florence, and Audiffredi in Rome, while other observations were attempted in Milan, Padua, and Venice.[90] Carcani's readings were equated to a parallax of 10.29″.[91] The British mathematician James Short had studied the published findings, determined the mean parallax, and proposed the figure of 8.5″, a value vigorously disputed by the French astronomer Alexandre Guy Pingré, who argued for 10.60″.[92] Ultimately, Carcani had employed the round value of 8″ for the two calculations he performed on the sundial, though it is not clear how he had settled on a value relatively close to that determined only subsequent to his study of the prosciutto: the data from the 1761 transit were still being debated in the academies and scientific publications when this third volume of the *Le Antichità* was produced.[93]

Carcani had factored all three of these values into his final calculation, the quantity of the obliquity of the ecliptic at the time in which the prosciutto

---

90. L. Pigatto, "The 1761 Transit of Venus Dispute between Audiffredi and Pingré," in D. W. Kurtz, ed., *Transits of Venus: New Views of the Solar System and Galaxy* (Proceedings of the International Astronomical Union, Colloquium No. 196, 2004) 74–86, including a list of all the observation stations in Italy. Ultimately a hybrid of Halley's and Delisle's methods of taking the readings had been employed. For a full list, see Aspaas 2012, 217–18, table 1. Carcani is missing from both Aspaas' and Woolf's catalogs of observers (Woolf 1959, 135–40).

91. R. Calanca, *Il transito di Venere sul disco del sole* (Mestre 2004) 69–103, including a table of readings from stations around the globe based on Pingré. See also R. Proctor, *The Transits of Venus* (London 1871) 27–66.

92. J. Short, "The Observations of the Internal Contact of Venus with the Sun's Limb, in the Late Transit, Made in Different Places of Europe Compared with the Time of the Same Contact Observed at the Cape of Good Hope, and the Parallax of the Sun from Thence Determined," *PhilTrans* 52 (1761–1762) 611–28; Short enhanced the data of this contribution in a second one the following year, arguing for a value between 8″ and 9″: "Second paper concerning the parallax of the sun determined from observations of the late transit of Venus in which this subject is treated more at length and the quantity of the parallax more fully ascertained," in *PhilTrans* 53 (1763) 300–45. For the dispute, see Pigatto 2004, 74–84. James Short (1710–1768), an astronomer and a Fellow of the Royal Society (elected 1737), was an optician famed for the quality of his reflecting telescopes, one of which may have been given to Paderni (see Chapter 1.g). See G. Turner, "James Short, FSR, and His Contribution to the Construction of Reflecting Telescopes," *The Royal Society Journal of the History of Science* 24.1 (June 1969) 91–108, who notes that many of Short's telescopes were used in the 1761 and 1769 efforts to record the transits of Venus and were acquired by astronomers throughout Europe, singling out the Swiss astronomer, Jean Bernoulli (see Chapter 3.a).

93. A value ranging from 8.43″ to 8.80″ was established as a result of the 1769 transit, when the techniques of observation took into account the problems encountered in 1761 (Ratcliff 2016, 18). S. Newcomb, "Discussion of the Observations of the Transits of Venus in 1761 and 1769," *Astronomical Papers Prepared for the Use of the American Ephemeris and Nautical Almanac* (1890) 401–2, recalculated all the available data from the 1761 transit and settled on a mean value of 8.79″. The currently accepted value of solar parallax is 8.794″.

Analysis 137

had been manufactured. That the plane of the equator was tilted in respect to that of earth's orbit around the sun had been known since antiquity, and the angle of the difference had been determined to a remarkable degree of accuracy. Even so, the precise value of this angle and, in particular, whether it was fixed or diminishing over time were questions still bedeviling astronomers in the eighteenth century. Ptolemy (*Almagest* 1.13), the second-century Alexandrian astronomer, reported measuring the sun's altitude at noon on the summer and winter solstices over many years and determined that the ratio of the height of a gnomon to the length of its shadow in his day was 11:83. This yields an obliquity of 23° 51′ 20″ (23.85°). Earlier, Hipparchus of Nicaea, from the second century BCE, and Eratosthenes of Cyrene, from the third century BCE, had found the same ratio of height to length.[94] Hipparchus, in turn, had stated that his readings at Byzantium were similar to those at Massilia (Marseille), which had been recorded in the fourth century BCE by the geographer Pytheas (Strabo, 1.4.4, 2.5.8).[95] The accuracy of these readings had been questioned by astronomers for centuries, but in 1719 the French astronomer Jacques-Eugene d'Allonville, Chevalier de Louville (1671–1732) had taken the novel approach of going to Marseilles to establish the obliquity at this latitude in his own day and comparing it with the ratio determined by Pytheas, recalculating it to take into account the three factors Pytheas would not have considered: the semi-diameter of the sun, solar refraction, and the solar parallax. Pytheas' angle of the sun at the summer solstice was 70° 47′ 41″ (70.79°), from which de Louville subtracted the combined refraction and parallax of 17″ and a semi-diameter of 15′ 49″ (0.26°) to yield 70° 31′ 35″ (70.53°). Further subtracting the latitude of Marseilles of 46° 42′ 12″ (46.70°) left the true obliquity of the ecliptic in Pytheas' time as 23° 49′ 23″ [= 23.82°]. This differed 21′ [= 0.35°] from the obliquity of 23° 28′ 24″ [= 23.47°] de Louville had

---

94. D. Shcheglov, "Hipparchus' Table of Climata and Ptolemy's *Geography*," *Orbis Terrarum: International Journal for the Historical Geography of the Ancient World* 9 (2007) 180–82; Wilson 1980, 63: the current value for the obliquity in the year 130 CE is 23° 40′ 46″ [= 23.67°].

95. Strabo 1.4.4: "The parallel through Byzantium is the same as that through Massilia; for as to the relation of the gnomon to the shadow, which Pytheas has given for Massilia, this same relation Hipparchus says he observed at Byzantium, at the same time of the year as that mentioned by Pytheas." A. Jones, "Eratosthenes, Hipparchus, and the Obliquity of the Ecliptic," *Journal of the History of Astronomy* 33.1 (2002) 15–19, questions whether Pytheas' calculation was as precise as often assumed; B. Goldstein, "The Obliquity of the Ecliptic in Ancient Greek Astronomy," *Archives Internationales d'Histoire des Sciences* 33.3 (1983) 3–14, believed that neither Pytheas nor Hipparchus had made precise calculations; J. Britton, "Ptolemy's Determination of the Obliquity of the Ecliptic," *Centaurus* 14 (1969) 29–41, reviews the potential errors in Ptolemy's calculations.

138 THE PROSCIUTTO SUNDIAL

obtained, allowing him to conclude that the value had diminished a full 21′ over the course of 2,075 years, or about 1′ every century.[96]

While de Louville's value for the rate of diminution continued to be debated—even La Caille would propose a rate that more than halved de Louville's—Carcani had remained firmly in de Louville's camp, specifically referencing his research in the Preface.[97] This may have been because of the perceived convenience of de Louville's round number for the rate of diminution and his table that associated a specific date of 360 BCE for Pytheas' original determination of the obliquity. De Louville's recalculation of Pytheas' reading was 23° 49′ 23″ [= 23.823°] while Carcani had determined an obliquity of 23° 46′ 30″ [= 23.775°] for the time when the prosciutto was manufactured, a difference of 2′ 53″ [= 0.048°], which, according to de Louville's rate of diminution, would date the prosciutto to the second half of the first century BCE. De Louville's rate was equally convenient for ensuring the date of manufacture agreed with the inscriptional evidence that required the prosciutto to date after the change in the name of the eighth month to Augustus. De Louville's rate has a secular diminution of 0.0105°, and Carcani had determined a total diminution of 18′ 12″ [= 0.303°] from the date of manufacture. The years from "0" to 1762 would yield 18′ 30″, meaning the difference of 18″, or about thirty years, would need to be added back on. The Accademia therefore settled on 28 CE, placing it in the middle of the reign of the emperor Tiberius (14–37 CE). This, however, is well after the traditional date of January 16, 27 BCE, for the Senate's awarding of the honorific title of Augustus. It was also subsequent to the so-called *Lex Pacuvia de mense sextilii* of 8 BCE, the legislation officially renaming the eighth month in honor of

---

96. J-E. d'Allonville de Louville, "De mutabilitate eclipticae dissertatio," *Acta Eruditorum* (1719) 281–94; de Louville provides a table with conversions of the values of the obliquity determined by earlier astronomers. This shows a difference of 3′ 10″ between the readings of Pytheas and Eratosthenes 160 years later.

97. N. L. La Caille, "Mémoire sur la Théorie du Soleil," *Memoires de l'Académie Royale des Sciences, Année 1757* (Paris 1762) 108–44, esp. 109–16: La Caille determined 23° 28′ 19″ in 1750 from Paris, for 47.3″ per century, but ultimately settled on a diminution of 44″ per century. La Caille noted that L. Ximenes at Florence had determined it had diminished 1′ 11″ in 245 years: Ximenes' (*Del vecchio e nuovo gnomone fiorentino e delle osservazioni astronomiche, fisiche et architettoniche fatte nel verificarne la costruzione, libri IV* [Florence 1757] 297–318) observations had established the obliquity as 23° 28′ 19″ in 1756 with a secular diminution of 30.6″. Lalande gives the value in the 1763 (1761) edition of the *Connaissance de Mouvemens Célestes* as 23° 28′ 21″ (23.47°); in his *Tables astronomiques* of 1792, he gives the value of 23° 45′ 27″ for the time of Pytheas, 23° 42′ 22″ for the year "0," and 23° 28′ 17″ for 1762. The present value of the obliquity is about 23° 26′ 11″ (23.43°) and diminishing at a rate of 0.013° every century.

Analysis 139

the emperor, which provides the more likely *terminus post quem* for the prosciutto.[98]

2.g) *The Accademia's Engraving of the Prosciutto*

The Accademia illustrated their philological and astronomical exegesis with an engraving of the prosciutto (Figure 2.10). This featured prominently at the top of the opening page of the Preface, rather than as a separate plate following the written discussion, as was the practice with the paintings. For this work they had commissioned two of the best artists working for the Regale Stamperia Palatina, the royal printing house in Naples that had been producing the volumes of *Le Antichità di Ercolano* since 1755. Giovanni Elia Morghen, a Florentine artist who had begun to contribute his work with the second volume of *Le Antichità* and who would have his hand in many of the antiquities reproduced in the Accademia's publications, was the "Regio delineatore," the royal illustrator, while Ferdinando Campana, a Roman whose career largely paralleled that of Morghen, was the "Regio incisore," the royal engraver of Morghen's drawing. Both signed this engraving in opposite corners at the bottom.[99]

The Accademia's engraving remains the most detailed and accurate representation of the entire instrument available. The divisions of the hour lines that can be obtained from the engraving are quite close to those inscribed on the prosciutto (Figure 2.11). It rendered the object at a 1:1 scale and depicted both sides in full, complete with the suspension ring at the top, the obverse bearing the dial shown on the left and the reverse on the right. The hatchings and cross-hatchings of the engraving capture the full shape and form of the porcine model on which the object was based, the tonal gradations highlighting its curved sides and dimpling along with the subtle undulation of the surface upon which the dial was etched. Represented as well are such details as the head of the femur bone, the bulging out at the stifle joint where the femur links to the fibula and tibia, and the narrowing of the shank at the hock, where the

---

98. Macrobius (1.12.35) attributes the change to a tribunician law put forth by Sextus Pacuvius and hence today referred to as the *Lex Pacuvia*. Censorinus (*de Die Nat.* 22) adds that the change occurred in the twentieth year of Augustus' reign, during the consulship of Caius Marcius Censorinus and Caius Asinius Gallus, which dated to 8 BCE and is the commonly accepted date for the change. Both Cassius Dio (55.6.6) and Suetonius (31.2) imply Augustus himself made the change.

99. Pannuti 2000, 151–78, esp. 161 (Campana), 170 (Morghen); Mansi 2008, 115–45, esp. 120–28 (engravers), 133, fig. 8 (prosciutto engraving); and Mansi 2015, 21–30.

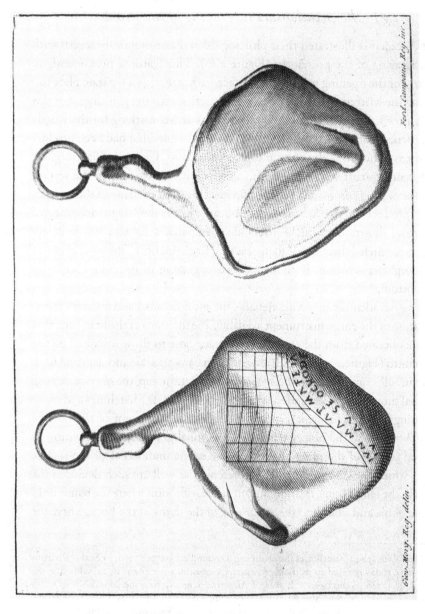

FIGURE 2.10 The *orologio di prosciutto* as drawn by Giovanni Morghen and engraved by Ferdinando Campana for the preface to the third volume of the Accademia Ercolanese's *L'Antichità di Ercolano* (Naples 1762).

Comparison of Accademia's Grid with the Prosciutto

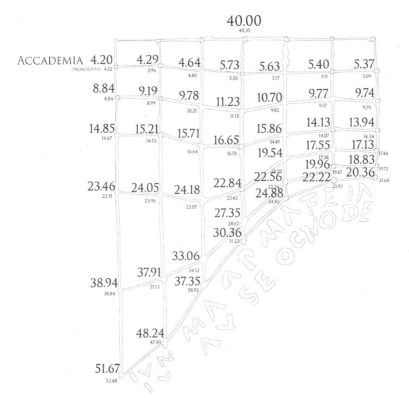

FIGURE 2.11 Comparison between the intersections of the horizontal horary lines with the vertical month lines on the Accademia's engraving (see Figure 28) and on the original prosciutto. Author's image.

leg likely was cut above the metatarsals to remove the trotters, or pig's feet. The reverse shows the characteristic manner in which the thick rind wraps around the sides from the front where it has been trimmed back to expose the meat around the head of the femur (Figure 2.12). The Accademia clearly wanted to ensure that their engraving was recognized for what it was: a prosciutto and not a carafe. Moreover, the artist's intent evidently had been to portray a prosciutto slaughtered from the variety of large, smooth-skinned, short-legged pig documented both archaeologically and in other works of art from the Vesuvian region.[100] The plump bronze piglet found in 1739 in Herculaneum offers a

---

100. Perhaps the *sus scrofa*. See M. Di Gerio and A. Genovese, "Studio zoo-archeologico sui reperti provenienti dalla Regio IX, Insulae 11/12, saggi B e C," *Rivista di Studi Pompeiani* 22 (2011) 143–49; M. MacKinnon, "High on the Hog: Linking Zooarchaeological, Literary, and

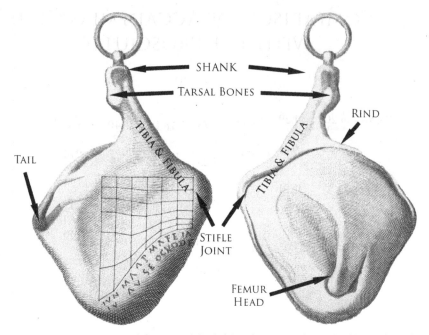

FIGURE 2.12 Anatomical features of the left hind quarter of a pig highlighted on the Accademia's engraving of the prosciutto. Author's image.

good example of just such a type, and it is odd the Accademia did not reference this earlier find in their lengthy discussion (Figure 2.13).[101]

The tail is shown here as projecting out from the surface, its tubular form following the angled outer edge of the prosciutto's hind quarter, as if captured in a swing to the animal's left. In the original the remaining stub is now mashed against the surface and bent slightly beyond the outer edge,

---

Artistic Data for Pig Breeds in Roman Italy," *American Journal of Archaeology* 105.4 (2001) 649–73, esp. 659–66, figs. 16–18. M. A. Anderson and D. Robinson, *House of the Surgeon, Pompeii: Excavations in the Casa del Chirurgo* (VI 1, 9–10.23), (Oxford 2018) 503 (Richardson), found that three-quarters of the pigs had been slaughtered before they were thirty months old. See also M. Fulford et al., "Towards a History of Pre-Roman Pompeii: Excavations beneath the House of Amarantus (I.9.11–12), 1995–8," *Papers of the British School at Rome* 67 (1999) 86–87, where the pig remains are compared in size to those found at other sites.

101. The piglet is MANN 4905 (11.5 cm H x 13.5 cm L), found on September 12, 1739, in Herculaneum, exact provenance unknown (*StErc* 530). The rectangular bronze pedestal with cloven hooves emerging from palmettes on which it is mounted is original. Chiseled into its side is the inscription HER · VOE · M ·L ·, which has been interpreted as a votive offering, perhaps to Hercules, though no satisfactory reading of the abbreviations has been offered. The Accademia published an engraving of it as the frontispiece to plate 17 in the fifth volume of *Le Antichità* (5.71 plate 17) and deciphered it as *Herculi Voefilus Marci Libertus*, but the *nomina* Voefilus or Voesius, an alternative they suggested, are unattested in the Vesuvian sites. See also MacKinnon 2001, 661 n. 15, fig. 16.

Analysis

FIGURE 2.13 Bronze piglet found in 1739 in Herculaneum, precise provenance unknown. Inscribed *HER · VOE · M ·L ·* 11.5 cm high × 13.5 cm long. MANN 4905. Su concessione del Ministero della Cultura—Museo Archeologico Nazionale di Napoli.

making it impossible to determine its original trajectory. The end of the tail is sheared off straighter in the engraving than found on the original, and a lighter stippling provides a clear contrast between the surviving portion of the tail and Carcani's reconstructed wax extension. This continues the trajectory of the original tail toward the upper right before it turns to point horizontally with a sharpened tip toward the upper left corner of the dial. The tip ends about 7.5 mm to the left of the dial. Despite Piaggio's claim, the original gnomon need not have been curled to any degree, the curling tail commonly associated with pigs not being typical in all breeds either in antiquity or today. Dots mark all the points of intersection between the horizontal and vertical lines on the grid in the same way that many, though not all, points of intersection are marked by round depressions on the instrument, but the carefully drafted lines of the engraving fail to capture the roughly furrowed lines found at points on the original. The lines of the dial are reproduced in their precise dimensions. Finally, the inscription is accurately reproduced here for the first time. It correctly portrays the staggered positioning of the calendrical abbreviations in respect to the grid, the irregular spacing between them, and, of course, their *boustrophedon* pattern. Despite their small scale, the individual letters are legible, sans serif, with greater clarity attained through the elimination of the middle bar in the "As" and the use of the open stroke in the "P." The forms of these letters, together with the irregular furrowing of the lines making up the grid, reflect the impressions etched by the artist with his stylus in the original beeswax model from which the final bronze cast was made.

2.h) Paderni and the Sundials of Pompeii

As it happened, the delay in publication meant the prosciutto was not the only sundial gracing the third volume of *Le Pitture*. On January 29, 1762, a canonical hemispherical sundial of marble was recovered in the excavations at Pompeii. This fortuitous discovery led to this sundial serving as a kind of bookend to the prosciutto, the decision having been made quickly that it should be engraved for inclusion in the volume, at that time nearing the final stages of production. Highlighting the hastiness with which it was added is the fact that it is the last engraving in the volume, tacked onto the end of the *Alcuni Osservazioni* ("Some Observations"), the notes detailing the small vignettes of paintings sprinkled throughout the text, and its treatment by the Accademia is uncharacteristically concise.

The sundial's likely provenance in Pompeii was the garden of a house adjacent to the site where a Pentelic marble statue of Diana in the archaicizing manner had been found two years earlier.[102] Despite it being the first sundial found in Pompeii, and the first sundial of this type from the Vesuvian sites, little notice of it was made in the excavation records. The weekly inventory simply states, "A piece of white marble has been found [measuring] 10 *oncie* (0.22 m) wide by 17 *oncie* and a half (0.38 m) [high], and it contains a sundial with 11 lines or rays and one parallel."[103] Paderni entered it in his *Diario* as "a semicircular concave quadrant of marble one *palm* and one *oncia* (0.28 m)

---

102. The Diana (MANN 6008) was found on July 19, 1760 (*PAH* I.i.114), during excavations in the so-called Irace Property in Pompeii, in the Casa di Diana, also known as the Casa di M. Spurius Saturninus e di D. Volcius Modestus (Regio VII.vi.3). The sundial probably came from the area of the main peristyle of the house next door, at Regio VII.vi.7, where a large painting of Venus emerging from her shell had been found (MANN 27704; *PAH* I.i.140 [March 6, 1762], *PdE* 4.11–15, plate 3); see Parslow 2007, 9–12.

103. *PAH* I.i.139 (January 30, 1762).

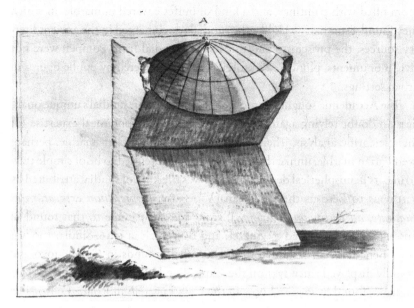

FIGURE 2.14 Marble sundial from house at Regio VII.vi.7, Pompeii, drawn by Camillo Paderni for his manuscript *Monumenti antichi rinvenuti ne Reali Scavi di Ercolano e Pompei* (1768), Plate 24. Courtesy the Bibliothèque de l'École Française de Rome (Ms. 26).

high and 10 [*oncie*] (0.22 m) wide."[104] He subsequently drew it for his manuscript, the *Monumenti antichi* (Pl. 24a), depicted as if it were still standing in its garden setting, with the caption, "Solar quadrant of marble found in the habitation pertaining to the above-mentioned garden" (Figure 2.14).

Paderni's contextual sketch was insufficiently accurate to serve as the basis for the Accademia's engraving of the sundial.[105] On the other hand, it illustrates the increasing recognition of the importance of documenting the full provenance of the finds. This sundial could be placed in the garden of a private house, near a shrine displaying a statue of Diana, while the prosciutto's similar origins in a domestic setting had long been forgotten. Even Winckelmann, when he toured the excavations at Pompeii with Paderni soon after its discovery, could place it in its context: "The only building of two stories found since they began to dig for antiquities is in this place, and it is now laid quite bare. I happened to be on the spot in the month of February 1762 with the

---

104. Paderni *Diario* (January 29, 1762).

105. *PdE* 3.337–39.

inspector of the cabinet [Paderni] while the laborers were disencumbering a room filled with paintings, and a kind of buffet covered in marble, in which they found a sundial."[106] Unlike the sundials documented in the ancient literary sources, the prosciutto and this marble sundial from Pompeii were not civic monuments, plundered in foreign lands, or erected by public figures in urban settings.[107]

The Accademia sought to emphasize the Pompeian sundial's unique qualities, no doubt relying again on Nicola Carcani's astronomical expertise for their scientific analysis. Their sundial was fashioned from marble, perhaps even Parian marble, unlike those in Rome, which tended to be of simple travertine. Its hemispherical design proved it was the kind of sundial attributed by Vitruvius to Berosus the Chaldean (Vitr. 9.8.1: *hemicyclium excavatum ex quadrato, ad enclimaque succisum*), so it was comparable to that found in Tusculum and published by Zuzzeri and Boscovich in 1746. Similar too was Pope Benedict XIV's sundial from the papal grounds in Castelnuovo and proudly displayed since 1751 on the Capitoline Hill in Rome, and that from Rignano, found in 1755 and now in the Casa Lucatelli in Rome. Le Roy had described one of marble on the southern slope of the Acropolis in Athens.[108]

Two views of the sundial were provided, one showing it in frontal perspective and the other in profile, the latter supplemented with lines and labeled according to specific points made in the text (Figure 2.15). These clearly illustrated the sundial's canonical form, its state of preservation, and its design, which the Accademia described. As is often the case with this type of ancient sundial, the delicate outer points at the top of the dial had been sheared off and the gnomon was no longer extant, but the socket into which it had been anchored survived, and the remainder of the sundial was intact. The dial consisted of the usual eleven horary lines, the central or sixth line marking the meridian and the upper edge constituting the horizon. A single arc of a circle intersecting the horary lines and running parallel to the outer edge of the dial represented the Equator, or the "inclination" of the Equator with respect to

---

106. Winckelmann 1762, #29; Mattusch 2011, 81.

107. See also R. P. Jacquier, "Lettre du R.P. Jacquier aux auteurs de la Gazette Littéraire," *Gazette Littéraire de l'Europe* (Paris 1764–1766) 295–99; Antonini 1790, plate 14: "Orologio solare antico esistente in Napoli"; G. H. Martini, *Abhandlung von den Sonnenuhren der Alten* (Leipzig 1777) 57–60, figs. III and IV, copied the engraving from the *PdE* as well as the Accademia's analysis; Gibbs 1976, 296 #3075 (present location unknown); it is missing as well from the catalog of sundials in the Naples museum studied by A. Pagliano, *Le Ore del Sole: Geometria e astronomia negli antichi orologi solari Romani* (Rome 2018).

108. J-D. Le Roy, *Les ruines des plus beaux monuments de la Grece* (Paris 1758) 15 n. 8.

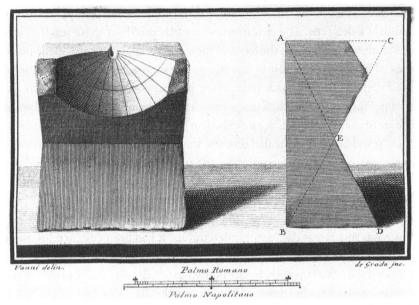

FIGURE 2.15 Marble sundial from house at Regio VII.vi.7, Pompeii. Drawn by Nicola Vanni and engraved by Filippo de Grado for the third volume of the Accademia Ercolanese's *L'Antichità di Ercolano* (Naples 1762).

the horizon, the latter being equivalent to the line formed by the sundial's base.[109] This example differed from others, noted the Accademia, in lacking the arcs corresponding to the Tropic of Cancer and the Tropic of Capricorn.[110] The gnomon, set into a vertical cutting on the top of the sundial and bent at a right angle, would have projected out horizontally, running parallel to the top of the sundial along the meridian line, its tip corresponding with the plane of the equatorial line inscribed below.[111]

What set their sundial apart from others, however, was the elevation of the pole, or the latitude, for which it had been designed to function. Determining this had not required factoring in solar refraction or the semi-diameter of the sun, nor had the obliquity of the ecliptic come into play here: its beauty lay in its simplicity and its conformity to basic trigonometric rules. Consequently, the Accademia could perform the same simple calculations that Boscovich

---

109. The "inclination," or declination of the sun, is the ecliptic. The day line representing the Equator represents the equinoxes.

110. The day lines representing the winter and summer solstices, respectively.

111. The length of the gnomon in such hemispherical dials is equivalent to the radius of the hemisphere.

148 THE PROSCIUTTO SUNDIAL

had carried out on the Tusculum sundial and could, in fact, be applied to any sundial of this type. The characteristic angular profile of these sundials was the result of cutting back the front of the stone parallel to, or along, the plane created by the shadow cast by the sun on the summer solstice. In the case of the Pompeian sundial, this front plane had been cut to correspond to the summer solstice, rather than projecting slightly beyond and having the line of the arc of the summer solstice inscribed on the stone, as was common. Because the top and bottom of the dial on these sundials represented the horizon, the angle created by the intersection of the upper edge of the sundial and the plane of the summer solstice revealed the latitude for which it was made.[112] Boscovich's Tusculum sundial was similar to the Pompeian example, the projecting points of its upper edge similarly broken off and its gnomon missing, but it was squatter at the base. He determined it had been designed for use at the latitude of 41° 43′, roughly that of Tusculum, and the Accademia had rounded the others up in their data pool to 42°.[113] The distinctive profile of the Pompeian sundial, the upper half containing the dial being equivalent in size to the lower one serving as the base, allowed the angle of the dial's front to be extended—and indeed the Accademia noted just such a line had actually been incised in the marble on the sundial's sides—to intersect the right angle created by the horizontal line of the sundial's bottom and its vertical back. By dividing the length (0.157 m) of the upper edge of the sundial by its vertical height (0.28 m), they determined the tangent of this right triangle (0.561) and then converted this into degrees. This yielded an elevation of the pole equivalent to 29° 18′ (= 29.3°), a latitude only slightly less than that given by Ptolemy in his *Geography* (IV.5) for Memphis in Egypt (= 29° 50′).[114] "It seems very likely that this marble was made for a pole in that illustrious kingdom of Egypt and from there transported into these parts," concluded the Accademia, in raising the intriguing possibility it was, in fact, foreign booty, like the plunder referenced in the literary sources that had been brought back by the Romans from conquered lands to enrich both public and private edifices. Alternatively, they mused, it may have been a copy of one originally designed for that city. Nevertheless, this was an exotic aspect that set their sundial apart from all other published examples, in the same way the prosciutto's design made it unique.

---

112. Gibbs 1976, 12–18, explains the underlying geometry succinctly.

113. The actual latitude of Tusculum is 41° 47′ while that of Rome is 41° 54′.

114. The actual latitude of Memphis is 29° 50′ 58″.

Analysis

If the hemispherical sundial from Pompeii had been celebrated for its unique qualities, a third sundial found the following year evidently had reignited tensions in the museum. This was a horizontal sundial found in a small room in the so-called Villa di Cicerone, a richly appointed suburban villa just outside the Porta Ercolano in Pompeii that the Bourbons had excavated only in part. In this case, virtually everything known about this sundial can be traced back to Paderni's documentation. In keeping with his practice, he inventoried the sundial's arrival at the museum in his *Diario* on May 11, 1763: "A horizontal quadrant in marble with its bronze gnomon secured with lead in said sheet of marble, 2 *palmi* and 2 *oncie* wide and in height 1 *palm* 9 ½ *oncie* [0.57 x 0.45 m]." Curiously, it is missing from the inventories of finds produced by the excavators themselves, though other objects recovered at the same time and place are listed.[115] Among these was a magnificent mosaic panel showing street musicians signed by the artist Dioskurides of Samos (MANN 9985), found in an adjacent space the previous week.[116] The sundial's precise provenance was securely identified only years later, when Francesco La Vega, Paderni's successor as *custode* of the museum, included it in an inventory of finds accompanying a plan of the Villa he drew sometime between 1778 and 1790. La Vega had relied on Paderni's *Diario* in drawing up his plan so his entry for the sundial describes it in similar terms: "A sundial on a sheet [of marble], of 2 *palmi* and 2 *oncie* and 1 *palm* and 9 *oncie*, with its bronze gnomon."[117]

---

115. This is not the same as the horizontal sundial first published in G. Fiorelli, *Giornale degli Scavi di Pompei* (Naples 1865) 14–16, and now in the Naples Museum (MANN 2476). At 0.53 m wide and 0.345 m high and 0.03 m thick, that sundial is remarkably similar in dimensions to Paderni's, but its provenance is unknown, it is missing its gnomon, and it was evidently reassembled from pieces found in the museum. The most important distinguishing feature is that it is labeled in Greek, including abbreviations in Greek for the signs of the zodiac whose day curves are incised as well (see Gibbs 1976, 331 #4007, who did not see the original). Paderni could not have overlooked both the Greek and the additional lines on the grid; his illustration in the *Monumenti* provides the visual proof. For this sundial, see also Pagliano 2018, 237–55, "Orologio a piano orizzontale L"; J. D. Hey, "On an Ancient Sundial Unearthed at Pompeii," *Transactions of the Royal Society of South Africa* 73.1 (2018) 56–72; and the exhibition catalog *Time and Cosmos in Greco-Roman Antiquity*, ed. A. Jones (New York and Princeton 2016) 82, 185 #7.

116. This is the same mosaic referenced by Paciaudi in a letter to Caylus. Another mosaic depicting a group of women seated around a table and also signed by Dioskurides (MANN 9987) was found on the opposite side of the room on March 8, 1764; Winckelmann witnessed its discovery and described it in a letter to Paciaudi dated March 24, 1764, but does not mention either the sundial or the "lustral basin" (the original letter is in the Winckelmann-Museum Stendahl [WG-H-15]; https://st.museum-digital.de/object/37276). The scenes are thought to be taken from plays by Menander, the *Theophoroumene* in the case of the musicians and the *Synaristosai* in the case of the women.

117. *PAH* I.2 Addenda 3.106, together with finds recovered between April 29 to May 11, 1763. La Vega had carried on the *Diario* once he became *custode* in 1781.

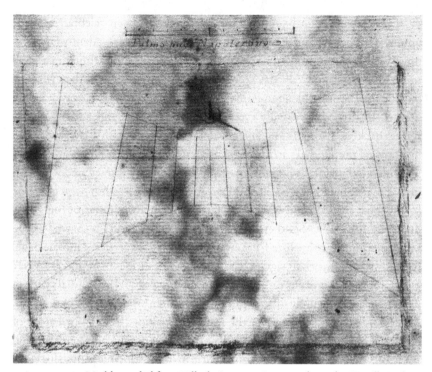

FIGURE 2.16 Marble sundial from Villa di Cicerone, Pompeii, drawn by Camillo Paderni for his manuscript *Monumenti antichi rinvenuti ne Reali Scavi di Ercolano e Pompei* (1768), Plate 27. Courtesy the Bibliothèque of the École Française de Rome (Ms. 26).

As with the hemispherical sundial, Paderni had featured this in his manuscript of 1768, the *Monumenti antichi rinvenuti ne Reali Scavi di Ercolano e Pompei* (Figure 2.16).[118] His scaled drawing of it appears together with that of a large bronze vessel he termed a "lustral basin," also found in the Villa. It is the most basic kind of horizontal, or planar, sundial, with the classic double axe-shaped dial filling the sheet of marble from one side to the other.[119] Hyperbolic curves at the top and bottom representing the summer and winter solstices, a straight line marking the equinox, and eleven horary lines

---

118. Paderni, *Monumenti antichi*, plate 27. See also N. Severino, *De Monumentis Gnomonicis apud Graecos et Romanos: Catalogo degli orologi solari Greco-Romani* (Rome 2011) 31–32.

119. Gibbs 1976, 39–42, where the sundial grid illustrating horizontal dials (fig. 22) is virtually identical to Paderni's, as is Gibbs 1976, 328 #4004G from Pompeii (Inv. 49725). The "lustral basin" (MANN 72990) is listed along with the finds for the week ending May 7, 1763, and described as "very precious and rare, and served for lustral water," with dimensions of roughly 1.00 m in diameter and 0.12 m deep (*PAH* I.i.150); in his *Diario* Paderni cataloged its arrival in the museum on May 10, 1763. This appears to have been the centerpiece of the eighth room of the Herculanense Museum; see, e.g., Allroggen-Bedel and Kammerer-Grothaus 1983, 115.

Analysis 151

constitute the dial; there are no other markings. Just above the top of the grid and centered on the meridian line is the vertical gnomon, a pointed bronze projection secured by a dollop of lead in a round, concave socket cut in the marble. Paderni represented it as casting a shadow toward the fourth hour on the summer solstice. The sheet of marble was shown in slight perspective and appears to be unusually thin, perhaps only 5.0 cm thick; other examples of this type from Pompeii tend to be substantially thicker.[120] By a curious coincidence, this sundial appears to be identical in form to the one Piaggio claimed Paderni had fabricated together with the marble carver. Piaggio described their example as having been incised in one of the thin sheets of Genova stone used as backing in framing the ancient paintings cut down from the walls of the ruins for display in the museum.[121]

It is Paderni's accompanying text that renders this otherwise ordinary sundial particularly intriguing. The luxurious quality of the lustral basin, the mosaic, and other finds signified to him that the owner of the habitation was wealthy. He ignored the dial itself and offered only three specific details: that it was made of white marble, measured 2 *palmi* and 1 *oncia* in width, and had been found with its bronze gnomon still anchored in place. It was precisely the fact that its ancient gnomon was intact, a rare exception to the usual condition in which these sundials were recovered, that entangled Paderni in another controversy, however.

Soon after its discovery the sundial had been consigned to Carcani for analysis to determine its design latitude. He must have completed his calculations shortly before succumbing to the plague that ravaged Naples in 1764 and the sundial had remained at his observatory at S. Carlo delle Mortelle. Paderni eventually grew anxious that it be returned to the museum and in early 1765 wrote a request to initiate the process for its reacquisition:

> Having come to Naples twice, I always forgot to speak to you on one point, namely that for a long time a marble sundial with an antique metal gnomon has remained out of this Royal Museum, which, by order of His Excellency dated November, 1763, I had to deliver to the deceased Provincial Prelate [Nicola] Carcani so that he could make explications of it, which I want to believe are done, but in order to find

120. Gibbs 1976, 328–31 #4004G–4006, all horizontal examples from Pompeii; #4004G is 0.092 m thick and 4005G is 0.43 m thick.

121. It is possible, however, that with the passage of time Piaggio had confused the ancient one for a modern one, given the sundial's subsequent history.

152   THE PROSCIUTTO SUNDIAL

said monument it is better that His Excellency gives the order [for its return] for which I pray you would be pleased to do at present.[122]

Carcani's analysis evidently had established that this sundial had been designed to function at a different latitude from Naples', and so the ancient gnomon had been removed and replaced with one that allowed the readings to conform to conditions in the museum. If Carcani himself had not altered it, Paderni may have performed the modification himself, under pressure from what he termed the "experts and intellectuals," but he came to regret the awkward position in which it had put him. Who might have persuaded him to replace the gnomon is unclear. Martorelli is a likely culprit, given the apparent outrage Piaggio ascribed to him over the incomplete state in which the prosciutto sundial had been displayed; he may have considered it a liability to draw scholarly attention to its modern inaccuracy. Piaggio, though bitter that the prosciutto had not been altered according to his specifications, may have advocated for the change but could not have executed it himself.

A fierce debate ensued about whether the sundial should be displayed in this altered form.[123] Paderni did not reveal who had been most offended by the modification, but it had been enough to trigger a royal decree ordering him to display the original gnomon alongside the revised sundial and to provide an explanation of why the change had been made.

> As a matter of fact, a few times I have been obligated to speak [to visitors] about this huge inconvenience, but the result is always wonder and amazement by the erudite, first because we do not need a sundial that marks the hours for us; second, because an antiquity found with nothing missing is always of great value; and third, especially, this must be preserved precisely because it does not correspond to today's hours, and the reason is clear.

Paderni, who was addressing Charles III in these comments, continues with a point Piaggio would make later in his *L'Orologio Solare*: that despite the seeming rarity of the ancient sundials that had already been recovered, more would certainly be found. Paderni, newly sensitive to the keen interest in ancient dials generated by the controversy over the prosciutto, assured the

---

122. ASN, *Segreteria di Casa Reale detta casa reale antica* Fascio 1541/35: letter of Paderni dated January 25, 1765, to Don Angelo Fernandez.

123. Forcellino 1999, 32–33, summarizes Paderni's version of the dispute.

Analysis

King that these, too, would stimulate additional scholarly enthusiasm within the international astronomical community that would reflect well on the royal excavations.

> We already know that such a monument was found in a city that was covered by Vesuvius. Naturally we must believe it was not the only sundial in that city, and as a consequence with time many will be found, as in fact were found two other smaller ones but lacking their gnomons. One of these is shown in Plate 24 [the Pompeian sundial also illustrated in the third volume of *Le Antichità*; the other is likely the prosciutto], and when another like the one above is found, with its own gnomon and corresponding exactly with the one found first [the prosciutto?], and especially by similar ones that could be found, how many pens will be set in motion to explain such a discovery?

The Prosciutto Sundial: Casting Light on an Ancient Roman Timepiece from the Villa dei Papiri in Herculaneum.
Christopher Parslow, Oxford University Press. © Oxford University Press 2024. DOI: 10.1093/9780197749418.003.0002

3

Enlightenment

THE ACCADEMIA DOUBTLESS considered their analysis of the prosciutto definitive. In the case of attempts to explore its parallels in the literary and historical sources this would prove largely true. As an archaeological and artistic object there were similarly no subsequent efforts to explain the artist's inspiration for its form or to associate its shape with a specific individual or understand its original context beyond its provenance in Herculaneum as it increasingly became simply a museum oddity, notable for its amusing shape but with its functionality unquestioned. Even the Accademia's detailed and accurate engraving of the prosciutto was often ignored in subsequent scholarship or distorted in bizarre ways ranging from crude renderings to complete inversions and striking variations on Paciaudi's carafe form. The Bourbon court's tight grip on access to the antiquities and the very limited distribution of this volume of *Le Antichità* meant that d'Alembert's discussion in the *Encylopédie* remained the more widely available, if erroneous, discussion of the prosciutto. As a result, the Accademia's astronomical interrogation of it went largely unchallenged in subsequent years, though the prosciutto's early position in the history of astronomical instruments helped sustain interest in it within the scientific community.

3.a) Lalande's "Moveable" Gnomon and the Follies of the Encyclopédie's *Successors*

The earliest treatment of the prosciutto following the publication of the third volume of *Le Antichità* appeared already in 1763 in an anonymous submission to the *Bibliotheque des Sciences et des Beaux Arts*, published in The Hague. The *Bibliotheque*, like its Italian counterparts the *Memorie* and the *Novelle Letterarie*, was a quarterly review of recent publications and news of interest to scholars, many of its entries, like this one, unattributed. Because of the limited publication run of *Le Antichità* and the fact that these volumes were distributed by the Bourbon court exclusively as gifts, this review would have played a crucial role in disseminating information about the prosciutto to a broad community, who would have been acquainted with it primarily through the entry in the *Encyclopédie*. Indeed, the alacrity with which this review appeared suggests it was written by a scholar with close enough ties with the Bourbon court to receive a copy of the volume who would not have wanted to jeopardize that relationship by being overly critical.[1] The article is essentially a summary of the Accademia's remarks on each of the volume's plates. It concludes first with a description of the engraving of the marble sundial from Pompeii, here ascribed to Herculaneum, and then turns back to the prosciutto sundial of the volume's preface:

> But a much more singular sundial, and one unique for its type, is the vertical and portable sundial of which the Academicians provide a description in the preface of the volume which we are paging through and which one sees represented in the vignette of this same preface. It is of bronze. It was found in the excavations in Portici in 1755. Until now no one has described or illustrated it as it should be. In truth the celebrated author of the article *Gnomonique* in the seventh volume of the *Dictionnaire Encyclopedique* gave a description but we see here that there are almost as many mistakes as there are words [*presque autant de fautes que de mots*]. Yet it was copied even in Italy itself, in a work titled *Monumenta Peloponnesia* and what is more serious is that the plagiarist, in order to disguise himself better, had the imprudence to add that he had seen in 1754 an illustration of the sundial in question, being ignorant no doubt that it had been disinterred only in 1755.
>
> One sees a ham suspended from a ring and represented on its two sides on one of which is the dial for which the bone or the tail of the ham serves as the style. The composition is as simple as it is clever. Its use is quite easy and its effect very exacting, but there is no way to give a fair idea without a drawing. We add only that among the many remarks both historical and astronomical concerning this precious sundial the wise Academicians make one that should fix the date of its construction, if it is well founded, as we must assume. They observe that, judging from the actual proportion of the shadows that can be

---

1. Such notices about recent publications in these journals, however, often took the form of informative summaries rather than critical analysis of the books' content.

seen on this dial, the obliquity of the ecliptic at the time it left the artisan's hands was 23° 46′ 30″, which is to say 18′ 12″ more than is found in 1762. Since according to the calculations of Chevalier de Louville, the inclination of the ecliptic only diminishes 21′ in 2000 years, this singular dial of which we speak was made around the year 28 of the Christian era.[2]

The anonymous author of another review of this volume took a different approach for their contribution to the *Nova Acta Eruditorum* of 1767. This journal had commenced publication in 1682 in Leipzig as the *Acta Eruditorum*, targeting a German-speaking audience, though the lingua franca was Latin, and while its format too was modeled after the *Novelle Litterarie* and the *Bibliotheque des Sciences et des Beaux Arts* its articles were primarily scientific in nature. The French astronomer Joseph Jérôme Lefrançais de Lalande was a contributor, for example, but while he had seen the prosciutto during his visit to Portici in just this period, he could not have authored this article.[3] Earlier fascicles of the *Nova Acta* had featured reviews of Bayardi's *Catalogo* and the first two volumes of *Le Antichità*. All were submitted anonymously, offered highlights together with more focused summaries of the books' contents, and emphasized specifically the finds' capacity to illustrate the ancient literary sources.[4] This review of the third volume opened with a long assessment of the Accademia's analysis of the prosciutto whose exclusive focus on literary matters reveals that it had been written by a philologist: nothing is said, for example, about the Accademia's use of the obliquity of the ecliptic to determine its date. After emphasizing the great value imparted by the rarity of its type the author provided a brief overview of the dial's design and calendrical inscription before noting the inadequate earlier treatments by the *Encyclopédie* and Paciaudi. Only two questions figured in the remaining discussion, both weighed through examination of the literary sources laid out over more than three pages of the review, far more space than had been devoted to any other object illustrated in this third volume. The first concerned what it was the

---

2. Anonymous, "Peintures antiques d'Herculanum etc, Tom. III," *Bibliotheque des Sciences et des Beaux Arts* 20.2 (The Hague 1763) 309–30, esp. 329–30. The fascicle covers the months of October to December 1763.

3. Lalande 1769, 7.126.

4. Anonymous, "Catalogo degli Antichi Monumenti de Ercolano," *Nova Acta Eruditorum*, August 1757 (Leipzig 1757) 443–45. [= Bayardi's *Catalogo*]; (January 1759) 1–11 [= *PdE* 1]; (April–May 1764) 41–80 [= *PdE* 2].

Enlightenment

philosopher in Bato's comedy had been carrying around with him in the reference from Athenaeus (*Deip.* 4.163b–c), the reviewer in this case agreeing with the Accademia that it had been a *clepsydra* of glass. The second was a topic closer to the heart of the journal's readers: whether the sundial's grid had been inscribed on a *petaso* or a *perna*. After examining the ancient Greek and Latin sources that distinguished each type, just as the Academicians had done, the author concluded by noting that the Romans valued the *perna*, derived from the *petaso* and requiring human agency to properly salt and age it, as a delicacy regardless of whether it was from the Carretanians or the Cantabians or, better, the Menapians—that is to say, a properly cured and smoked *schinken* from the region of contemporary Westphalia. What rendered this contribution particularly valuable, given the journal's broad distribution, was that it included a reasonably accurate engraving of the prosciutto that was a very close copy of the Accademia's and made the image available to a wider audience (Figure 3.1). The engraving was made by Johann Gottfried Krügner, Jr. (1714–1782), an accomplished artist working for publishers in Leipzig known, like his more famous father, for engraved portraits.[5] His prosciutto is sized a bit larger with a somewhat more rugged shape along its right side but is otherwise appropriately stippled to capture its undulating surface and edges. The reconstructed gnomon is thicker than the Accademia's, but its shadow falls in the same manner. While the latitude lines around the summer solstice on the left extend slightly longer than on the original, the dial is otherwise almost an exact copy, reproduced at the correct scale, though no dimensional scale was provided and no dimensions were given in the review. The variations come in the inscription, where the letters bear serifs, the "A"s have been given middle bars, and the open stroke of the "P" in Aprilis is shown closed.[6]

Further evidence of the misconceptions about the prosciutto caused by the Accademia's tight-fisted control on the dissemination of information is documented in the work of the Swiss scholar Francois Seigneux de Correvon. In a two-volume work published in 1770 he took a similar approach to that of the *Bibliotheque* and the *Nova Acta* in offering a critical review of the Accademia's publication rather than a first-hand description. His book's

---

5. Thieme-Becker, *Allgemeines Lexikon der bildenden Künstler von der Antike bis zur Gegenwart* (Leipzig 1978) 22.1, s.v. *Krügner, Johann Gottfried (1684–1769)*; he had provided illustrations for the *Acta* between 1727 and 1735.

6. Anonymous, "*Le Pitture Antiche d'Ercolano e Contorni*, etc. id est Picturae Antiquae Herculani, in aes incisae, cum explicationibus: Tomus III…" *Nova Acta Eruditorum*, March–April 1765 (Leipzig 1767) 280–335.

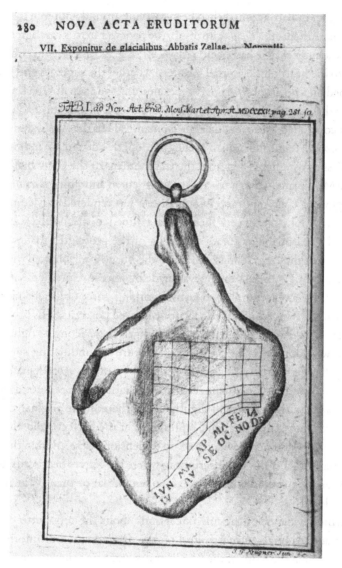

FIGURE 3.1 Engraving of the prosciutto by Johann Gottfried Krügner, jr., for the *Nova Acta Eruditorum* of March–April 1765 (Leipzig 1767).

conceit consisted of letters written to an anonymous, and presumably fictitious, "monsieur" who had tasked him with reviewing the published sources on the excavations and their findings, a process that involved directly translating long passages of those sources. In the process of commenting on the Accademia's explication for each plate in the first volumes of *Le Pitture di Ercolano*, he eventually reached the Preface to volume three and its discussion of the prosciutto, "an object worthy of curiosity relative to the arts." He dutifully summarized the Accademia's analysis, emphasizing the dial's coordination with the signs of the zodiac and their associated degrees of solar elevation, as well as the staggered arrangement of the corresponding calendrical labels. He cataloged the points on which the Accademia had sought to refute the authors of the *Encyclopédie* and even reprinted their incorrect transcription of the inscription. He concluded by quoting a critique of the Accademia's analysis he had received from a "very famous philosopher."

"This description," began the philosopher, "although doubtless very judicious, nevertheless seems to me little suited to satisfying amateurs for the following reasons." Seigneux de Correvon must have provided him with a translation of the Italian in which he had translated the pig's tail as the shank (*manche*), and so the author could not imagine how the shank, being on roughly the same plane as the rest of the dial, could serve as the gnomon to cast a shadow when suspended vertically, especially since the Accademia had not stated the gnomon/tail's height. Second, he correctly observed that the Accademia had only discussed the shadow lengths in each zodiacal sign and had not explained whether hourly readings also could be taken. Perhaps because Seigneux de Correvon had not translated the text in full, he incorrectly believed they had not identified the latitude for which it was designed and wondered in which plane it needed to be suspended in order to function, in the evident belief that azimuth played a role. Finally, he agreed with the Accademia on the need to stagger the position of the calendrical abbreviations in relation to the month lines.

If Seigneux de Correvon had provided his philosopher friend with both an incorrect translation and a copy of the engraving that appears in this volume, then the source of his confusion concerning the gnomon becomes clear. The engraving is a reasonably close copy of the Accademia's as far as the shape, the grid, and the inscription go, but it lacks any indication of the tail on the left side, in contrast with, for example, Krügner's in the *Nova Acta* (Figure 3.2). For this reason, the philosopher had wondered how the prosciutto suspended vertically from the ring attached to its shank, without anything projecting from the surface, could have cast a shadow. His transcription of the inscription includes periods after the abbreviations for IVN and AP, while August is abbreviated AI. Inexplicably, this engraving does not appear together with this commentary but more than one hundred pages later, keyed specifically to a separate letter discussing not the prosciutto but the marble sundial found in 1762 in Pompeii that had served as the coda to the

160 THE PROSCIUTTO SUNDIAL

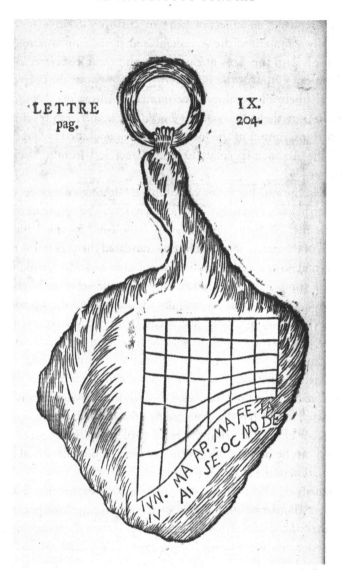

FIGURE 3.2 Engraving of the prosciutto in Francois Seigneux de Correvon's *Lettres sur la decouverte de l'ancienne ville d'Herculane et de ses principales antiquities* (Yverdon 1770).

third volume. Moreover, this is the only engraving in the entire two-volume work.[7]

---

7. F. Seigneux de Correvon, *Lettres sur la decouverte de l'ancienne ville d'Herculane et de ses principales antiquities*, 2 vols. (Yverdon 1770) 2.99–109 (Letter VI); the engraving follows page 204 of Letter IX. Its placement here may have been the typesetter's fault since an "Avis au Lecteur" in volume one alerts the reader to the fact that all the letters were accidentally dated to the same year, 1750, though they had been "written" over several years.

Enlightenment

By now the prosciutto had taken its place among the antiquities displayed in the Herculanense Museum at Portici, where it was seen and documented in the correspondence and descriptive catalogs of foreign travelers. Depending upon how they enumerated the museum's rooms, these visitors placed the sundial in either the tenth or the ninth room, a relatively small space roughly 38 m square, where it would have been overshadowed by the room's centerpiece: the exuberant life-size bronze statue of a drunken satyr reclining on a wineskin found in the grand peristyle of the Villa dei Papiri (MANN 5628). Thanks to its small scale and silvered exterior the prosciutto was displayed in one of the glass-covered wooden cases arranged along the walls (Figure 3.3). These held the other precious small finds recovered in the excavations, ranging from silver vases and engravings to cameos, rings, bracelets, earrings, and gemstones. Exhibited here as well was the large gold coin bearing Augustus on the obverse and Diana on the reverse (MANN 3692) that the Accademia had featured in the preface to the second volume of *Le Antichità* (see Figure 2.1).

The prosciutto must have been one of the "cadrans solaires" the Abbe Francois Gabriel Coyer had seen in February 1764 and included in a general inventory of the antiquities on view.[8] The French plant physiologist Auguste-Denis Fougeroux de Bondaroy considered it of sufficient interest to include in his 1770 book of "recherches" on the ruins of Herculaneum, where it was assigned to the ninth room:

> A small sundial with the shape of a ham and which one suspends from a ring; the animal's tail serves as the gnomon. A description of this sundial under the entry *Gnomonique* was given in the *Encyclopédie*. One may also consult what was said in the Preface to the third volume of the Antiquités of Herculanum; *Le Pitture antiche*, etc., already cited.[9]

Whether or not the prosciutto was mentioned in these travel journals likely depended on who had guided the author through the museum—Piaggio implies Paderni would tap the doorkeeper for this duty when he was

---

8. F. G. Coyer, *Voyage d'Italie et Hollande*, 2 vols. (Paris 1775) 1.230: Letter 29 from Naples, February 4, 1764.

9. A. D. Fougeroux de Bondaroy, *Recherches sur les Ruines d'Herculanum et ses lumieres qui peuvent en résulter relativement à l'état présent des Sciences et des Arts* (Paris 1770) 86.

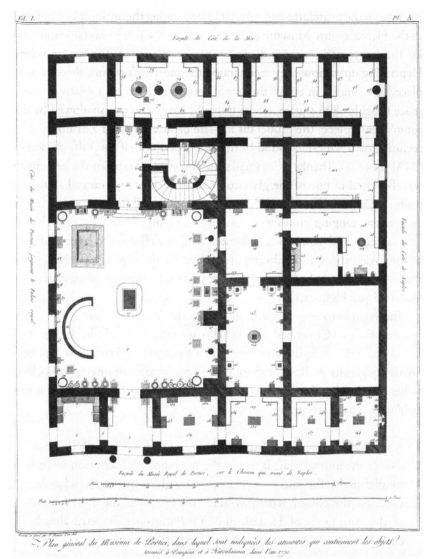

FIGURE 3.3 Plan of the first floor of the Herculanense Museum at Portici in 1770. Engraving by Francesco Piranesi, *Antiquités de la Grande Grèce, aujourd'hui Royaume de Naples...*, vol. 1 (Paris 1807), Plan A. In this period the prosciutto sundial was displayed in one of the wooden cases along the walls of the fourth room from the left along the bottom.

otherwise engaged—and what finds the guide had highlighted. Visitors were expected to tip their guide, and this influenced what they saw.[10] The prosciutto

---

10. Jean-Marie Roland De La Platière, *Lettres ecrites de Suisse, d'Italie, de Sicile, et de Malthe*, 4 vols. (Amsterdam 1780) 4.255: "One gives the guard of this museum 6 to 8 *carlins*, sometimes 10, depending on the number of people, the length of the visit, the zealous and honest manner in which he opened the cupboards, let one look more closely, even touch the objects when

Enlightenment

is strangely absent from the detailed discussion of the contents of the museum in Heinrich Cramer's *Nachrichten zur Geschichte der Herkulanischen Entdeckungen* from 1773 and the room-by-room description of the museum offered by Jean Bernoulli in his *Beschreibung des Herkulanischen Musäums nach den Zimmern* of 1775. Bernoulli (1744–1807) was the youngest member of a family of famous Swiss mathematicians and astronomers. Appointed royal astronomer in Berlin at the tender age of nineteen, he had published a modest number of his own papers as well as a three-volume collection of important works by other astronomers entitled the *Recueil pour les Astronomes* (Berlin 1771–1776). His catalog of the precious antiquities in what he had labeled the museum's tenth room is otherwise among the most comprehensive from this period, including the sardonyx cameo set in a gold ring that Charles III had famously removed from his finger before departing for Spain (MANN 25181). He had described in some detail the Egyptianizing base (MANN 1107/76384) in the eleventh room that the Accademia had forsaken in favor of the prosciutto, but the prosciutto itself evidently failed to pique his interest (see Figure 2.2).[11]

A remarkably detailed description of the antiquities displayed in the museum in this period was offered by Lady Anna Miller in a letter addressed to a friend and dated February 9, 1771. Like others, however, she complained that "my time and memory both fail me, it being with the utmost difficulty I contrived to take a few notes in my pocketbook without being observed." Her museum guide—presumably Paderni—had soundly rebuked her for touching a folded piece of ancient purple cloth that had promptly yielded under the pressure in a puff of colored dust, leaving traces on the tip of her finger.[12] He had "insisted peremptorily on my not carrying off an atom, 'For,' said he, 'it is a curiosity no monarch upon earth can boast the possession of besides my master, the King of Naples.'" As she moved into the tenth room one object

---

desired, and answer all the questions." The prosciutto is missing from his lengthy catalog of objects in the museum (4.233–41). It is likewise missing from the otherwise detailed account in Mariana Starke's letter of September 1797. She notes that a letter of permission from the king to visit the museum was still required; that the museum could be visited from nine to noon and three to five or six every day except holidays; that each floor of the museum had its own custodian who each received six *carlins*; and that it was still prohibited to take notes (M. Starke, *Letters from Italy, between the Years 1792 and 1798*...(London 1800) 96–129 (Letter 21, from Naples). See also Cantilena and Porzio 2008, 88.

11. J. J. Bernoulli, "Beschreibung des Herkulanischen Musäums nach den Zimmern," *Neue Bibliothek der schönen Wissenshaften und Freyen Künste* 17.1 (1775) 78–79, 83–85.

12. Evidently she had not heeded the cautionary notice posted by Paderni, "Look but don't touch, or accept the insults from the custodian" (*Si vede e non si tocca, altrimente dal custode riceveranno affronto*). Bassi 1907, 668, as recorded by Piaggio.

164 THE PROSCIUTTO SUNDIAL

caught her attentive eye: "I observed a very curious quadrant engraved on silver, in the shape of a ham; the tail of the hog forms the style."[13] Only a few years later the German writer Johann Jacob Volkmann (1732–1803) described it in terms so similar to Lady Miller's that both were seemingly parroting the boilerplate explication provided by their guide: "A sundial drawn on a piece of silver that has the shape of a ham and the tail of the animal serves as the gnomon."[14]

Fougeroux de Bondaroy may have been guided in his exploration of the museum by the earlier published account of his compatriot Joseph Jérôme Lefrançais de Lalande (1732–1807). Like his mentor, Joseph-Nicolas Delisle, Lalande was a prominent French astronomer and ultimately assumed the title of Chair of Astronomy (1774) at the Collège de France after Delisle's death. He was a member of the French Académie des Sciences (1753), the Berlin Akademie der Wissenschaften (1752), and the Royal Society of London (1763), his election to the latter having been supported by La Condamine (Figure 3.4).[15] His two-volume work *Traité d'Astronomie* (Paris 1764) was a major contribution to the field and for much of his career he served as editor of the *Connaissance de Mouvemens Célestes pour l'Annee*, a compendium of astronomical tables and formulae relied upon by astronomers throughout Europe; Carcani likely had used it in his calculations for the semi-diameter of the sun. Lalande had assisted Delisle in the international effort to document the transit of Venus in 1761 and then spearheaded the similar, and more successful, effort in 1769, mapping out the observation points and then coordinating the collection and processing of the resulting data.[16] He had traveled

---

13. A. R. Miller, *Letters from Italy, Describing the Manners, Customs, Antiquities, Paintings, &c. of that Country, in the Years MDCCLXX and MDCCLXXI, to a Friend Residing in France*, 2 vols. (London 1777) 2.59–99 (Letter 36, February 9, 1771): 63–84 (museum), 74 (prosciutto sundial). Lady Miller, who recounts her repeated efforts to elude the guards overseeing the excavations at Pompeii, also confessed that she had damaged the wall of a house when she climbed a ladder and leaned through an opening for a look inside. In a later letter from Rome she singled out Zuzzeri's Tusculum sundial in the Museum Kircherianum as being "esteemed a very great curiosity: by this dial it appears that the Romans reckoned twelve hours to the day, including one hour of twilight" (2.242, Letter 44, Rome, May 1, 1771).

14. J. J. Volkmann, *Historisch Kritische Nachrichten von Italien*, 3 vols. (Leipzig 1771) 3.287.

15. Royal Society of London: *Archives* EC/1763/12. He was elected to the American Academy of Arts and Sciences in 1781.

16. Dumont and Gros 2013, 131–44. See also *Biographical Encyclopedia of Astronomers* (New York 2014) 1264–66, s.v. *Lalande, Joseph-Jérôme* (S. Dumont): he determined a solar parallax of 8.5 to 8.75″ and a diminution of the obliquity of the ecliptic at 38′ per century. Lalande published an annotated bibliography of known scholarship on astronomy and arranged chronologically from roughly 480 BCE through 1802: *Bibliographie astronomique,*

Enlightenment

FIGURE 3.4 Portrait of Joseph Jérôme de Lalande (1732–1807). Painting from 1802 by Joseph Ducreux (1735–1802). Château de Versailles (Inv. Dess 1060). https://commons.wikimedia.org/wiki/File:J%C3%A9r%C3%B4me_Lalande.jpg.

through Italy in 1765 and 1766 and collected his observations in an eight-volume work entitled the *Voyage d'un François en Italie, fait dans les années 1765 et 1766* (Paris 1769).[17] His description of the museum, which he toured in 1765, followed no clear organizational principle and did not correspond to the individual rooms but is detailed and perceptive. Like others before him,

---

avec l'histoire de l'astronomie depuis 1781 jusqu'a 1802 (Paris 1802); the 1749 entry for the Zuzzeri/Boscovich sundial from Tusculum on page 430 cross-references the preface to the third volume of *Le Antichità* and Jacquier's 1765 discussion of the marble sundial from Pompeii (see Chapter 3.a). A star in the constellation Ursa Major is named for Lalande (Lalande 21185), who discovered it in 1801.

17. The full title is *Voyage d'un François en Italie, fait dans les années 1765 et 1766: Contenant l'historie et les anecdotes les plus singulieres de l'Italie, et sa description, les mœurs, les usages, le gouvernement, le commerce, la littérature, les arts, l'histoire naturelle et les antiquités, avec des jugemens sur les ouvrages de peinture, sculpture et architecture, et les plans de toutes les grandes villes d'Italie*, 8 vols. (Paris 1769). He had seen as well the *osservatorio astronomico* installed by N. Carcani in the Collegio Reale at San Carlo alle Mortelle; Carcani had died a short time earlier: Lalande 1769, 6.143–44.

THE PROSCIUTTO SUNDIAL

Lalande found the restrictions placed on visitors to the museum a challenge, writing, "Besides, one should not hazard a description and a very extensive criticism of these monuments, it being prohibited for anyone to write in these cabinets [the Museum], which means that one can only relate the various peculiarities from memory," but nevertheless he had succeeded in thwarting them.[18] Fougeroux de Bondaroy owned a copy of Lalande's *Voyage*, which he annotated, and his own published account shows parallels with Lalande's: a brief description of the sundial and references to earlier publications of it that provided more detailed information.[19]

Lalande's account of the prosciutto appeared near the middle of his narrative, in what was becoming the stock description, sufficiently succinct and accurate but virtually impossible for a reader to correctly envisage without an accompanying illustration:

> A small sundial inscribed on a piece of silver in the shape of a ham, the animal's tail serves as the style. It was engraved in the third volume of the *Antichità di Ercolano*, p. 337. M[onsieur] de la Condamine spoke of it in the Mémoires of the Académie for 1750, p. 370.[20]

As it happened, this renowned French astronomer had dwelled longer over the collection of ancient phalluses than on this early sundial, though in fact he had combined these two passions in a puzzling observation that the gnomon on one small sundial was shaped like a phallus.[21] Yet if the brevity of his description is disappointing, its errors are particularly jarring. The page reference for *Le Antichità* is actually to the Accademia's discussion of the marble sundial from Pompeii, which he otherwise did not describe. As for the reference to the *Mémoires*, La Condamine's report to the Académie in which he recounted seeing the prosciutto soon after its arrival in the museum was read in 1757 and published in 1762; the prosciutto had not yet been found in 1750,

---

18. Lalande 1769, 7.152–53.

19. American Philosophical Society, Phildelphia PA, *Auguste-Denis Fougeroux de Bondaroy Collection*, Mss.B.F8245: Fougeroux de Bondaroy's copy of Lalande's *Voyage d'un François en Italie*, though his annotations from Portici are limited to the remains of minerals used in the manufacture of ancient paints and to the papyrus.

20. Lalande 1769, 7.126.

21. Lalande 1769, 7.110–133 (museum as a whole), 7.116–118 ("Priapi"), 7.117 ("un petit cadran dont le style étoit de même forme"). It is not known to which sundial he is referring.

Enlightenment 167

an error that recalls Paciaudi's damning blunder of dating its discovery to 1754. Lalande's page number, at least, is correct.[22]

Lalande contributed both directly and indirectly to updated entries on the prosciutto sundial for new editions of Diderot and d'Alembert's *Encyclopédie*. The popularity of the original *Encyclopédie* had sparked enthusiasm for revised and copy-cat editions even as the last volumes of the original version, published between 1751 and 1765, were coming off the press. The revisions ultimately took the form of three different editions of encyclopedias modeled on the original. One, the so-called Yverdon *Encyclopédie*, edited and published by F. B. De Felice in Yverdon, Switzerland, consisted of fifty-eight volumes published between 1770 and 1780. Another was the series of five volumes of *Supplement* published between 1776 and 1780, while the third was a substantially expanded edition with the general title of *Encyclopédie Methodique* whose 166 volumes, grouped according to subject, appeared between 1782 and 1832.[23] The entries of relevance to the prosciutto were *Cadrans Solaires, Gnomon*, and, especially, *Gnomonique*, though the text of the latter two were interchangeable depending on the edition. In some editions the entries are amended revisions to d'Alembert's description of the prosciutto in the original *Encyclopédie*, in others the entry is entirely reworked, but all appear to have willfully ignored the scientific analysis of the prosciutto published by the Accademia.

D'Alembert's description of the prosciutto had been included wholesale in the entry *Gnomonique*, while the earliest of the revised descriptions of the prosciutto appeared in volume 21 of the Yverdon *Encyclopédie* under the entry *Gnomon*. After cataloging the sundials of Zuzzeri and two others of Parian marble and travertine installed in the Vatican by Pope Benedict XIV, the author observes that one of these sundials had been designed for the latitude of Memphis and brought by the Romans from Egypt. In a sign of the confusion to come, this was, however, the latitude of the marble sundial from Pompeii (see Figure 2.15), which the author then describes:

---

22. A later biographer reported that his *Voyage* was criticized in Italy for being fraught with errors; see *Serie di vite e ritratti de' famosi personaggi degli ultimi tempi*, 3 vols. (Milan 1818) vol. 3, s.v. *Giuseppe Girolamo di Lalande*.

23. K. Hardesty, *The Supplément to the Encyclopédie* (The Hague 1977) 1–18, clarifies the complex history of the revisions. Different publishing houses, in Paris, Switzerland, and the Netherlands, produced editions of the *Supplement* and the *Encyclopédie Methodique*, with slight variations, though the relevant entries are more or less consistent between editions.

168 THE PROSCIUTTO SUNDIAL

> In 1762 was found in the excavations at Civita [Pompeii] an ancient
> sundial of marble made for an elevation of the pole of 42 degrees [*sic*],
> containing simply part of the arc of the circle corresponding with the
> equator, in contrast to the other sundials which included in addition
> to this arc the semicircles of the two Tropics. The style of one of the
> sundials we have just described has the shape of a Priapus.

It was, in fact, the Pompeian sundial that the Accademia had determined was
designed to work at the latitude of Memphis while they had stated that the
Tusculum and Vatican examples were designed for 42°. The reference to the
phallic gnomon from Lalande's *Voyage* is further evidence of the variety of
sources cobbled together for this entry. The author then looked to the prosciutto to "prove" that the ancients did not limit themselves to such hemicyclium sundials, but had made portable ones as well:

> On June 11, 1755, was found in the excavations of Herculaneum or
> Portici, a small dial of silver-plated bronze, which resembles quite
> exactly a ham suspended perpendicularly by means of a ring, that is to
> say, represented on it are the concavities, the convexities, in a word the
> unevenness of the surface of ordinary hams. On one of the surfaces
> are drawn seven perpendicular lines, where are marked the length of
> the shadow for each month in the different hours of the day, which are
> designated by curved lines which intersect the perpendiculars. The
> lowest curved line indicates noon, etc. Below this line we see the first
> letters of each month; for example, *IA. FE. MA.,* etc. that is to say,
> *januarius, februarius, martius,* etc. The shortest of the perpendicular
> lines marks the incidence of the shadow in all the hours of the 21st of
> December and the longest of the perpendicular lines indicates the
> length of the shadow in all hours of the day on the 21st of June.

These opening statements demonstrate how the details cited in the secondary sources, though at times contradictory, were starting to coalesce into
a fuller picture of the prosciutto. Through slight modifications and significant
omissions, the author had fused the Accademia's explication with d'Alembert's.
He would have known the exact date of its discovery, for example, only by
consulting the Preface, where its specificity had served to highlight Paciaudi's
egregious error. The stated provenance as "Herculaneum or Portici" is an early
example of the confusion between where it was found with where it was
exhibited, a blurring that would endure through modern times as the sundial

Enlightenment

acquired the moniker "Portici ham" and its provenance in the Villa dei Papiri was forgotten entirely. That the ruins of the Villa lay closer to modern Portici than those associated with the ancient city of Herculaneum contributed to this process.[24] The description of its form clings to the *Encyclopédie*'s perceptive observation of its undulating surfaces while acknowledging the Accademia's point that this was simply an accurate reproduction of a cured ham. The explanation of the dial's design is too truncated to allow the reader to mentally reconstruct how it worked, but at least the author abandoned the *Encyclopédie*'s lunar calendar and correctly associated it with the zodiacal one detailed in the Preface. He wisely dispensed with D'Alembert's boxy representation of the inscription along with any mention of the *boustrophedon* arrangement. As a result, the treatment of the inscription is superficial and ignores the manner in which it functioned together with the dial.

The tail is entirely ignored, not simply because the tail is routinely absent from a prosciutto but because the author believed this sundial functioned like an altitude or cylinder dial (Figure 3.5).[25] As the name indicates, the dial on this type of sundial wrapped around a cylinder, its dial similar to that on the prosciutto, though in most versions it was mirrored around the line representing the summer solstice so that if unwound onto a flat plane it would be twice the size and represent the full twelve months of the year. The distance between the month lines depended only on the cylinder's diameter. The length of the gnomon, along with the projection of the horary lines, depended on the latitude in which it was intended to work, since accurate readings could be taken off this type of sundial only within a few degrees of the latitude for which it was designed. The gnomon extended out from the top of the cylinder so that it aligned with the horizon line at the top of the grid and was rotated to line up with the vertical line corresponding with the current month. The cylinder, suspended from a string, was turned so the gnomon's shadow was cast on the dial.[26] Since there was no cylinder and no indication of where

---

24. Similarly, it is likely that many readers would not have been aware that the modern toponym of Civita given as the provenance of the marble sundial from 1762 was ancient Pompeii.

25. Also known as an elevation dial, a shepherd's dial or a pillar dial. For the common types of sundials in the author's day, see B. Chandler and C. Vincent, "A Sure Reckoning: Sundials of the 17th and 18th centuries," *Metropolitan Museum of Art Bulletin* 26.4 (December 1967) 154–69.

26. For cylinder or "altitude" dials, see A. Mills, "Altitude Sundials for Seasonal and Equal Hours," *Annals of Science* 53 (1996) 75–84, and M. Arnaldi and K. Schaldach, "A Roman Cylinder Dial: Witness to a Forgotten Tradition," *Journal for the History of Astronomy* 28 (1997) 101–17: on the cylinder dial made of bone from Este (Museo Nazionale Atestino di Este 15397) dated to the first century CE and designed for a latitude of 45° (see Figure 3.36).

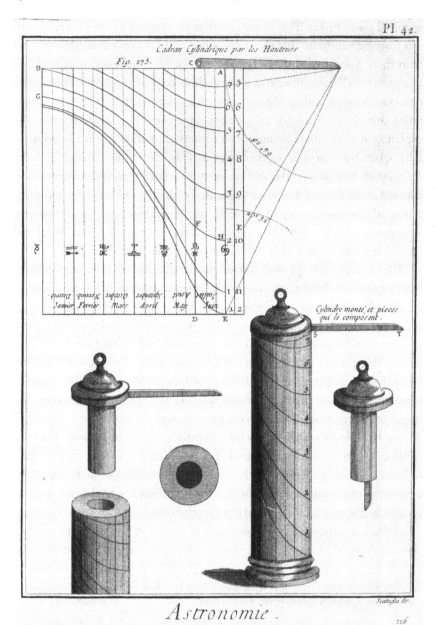

FIGURE 3.5 Cylinder or "shepherd's" altitude dial engraved for the *Encyclopédie Méthodique*. From the *Planches de l'Encyclopédie Méthodique: Mathematiques* (Padua 1790) 116, Plate 42.

Enlightenment

such a gnomon could be attached at the top of the prosciutto's dial, the author proposed an alternative solution:

> One must add a small machine, which served as a style or cursor along the horizontal line which is at the top of this dial; this style must move forward or backward in each month, so that it marks by the incidence of its shadow, or its luminous point, the present time.[27] But this style was not recovered, and we do not even understand how we could make it run in a reliable manner on this ham.

The author may also have had in mind a specific type of altitude dial, available in his day, with a grid virtually identical to the prosciutto's inscribed on a flat plaque that rotated on a vertical pole (Figure 3.6). A slidable gnomon attached to the top and corresponding with the horizon line could be moved from side to side to align with the appropriate month so its shadow was cast down to mark the hours. The problem was that there was no indication how such a "machine" could be attached to the prosciutto.[28]

Despite the similarities in their design the author concluded that the prosciutto was flawed in comparison to such modern altitude dials. These lacked the awkward undulating surface of the prosciutto and were more convenient in having the twelve hours of the day inscribed along the vertical line of the summer solstice beside the horizontal horary lines.

> It is obvious that this small dial is formed on the same principle as our cylindrical dials; but ours are more accurate and more convenient, (1) because they are drawn on a smooth surface; (2) we mark the hours outside the perpendicular line which the sun travels on June 21, etc.

Inexplicable is how the author could have missed the detailed explanation provided by the Accademia, not just in their written text but also illustrated by their engraving, of how they had reconstructed the gnomon to reanimate

---

27. The "luminous point" is likely intended to encompass gnomons fashioned from a pinhole in a disk at the end of the gnomon that highlighted the point on the dial with a circular ellipse of shadow, rather than simply a straight gnomon.

28. Often called a "flag dial"; see, e.g., A. Mills, "Altitude Sundials for Seasonal and Equal Hours," *Annals of Science* 53 (1996) 79–80, fig. 4, and R. Rohr, *Sundials: History, Theory, Practice* (Toronto 1970) plate 26, for the nineteenth-century flag dial with a grid identical to the prosciutto's from the Musée de la Vie Wallonne, Liège (Inv. No. 4000558).

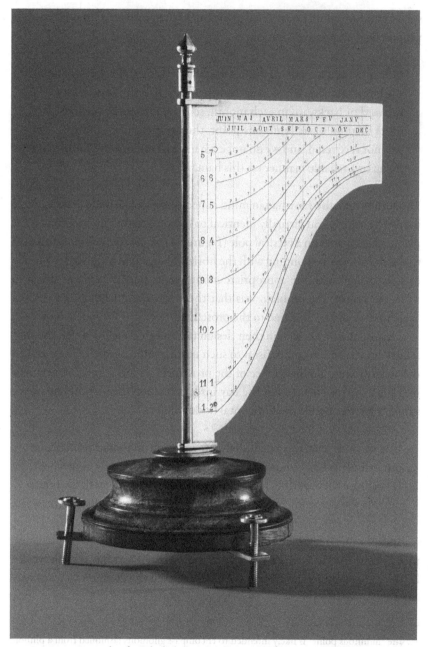

FIGURE 3.6 Example of a "Flag" dial with a "moveable gnomon." The gnomon is missing but was attached at the top of the dial and would slide left and right above the desired vertical month line. The similarities with the prosciutto's dial are self-evident. A plumb bob and a second grid on the back allow it to function as a clinometer. Copyright of the Musée de la Vie Wallonne, Liège (Inv. No. MVW-4000558).

Enlightenment 173

the prosciutto. He evidently had concluded that some detail even more crucial than the gnomon was missing in knowledge about the prosciutto and, in a diatribe awkwardly out of place in an entry for an encyclopedia, blamed this on the notorious restrictions placed by the Bourbon court on viewing and studying the finds.

> We observe, in passing, that in order to put readers in a position to pronounce a firm judgment between the admirers and the censors of this dial, it would be hoped that the Academicians of Naples would have a quantity of models of this machine made of plaster or lead and a quantity of other figures in relief or in bas-relief which they consider marvels, although several foreigners dare to suppress them. The scholars of all the Academies, seeing these models, could finish off the disputes and make a number of discoveries. But one dares to predict that the more than human jealousy of the Neapolitan scholars will never allow the King of Naples to employ so simple and judicious a means. They forbid strangers to stop their eyes for a moment on the curious objects extracted from Herculaneum and contained in the museum of Portici: it is prohibited for guides to allow even the inscriptions to be copied. However, it is obvious that the combined enlightenment of foreigners could dispel many prejudices of the Neapolitans.[29]

The entry was not a revision penned by d'Alembert, or at least he did not take credit for any of it. Credit is assigned at the end to a certain "V. A. L." This is not the abbreviation of a pseudonym for Lalande; he took credit with the initials "D. L." for the material covered in the first part of this entry, to which V. A. L. had amended his comments. The author has been identified as Paul-Joseph Vallet (1720–1781), a French jurist and magistrate who served as Lieutenant Générale de Police in Grenoble. Of a "difficult character," he also had assumed the role of "encyclopédiste" and has been credited with between

---

29. *Encyclopédie, ou Dictionnaire Universel Raisonné des Connaissances Humaines*, F. De Felice ed. (Yverdon 1773) 21.718–19, s.v. *Gnomon*. The entry is identical in the *Nouveau Dictionnaire pour server de Supplément aux Dictionnaires des Sciences des Arts et Des Metiers* (Paris and Amsterdam 1777) 3.240–241, s.v., *Gnomon*. The entry for *Herculanum* in this same Yverdon 1777 edition is essentially a reprint of the description of the site and the museum from Lalande's *Voyage*, including the reference to the prosciutto sundial, though it is credited to Claude Courtépée; *Nouveau Dictionnaire* (Paris and Amsterdam 1777) 3.349–58, s.v. *Herculanum*.

50 and 847 entries in the Yverdon edition.[30] Like other contributors to the revised editions of the *Encyclopédie*, Vallet may have sought the cover of anonymity in his submissions because of concerns about plagiarism. His suggestion that a cast be made of the prosciutto so that models distributed among scientists might lead to a better understanding of how the sundial functioned may have been influenced by a remark Lalande had made in his *Voyage*, as had Vallet's complaint about the Bourbon restrictions. Lalande had mused, after describing the marble equestrian statue of M. Nonius Balbus that confronted visitors in the courtyard of the museum, "It is to be hoped that we obtain from the King the permission to mold it [the equestrian statue of Balbus] to have models in our Schools. What benefits would not be derived from the study of these monuments by our sculptors who, for the preeminence of their talents, are chosen by the cities of France to execute the equestrian statues which they devote to the glory of our Kings."[31]

Lalande took credit for the entire entry for *Gnomonique* in the grand *Encyclopédie Méthodique*, published in 1785. The text is an odd amalgamation of the earlier entries of d'Alembert and Vallet together with Lalande's own additions. The result is a discussion of the prosciutto strikingly rife with errors and contradictions, including the resurgence of d'Alembert's dentilated gnomon. While Lalande acknowledged the Accademia's publication he incorporated none of its findings. After quoting Vallet word-for-word on the marble sundial from Pompeii, including his own reference to the sundial with the Priapic gnomon, Lalande continued with a description of the prosciutto almost identical to d'Alembert's from twenty-eight years earlier. He added to the earlier cross-references in his *Voyage* by referring the reader to the engraving in the recent treatment of it by a German astronomer, Georg Heinrich Martini, not to the more accurate engraving by the Accademia.

---

30. G. Salamand, "Paul-Joseph Vallet, Lieutenant de Police, et 'L'Encyclopédie d'Yverdon,'" *Les Affiches de Grenoble et du Dauphiné: Hebdomadaire de chroniques juridiques, commerciales, économiques, financières et d'échos régionaux* 4701 (Octobre 2014) 39, gives 847 entries submitted "en secret." Hardesty 1977, 146, credits him with fifty articles on the fine and mechanical arts. His publications included the *Method for making rapid progress in science and the arts* (*Méthode pour faire promptement des progrès dans les sciences et dans les arts*, Grenoble 1767), a manual on the principles and application of analysis. His familiarity with dialling may have been derived from locally sourced practical guides like F. Bedos de Celles' *La Gnomonique Pratique, ou L'art de tracer les cadrans solaires avec la plus grande précision*... (Paris 1760) 311–63, Pls. 28–29 (Cadrans Portatifs).

31. Lalande 1769, 81–82.

Enlightenment

The ancients, like us, made portable sundials. In 1755, in the excavations of Herculaneum, at Portici, was found a sundial of silver-plated bronze, which resembles almost exactly a ham suspended perpendicularly by means of a ring. *Le Pitture di Ercolano*, Tom[e] III, pag[e] 337, not[e] 130 [= again, the Pompeii marble sundial]. The figure is in *Martini*.[32] On it one sees the concavities, the convexities, in a word the inequalities of the surface of an ordinary ham. It has on one side a style a little long and dentilated, which is about a quarter part of the diameter of this instrument. One of the two surfaces, which one can consider the upper surface, is entirely covered in silver and divided by twelve parallel lines which form an equal number of small squares a bit hollow. The last six squares, which are ended by the lower part of the circumference of the circle, contain the letters of each month, arranged in the following manner [inserted here is the original reproduction of the inscription in d'Alembert's 1757 entry in the *Encyclopédie*]. The manner in which these months are arranged is remarkable because it is boustrophedon, the first figure going from right to left and the second from left to right to correspond to the months, like April-September when the sun finds itself at just about the same height in certain corresponding days. But this correspondence occurs only in the first two halves of each of these months: in the last fifteen days of April, the sun is higher than in the last fifteen days of September. It's the same for the others.

Lalande did not explain that it followed a zodiacal rather than a lunar cycle. He continued not with his own observations, as an astronomer with first-hand knowledge of the prosciutto, but with the analysis of Vallet, the police inspector from Grenoble, despite the fact that it contradicts in part what he had just written, in particular concerning the gnomon:

Thus, we have marked the length of the shadow for each month in the different hours of the day, which are designated by curved lines which cut the perpendiculars. The lowest curved line indicates noon, etc., below this line we see the first letters of each month, for example IA. FE. MA. etc., that is to say *januarius, februarius, martius*, etc. The shortest of the perpendicular lines marks the end of the shadow in all the hours of December 21, and the longest of the perpendicular lines

---

32. G. H. Martini, *Abhandlung von den Sonnenuhren der Alten* (Leipzig 1777); see Chapter 3.a.

THE PROSCIUTTO SUNDIAL

denotes the length of the shadow in all the hours of the day on the 21st of June. One probably added a kind of style or cursor along the horizontal line which is at the top of this dial, as in figure 275, and made this style move forward or backward in each month, so that it marks the hour by the incidence of its shadow, or of its luminous point. But this style was not recovered, and we do not see how we could make it move in a solid way on this ham. This small dial is formed on the same principle as our cylindrical dials, but ours are more correct and more convenient because they are marked on a solid surface. (M. Valet [sic], de Grenoble, Supplement de l'Encycl[opédie] Tom[e] 3).[33]

Figure number 275 to which Lalande refers is one of the plates illustrating this volume of the Encyclopédie Méthodique for the section devoted to astronomy.[34] The plate consists of an exploded view of a portable cylinder dial and details of its individual parts (see Figure 3.5). Its cap with the gnomon attached is shown separately with the gnomon extended as well as folded down into the cap to slide neatly inside the hollow cylinder for safe transport. The dial has been flattened and the months labeled across the bottom so that they fall between the vertical lines that themselves are labeled with the corresponding zodiacal signs. The hours are listed at the far right, just outside the line of the summer solstice, and the angle of the gnomon's shadow at noon on the summer solstice for the latitude of Paris. The similarities between this grid and that on the prosciutto sundial are vividly on display.

Lalande's other reference was to the first formal study of ancient sundials to feature the prosciutto after the Accademia's analysis of it. This was Georg Heinrich Martini's Abhandlung von den Sonnenuhren der Alten, published in 1777. Martini (1722–1794) was not an astronomer but the rektor of the Nikolaischule in Leipzig at the time and a philologist who dabbled in ancient material culture: two years later he published his book Pompeii, As It Were, Resurrected.[35] His book on sundials followed in the footsteps of earlier philological treatments such as Salmasius, to whom he refers frequently, and began with an historical survey drawn from the literary record. He then cataloged the known sundials in his day, from Zuzzeri's Tusculum sundial to Le Roy's

---

33. Encyclopédie Méthodique: Mathematiques (Paris 1785) 2.145–47, s.v. Gnomonique (Lalande). It is possible that this entry, though credited to Lalande, was concocted by the editor of the volume and the errors not entirely attributable to Lalande.

34. Planches de l'Encyclopédie Méthodique: Mathematiques (Padua 1790) 116, Plate 42.

35. G. H. Martini, Das gleichsam auflebende Pompeji (Leipzig 1779) 307–8, is a passing reference to the marble sundial from Pompeii found in 1762.

Enlightenment

example from Athens. The Pompeii sundial from 1762 served to introduce Roman sundials in the literary and historical record including such highlights as Vitruvius' parade of famous sundial designers to Petronius' trumpeter in the triclinium. Mention of Vitruvius' *viatoria pensilia*-type presented the opportunity to discuss Bato's portable sundial, the history of *clepsydras*, Baldini's odd portable sundial, and, finally, the prosciutto. Martini relied almost exclusively on the Accademia's treatment of it, though noted that they had "woven a variety of untimely scholarship and extensive disputes into their description." Ignoring entirely, or ignorant of, the confusions introduced by the French scholarship, Martini offered what was becoming the standard presentation of the prosciutto: a detailed description of the dial and its inscription, line-by-line, and an explanation of how to use it to read the passage of the sun through the ecliptic. He accepted the Accademia's conclusions that it was accurate to within a few minutes, designed for an elevation of the pole of 41° 39′ 45″, and made in 28 CE, but offered no new analysis of his own. He knew of the anonymous review of the third volume of *Le Antichità* in the *Nova Acta Eruditorum*, which was published in Leipzig, and so likely would have seen the engraving of the prosciutto there. His own, however, more closely copied the Accademia's (Figure 3.7). He slightly altered the body's shape and reduced the size of the dial but added the *Nova Acta*'s serifs to the lettering of the inscription, slanted the letters of the abbreviations slightly to the right, and did not center them on the grid.[36] Nevertheless, in the eyes of Bernoulli, his book was "full of inquisitive and sound erudition."[37]

Martini's treatment would later be summarized in Christoph Gottlieb von Murr's *Abbildungen der Gemälde und Alterthümer...*, one of the many pirated reprintings of engravings and commentary from the volumes of *Le Antichità* intended to feed the appetites of artists and scholars who could not obtain the originals. Von Murr (1733–1811) was a German historian, mathematician, philosopher, and archaeologist, but he brought few of these skills to his treatment of the prosciutto. The text is essentially a condensed translation of the original, limited to a synopsis of how the vertical and horizontal lines of the grid were read, the latitude for which it was designed, and the *terminus post quem* of its construction. As in the Preface to the third volume, an engraving of both sides of the prosciutto features prominently on the first page. The

---

36. Martini 1777, 134–38.

37. In his *Bibliographie astronomique* (Paris 1802) 559, Lalande cataloged Martini's book and Bernoulli's notice of it in the *Nouvelles Litteraires de Divers Pays* (Berlin 1777); there are no other historical surveys of sundials in Lalande's catalog between 1762 and 1777.

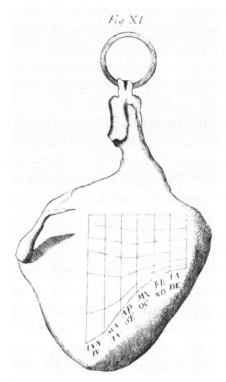

FIGURE 3.7 Georg Heinrich Martini's prosciutto. Figure XI from his *Abhandlung von den Sonnenuhren der Alten* (Leipzig 1777).

engraving is by Georg Christoph Kilian (1709–1781), a prolific German engraver and printmaker, and while this is virtually identical to the Accademia's in its size and reproduction of the inscription, it lacks the detailed stippling, and Kilian slightly altered the position and size of the gnomon (Figure 3.8).[38]

Even more popular than von Murr's reproduction of engravings from *Le Antichità* were the eight volumes of Pierre-Sylvain Maréchal and François-Anne David's *Les Antiquités d'Herculanum: avec leurs explications en François* published in Paris between 1780 and 1789. Maréchal (1750–1803) was a poet

---

38. C. G. von Murr, *Abbildungen der Gemälde und Alterthümer in dem königlich Neapolitanischen Museo zu Portici*...(Augsburg 1793) 3.1–3. Although the engraving of the prosciutto features prominently on the first page, the text begins with the 1762 Pompeii marble sundial, the engraving of which is shown on page 3. Murr, together with Kilian and Johann Balthasar Probst, had published both a German and an Italian version of their reproductions of the first volumes of *Le Pitture*. Kilian was given primary authorship of the Italian version, entitled *Li corntorni delle pitture antiche d'Ercolano, con le spiegazioni incise d'appresso l'originale*, 7 vols. (Augsburg 1778), where the engraving of the prosciutto, together with the 1762 marble sundial from Pompeii, appear together as the first engraving on page 24 in volume 3.

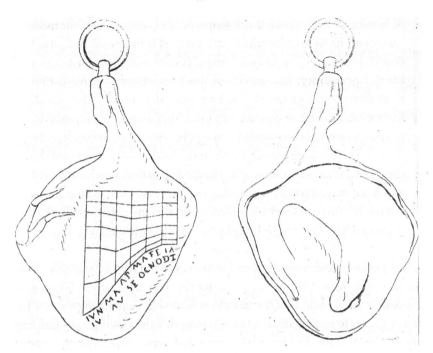

FIGURE 3.8 Georg Christoph Kilian's engraving as published in C. G. von Murr's *Abbildungen der Gemälde und Alterthümer in dem königlich Neapolitanischen Museo zu Portici...* (Augsburg 1793).

and philosopher while David (1741–1824) was the artist responsible for the engravings. Maréchal counted Diderot among his acquaintances, but he did not contribute to the *Encyclopédie* and his credentials for penning this work remain unclear, other than a desire to cash in on the popular thirst for reproductions of the antiquities. The dilettante and derivative approach to the material as a whole is evident in their treatment of the prosciutto. It makes its first appearance in the introductory essay of the first volume, where it is briefly described among the contents of the ninth room of the Herculanense Museum. This is a condensed version of Fougeroux de Bondaroy's room-by-room catalog, with no new or first-hand material added, and at points copied word for word, as is the case with the prosciutto. A full explication accompanies the discussion of the sundial from Pompeii that features as the seventh plate of the fourth volume, published in 1781. This further revises Lalande's entry for *Gnomonique* in the most recent editions of the *Encyclopédie*, right down to its contradictory explication, but with d'Alembert's inaccuracies removed.

Although this volume is primarily devoted to the ancient paintings of Herculaneum, in order to follow step by step our wise Academicians of Portici, we will publish the description of a bronze, one of the most curious of the Royal Museum. It was removed from the excavations of Portici, June 11, 1755. An ancient sundial is a very rare monument. Ours has this peculiarity: that it is of an absolutely different construction from all the other gnomons that have come down to us. They are all horizontal, or fixed, or concave, etc. Ours, on the contrary, is portable, vertical, much less complicated and quicker and easier to use. Its construction is also more perfect, in that it marks all the movements of the sun through the ecliptic during all the months of the year, the names of which we see inscribed, a singular and precious circumstance. This bronze, the original of which is identical to this engraved copy, has the shape of a ham that can be hung by the handle with a moving ring.

The disposition of the horizontal and vertical lines and their relationship to the zodiacal months and the hours of the day are then explained in detail, with the correct abbreviations of the months indicated by capitalizing the first letters of the Latin names, arranged in two rows to reflect their proper ordering but dispensing with the boxed grid. The initial steps in operating the sundial are then spelled out. Nothing is said about how to read the shadow cast by the gnomon on the dial, however, evidently because of concerns that it was no longer accurate because of the diminution of the obliquity of the ecliptic:

> To make use of this antique dial, one first had to hang it from its ring and by its own weight it remained in a vertical equilibrium. Then it was rotated toward the sun, only on the side of the gnomon. But if one wanted to use it today, it would be necessary, by astronomical calculations, to assess the differences that the planetary system offers us since the time of the construction of this dial, which we can trace back to the year 28 of Jesus Christ.

As Lalande would do in his revised entry in the *Encyclopédie Methodique*, Maréchal and David's text then reverted to a different description of the dial that ended with Lalande's notion that a moveable gnomon needed to be added to it.[39] Following a brief explanation of the difference between portable sundials and *clepsydras*, other topics of relevance are addressed, such as the belief that Romans ate ham as a second course "to awaken the appetite and induce drinking" and the notion that the prosciutto shape referenced the

---

39. S. Maréchal and F. David, *Les Antiquités d'Herculanum: avec leurs explications en François* (Paris 1780–1789) 4.14–17.

Enlightenment

artist's or the owner's name or "perhaps a friend of the dear man who was quite comfortable remembering his favorite tastes at all hours of the day," certainly to be distinguished from Plautus' parasite who rebelled against sundials dictating when he could eat.[40]

The accompanying engraving coupled the prosciutto with the marble sundial from Pompeii to which space was devoted in the adjoining text in much the same terms as in the *Encyclopédie*, right down to the reference to a Priapic gnomon on an unspecified sundial (Figure 3.9). As in von Murr's reproduction, both sides of the prosciutto are shown, though with the same degree of stippling in the Accademia's version to capture the full contours of its form, at nearly the identical scale, but with the gnomon slightly wider and longer to extend closer to the dial. Remarkably, despite the wide availability of printed versions, the image was flipped a full 180° so the gnomon lies on the right rather than the left. While this would not have rendered the sundial useless, since theoretically it could have continued to work in the reverse, it nevertheless serves as further evidence that the prosciutto was increasingly interesting more as a curiosity than as a sophisticated astronomical instrument. To add insult to injury, the inscription was eliminated.[41]

Maréchal and David's folly paled in comparison to that of the Italian artist and architect Carlo Antonini (c. 1740–1821). In addition to his views of Rome in the manner of his more accomplished and renowned compatriot Giambatista Piranesi (1720–1778), Antonini engraved examples of the ancient decorative arts as a reference for artists and architects.[42] He collected these in a four-volume work, the fourth volume of which was dedicated to candelabra and sundials. Text in these volumes was limited and, in this case, Antonini's introduction to the eighteen ancient sundials engraved here was a mere six pages.[43] In addition to explaining the difference between the "natural" and "civic" hours in antiquity and the differing lengths of the days according to the time of year, he dwelled on the design of the hemicyclium-type that made up the majority of his examples. This type, he noted, was often displayed on top of a column, and he supported this with an engraving of a

---

40. Maréchal and David 1781, 4.17–18.

41. The lack of the inscription may have caused the printer to flip this image inadvertently since the Pompeii sundial is similarly reversed, but David appears to have been both engraver and printer. The mistake appears in all the editions available for consultation.

42. W. Y. Ottley, *Notices of Engravers, and their works...* (London 1831), s.v. *Carlo Antonini*: "(c. 1780) An engraver of moderate talents who resided at Rome, and whose prints do not appear to be numerous." Little else is known of Antonini.

43. Later scholars would credit Antonini with being the first to offer engravings of this entire collection of ancient sundials.

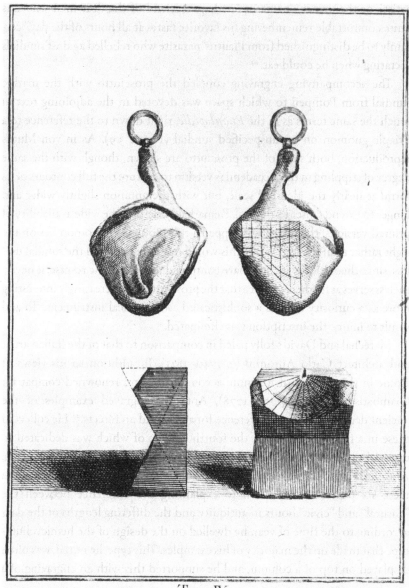

FIGURE 3.9 S. Maréchal and F. David's engravings of the prosciutto and the marble sundial from Pompeii, from the fourth volume of their *Les Antiquités d'Herculanum: avec leurs explications en François* (Paris 1780–1789). The prosciutto has been reproduced in reverse, placing the gnomon/tail on the right instead of the left.

Enlightenment 183

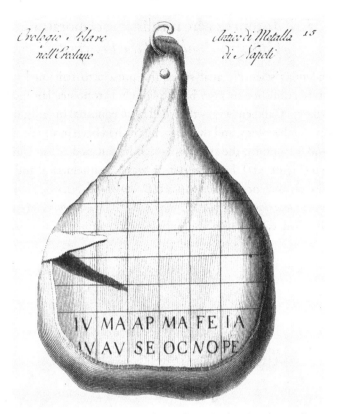

FIGURE 3.10 C. Antonini's prosciutto, plate 15 from the fourth volume of his *Manuale di varj ornamenti tratti dalle fabbriche, e frammenti antichi: Per uso, e commodo de' pittori, scultori, architetti, scarpellini, stuccatori, intagliatori di pietre, e legni, argentieri, giojellieri, ricamatori, ebanisti, Volume 4 che contiene la Serie de' Candelabri, e orologgi Solari Antichi* (Rome 1790).

relief from a vase he had seen discussed in Paciaudi's *Monumenta Peloponnesia*. The principal topic of Paciaudi's book had been the Columna Dianiae, which he believed consisted of a sundial on a column with an inscribed base. Antonini observed that Paciaudi also had illustrated "a bronze sundial of rare form, excavated in Herculaneum near Naples, which we are pleased to include in this collection." Given the passage of more than a quarter century since the Accademia's publication along with all the subsequent scholarship, it is stunning that Antonini's engraving was a replica of Paciaudi's "carafe," with cleaner lines, no "magical characters" in the grid, and a neater but still flawed inscription, yet certainly not the prosciutto sundial (Figure 3.10).[44]

---

44. C. Antonini, *Manuale di varj ornamenti tratti dalle fabbriche, e frammenti antichi: Per uso, e commodo de' pittori, scultori, architetti, scarpellini, stuccatori, intagliatori di pietre, e legni,*

3.b) Totum ex aere nitidissime elaboratum: Calkoen's Analysis of the Prosciutto

The Accademia's scientific analysis of the prosciutto remained unchecked until a study published in 1797 by the Dutch astronomer Jan Frederik van Beeck Calkoen. Calkoen (1772–1811) had been educated broadly, in philosophy, history, archaeology, and theology, but he had been most taken by mathematics and astronomy, the study of which he pursued at the University of Groningen (Figure 3.11). His dissertation was a "mathematical and antiquarian" treatise on ancient sundials entitled *De horologiis veterum sciothericis cui accedit theoria solariorum horam azimuthum et altitudinem solis* (Amsterdam 1797). This had earned him the appointment, in 1800, as extraordinary

FIGURE 3.11 Portrait of Jan Frederik van Beeck Calkoen (1772–1811). From *Ter Gedachtenisse Van Wijlen Jan Frederik Van Beek Calkoen* (Utrecht 1846).

argentieri, giojellieri, ricamatori, ebanisti, Volume 4 che contiene la Serie de' Candelabri, e orologgi Solari Antichi (Roma 1790) iii–viii, Pl. 15: "Orologio solare antico di metalla nell' Ercolano di Napoli." It is not clear why December has been abbreviated "PE." Antonini also illustrated (Pl. 14) the marble sundial from Pompeii, though he ascribed it to Naples.

Professor of Natural Philosophy and Mathematics, specializing in astronomy and hydraulics, at the University of Applied Sciences in Leiden, and subsequently in Utrecht.[45] Widespread knowledge of his treatise would have been bolstered by Lalande's inclusion of it in his 1802 *Bibliographie astronomique*, but its detailed assessment of the prosciutto appears to have gone largely unnoticed at the time.[46]

Calkoen had divided his thesis into four chapters. The last of these, the most theoretical and mathematical, was on the role of the azimuth and altitude of the sun in the design of different types of sundials oriented to specific directions. The other chapters followed the more traditional philological, or "antiquarian," pattern but differed in their use of mathematical equations and geometrical drawings to explain and evaluate the historical evidence. For example, in the first chapter on the history of the division of the days and the measurement of time in antiquity prior to the development of sundials he applied astronomical calculations to Palladius Rutilius' table of temporal shadow lengths and established to his satisfaction that they were reasonably accurate. This naturally led to a review of the "twelve-foot dinner" and the works of Salmasius and Petavius. In chapter 2, on the history of Greek and Roman sundials, he contrasted the gradual adoption of formal sundials in urban settings and the villas of the elite with the simple vertical gnomons and the pacing off of shadows that continued to be used in rural contexts long thereafter. The third chapter was dedicated to the forms and fabrics of ancient sundials, beginning with Vitruvius' catalog of sundial designers and followed by Zuzzeri's sundial and the other examples of the hemicyclium form known from Rome. Included here, naturally, was the 1762 marble sundial from Pompeii. Vitruvius' *viatoria pensilia* cued an inquiry into Bato and his character's timepiece, which Calkoen believed was not a *clepsydra* but more likely something akin to portable devices such as the "bronze sundial which is not fixed" referred to in the *Digest* (33.7.23). Anachronism aside—Bato's Cynic philosopher was the character in a third-century BCE New Comedy, and the *Digest* quote is from the second century CE—this illustrates the critical approach Calkoen had taken to the subject.

Calkoen's admiration for the prosciutto's form and function is reflected not only in how he wrote about it but in the fact that he devoted a full

---

45. K. J. R. Harderwijk, ed., *Biographisch woordenboek der Nederlanden* (Haarlem 1858) 3.7–8, s.v. *Calkoen (Jan Frederik van Beeck)*. His scholarly work spanned both of these fields and beyond.

46. Lalande 1802, 641. Unlike many other entries, Lalande did not annotate Calkoen's study.

seventeen pages to its analysis, more space than to any other subject. Initially observing that "the whole has been most handsomely worked from bronze" (*totumque ex aere nitidissime elaboratum*), he concluded that "whoever cherishes the arts and sciences will certainly acknowledge the elegant and convenient design of this instrument and venerate its ancient inventors." In Calkoen's view, the French had reaped the fame for first recognizing the dial and its calendrical abbreviations as a sundial, but the Accademia had corrected the inaccurate description of it in the *Encyclopédie* by publishing a detailed discussion illustrated by an accurate, properly scaled engraving. The Accademia, however, had treated it more as antiquarians than as mathematicians, failing to provide crucial information on the position and length of the gnomon and not taking into consideration the surface irregularities which prevented it from presenting a fully flat plane from which precise readings could be taken. Calkoen believed that the singularity of the prosciutto demanded a full explanation of how it worked, how the hours were read over the course of the zodiacal year, because the Accademia's treatment of its operation had been so perfunctory. His critical analysis ranks among the best available.

Calkoen illustrated his study of the prosciutto in several ways. First, he offered two views of the dial side of the prosciutto, one showing it as it was found, with its gnomon broken, and the other with his proposed reconstruction of the gnomon (Figure 3.12). These are very close copies of the Accademia's version, though their shading has been kept to a minimum and the broken gnomon curves in an exaggerated manner as if in the initial stages of forming the characteristic curl of a pig's tail. He provided no dimensional scale and nowhere states how large the object is. Two additional plates contained four graphic renderings of the geometry underlying his mathematical analysis. Finally, in addition to copying the inscription on his representations of the prosciutto, he included a transcription in his text (Figure 3.13). This shows the two lines of correctly abbreviated and properly paired months but with the letter forms oddly distorted as if to capture their worn, ancient forms. The letters are generally sans-serif, with most of the "A"s lacking their middle bar but with those in May, April, and August also shown with only the upper half of their left stem. The "C" in October is more open than normal and oddly askew, while the "D" of December looks like a "P" with the lower half of its *hasta* broken off. On the other hand, the "P" in April is not shown in the open form that earlier scholars had interpreted as a mark of its antiquity.

The Accademia had provided dimensions for the length of the vertical lines, but none for the horizontal lines, and had stated two dimensions resulting

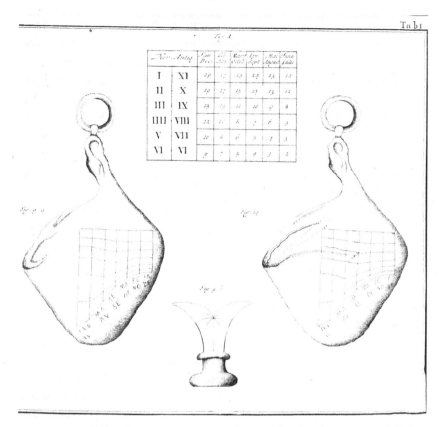

FIGURE 3.12 Calkoen's prosciutto, as extant (Figure 9a) and with gnomon and shadow reconstructed (Figure 14). Plate 1 from his *De horologiis veterum sciothericis cui accedit theoria solariorum horam azimuthum et altitudinem solis* (Amsterdam 1797).

IVN MA AP MA FE IA
IV AV SE OC NO PE

FIGURE 3.13 Calkoen's transcription of the prosciutto's inscription, from his *De horologiis veterum sciothericis*... (Amsterdam 1797) 72.

from their own inquiries: the distance from the tip of the gnomon to the equinoctial shadow on the fourth line and the distance from the tip to the seventh line representing the winter solstice, both along the first horizontal marking the horizon. They expressed these in units derived from the length of the equinoctial line, whose actual 31.23 mm length they had converted into 1,000 units, an even number to which the other dimensions could be related as ratios, the measured length of each also expressed in units.[47] The equinoctial shadow from their reconstructed gnomon had been 881 units and that of the solstitial shadow 1,482 units, or 28.23 mm and 47.50 mm, respectively. They had also concluded that the sundial was designed for a latitude of 41° 39′ 45″, which Calkoen rounded up to 41° 40′, when the obliquity of the ecliptic was 23° 46′ 30″, which Calkoen intended to prove was too great.

To these known quantities, Calkoen first added three of his own. The first was the altitude of the celestial equator, which he derived from the Accademia's stated latitude as 48° 20′—that is, the latitude of 41° 40′ subtracted from 90°. The second was the length of the upper horizontal line, literally the horizon line, which he would have had to obtain by measuring it off the Accademia's scaled engraving since it was otherwise not available. To maintain consistency, he expressed this as 1,620 units. Provided his measurement of the equinoctial vertical was 31.23 mm, giving the value of 0.0312 mm to each unit, this would equal 51.92 mm. This is much too long, however, and must be a typographic error: 1,260 gives 40.35 mm, which is equivalent to the actual length. He also established the distance between each of the vertical lines as 120 units, or 3.84 mm. This is too short, however, especially for his longer horizontal line, and again must be a misprint for 210 units, or 6.72 mm, a figure he used in his calculations at a later point as well and is close to the average distance between the verticals on the prosciutto. Six multiples of 210 units yield 40.38 mm.[48]

Calkoen then sought to determine the height of a gnomon necessary to cast a shadow of the desired length along each of the vertical lines on the prosciutto's dial, the values of which the Accademia had also expressed according to units. To do so, he needed to determine the altitude of the sun at noon as it entered each zodiacal sign. This provided him with the tangent of the right angle created by the height of the gnomon and the length of the vertical month line. By knowing the tangent and the length of the adjacent line of the right triangle created by the gnomon and its shadow, he could calculate the

---

47. Neither Calkoen nor the Accademia specifically states that the value of their units is 0.0312 m; this is the conversion of these units used throughout this study based on the length of the equinoctial line as measured off the prosciutto. Subsequent scholars used the Accademia's unit values rather than actual metrical measurements since they had not taken measurements off the prosciutto itself.

48. Calkoen's text is marred by several numerical mistakes, in some cases simple typographical errors but in other cases clearly miscalculations.

Enlightenment

height of the gnomon since, as Ptolemy had shown in the second century CE, the ratio between the height of the gnomon and the length of the shadow is fixed, just as they are in all right triangles. Given the Accademia's latitude of 41° 40′, Calkoen determined the altitude of the sun at noon on the equinoxes by subtracting the latitude from 90°, yielding 48° 20′. The sun travels along the ecliptic, the projection of the earth's orbit into the sky, which for Calkoen's purposes was inclined by 23° 46′ from the celestial equator, the projection of the earth's equator into the sky.[49] He then added the value of the ecliptic to the altitude of the sun to determine the altitude on the summer solstice while he subtracted it for the winter solstice. Because the sun inclines and declines throughout the year, the value of the altitude likewise inclines or declines, being 20° 26′ greater than the maximum altitude on the vernal equinox in Leo and Gemini but only 11° 38′ greater in Virgo and Taurus, while it is 11° 38′ less than the autumnal equinox in Libra and Pisces and 20° 26′ less in Sagittarius and Aquarius.[50] By subtracting these values in degrees from 90°, Calkoen determined the altitude of the sun at noon as it entered each of the zodiacal signs and the corresponding heights of the gnomons: a height of 889 units for the equinoctial line and 1,503 for the winter solstice. While the first of these was within the margin of error for the Accademia's 881 units, the solstice number was higher than their 1,482.[51]

Because these were the heights for seven gnomons serving each of the seven verticals, Calkoen had to determine whether they were valid as well for shadows projected from a single fixed gnomon since he knew the prosciutto's tail had served this function. His thesis was that a gnomon aligned with the dial's first vertical line would cast a shadow of the same length as one set directly on each vertical since all that was necessary was that the tip's shadow fall at some point along the appropriate vertical by rotating the suspended object. Calkoen assumed the gnomon had the height of the gnomon for the first vertical line representing the summer solstice that he had determined needed to be 533 units high (17.07 mm).[52] He then calculated the lengths of

---

49. In using the value of 23° 46′ Calkoen had dropped the seconds from the value used by Carcani.

50. Calkoen 1797, 75–76; these are the values he used in determining the resulting tangents.

51. His calculated heights, from left to right, are 533, 599, 719, 889, 1078, 1305, 1503, equivalent to 17.07, 19.18, 23.02, 28.46, 34.52, 41.79, 48.13 mm.

52. He erred in his calculation of this first vertical line, which he subsequently used for the height of his gnomon. He applied an altitude of the sun at the summer solstice at this latitude of 17° 54′ (17.90°; 48.33° + 23.77° = 72.10°; 90° − 72.10° = 17.90°), which produces a tangent of

the lines required to link the gnomon tip with the corresponding vertical line at its intersection with the upper line marking the horizon. He had spaced the vertical lines apart 210 units (6.72 mm) along his upper horizon of 1,260 units (40.35 mm long). Using the gnomon height (533 units) and the individual distances from the gnomon to the verticals along the horizon line to establish a right angle, he determined the length of each hypotenuse.[53] In comparing them with the hypotenuses created by a gnomon casting a shadow on each vertical, he found they varied by between 27 and 134 units (Figure 3.14).

CALKOEN RECALCULATED

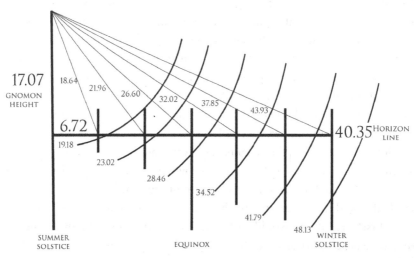

FIGURE 3.14 Calkoen's geometry recalculated using the actual length of the equinoctial line (31.23 mm). Author's image.

0.323 and a gnomon height of 544 (= 17.43 mm), or the tangent times the length in units of the vertical line (0.323 × 1686 = 544). His 533 units would require the tangent to be only 0.316 = 17° 32′ (17.54°); 533 ÷ 1686 = 0.316. Because he went on to use this value in subsequent calculations, those too are in error.

53. His points along the horizon were equally spaced 210 units apart, so y = x(210), where x corresponded with the number of the vertical line from left to right. These values were squared and added to the square of his gnomon height (533 units), so $y^2 + 533^2$. The square root of the sum provided the length of the hypotenuse. The resulting sums, from the second to the seventh vertical, were 572, 687 (for 678), 825, 979 (for 995), 1,178 and 1,369, equivalent to 18.32, 21.99 (for 21.71), 26.42, 31.34 (for 31.86), 37.72, 43.84 mm. Given his erroneous initial height for the gnomon, these should be 582, 686, 831, 1,000, 1,182, 1,372, equivalent to 18.64, 21.96, 26.60, 32.02, 37.85, 43.93 mm; these are used in Figure 3.14.

Enlightenment

As proof that this single gnomon would correctly cast its shadow along each vertical when set perpendicular to the plane of the dial, he demonstrated that lines equivalent in length with the shadow lengths he had calculated earlier intersected each vertical slightly below the horizon line. For example, the diagonal from the gnomon intersecting the equinoctial line at the horizon was 825 units (26.42 mm) long while the line of 889 units he previously had determined for this vertical's meridian shadow intersected it below the horizon line. This was the case for all the shadow lines, their intersections with the verticals occurring increasingly below the horizon line moving across the dial toward the winter solstice. Finding no evidence for gnomons along the horizon line, he concluded the tail must have extended as far as the first vertical month line.

Only the tip needed to cast a shadow on each vertical line, not the entire gnomon, a point Piaggio had argued as well. Consequently, Calkoen reconstructed the tail's shape mathematically in order to position its tip at the proper point in space to cast the desired shadows. He needed to position the tip 533 units above and perpendicular to the plane of the dial, above the intersection of the horizon line and the first vertical. For the length of the base of his right triangle, running from the stub of the tail at the left side of the prosciutto to the upper left-hand corner of the dial, he posited a length of close to 1,000 units (31.23 mm), "from a measurement I myself established, since the Accademia did not supply it." This is rather short, the true length being closer to 1,217 units (39 mm). From this he concluded that the length of the gnomon-tail was 1,224 units (39.19 mm) and this created an angle of 29° from stub to tip. In fact, however, the hypotenuse of the resulting right triangle is only 1,110 units (35.54 mm) long; a base of 1,099 units (35.19 mm) would be required to obtain 1,224, and this would produce a 25.8° angle.[54] His engraving shows a long, thin tail, springing from near the prosciutto's left side and curving up slightly to end near the dial's corner. It is straight and not curled at any point. "And so," he concluded, "this picture sets the sundial before our eyes in its entirety, as it was in antiquity."

He next explained how the sundial was used to tell time by referencing a shadow produced by the reconstructed gnomon on his engraving.[55] The tail side of the prosciutto needed to be rotated toward the sun and the tip of the shadow aligned to the vertical corresponding with the name of the month inscribed below. The horizontal line closest to the shadow's tip determined the hour, depending on whether it was before or after midday. For his example

---

54. The true length of the gnomon/tail was ca. 42.65 mm (1331 units).

55. Calkoen 1797, Plate 1, fig. 14 (see Figure 3.12).

192 THE PROSCIUTTO SUNDIAL

he illustrated an hour on April 15, so the tip fell to the left of the equinoctial line and just below the third horizontal, meaning the hour was just past the second or approaching the tenth hour.

Calkoen then turned to the matter of how the Accademia had determined the latitude for which the sundial was designed. They had found the latitude of 41° 39′ 45″ by dividing the length of the shadow (889 units) by the length of the equinoctial line (1,000 units), though Calkoen noted that La Caille's most recent tables refined this to 41° 38′ 18″.[56] The Accademia had observed that a slight imperfection in the surface along the equinoctial line had caused a deviation of "no more than two or three minutes" at the third horizontal marking the second and tenth hours. "But, if I may interject my judgement," continued Calkoen, "Although this sundial was exceptionally designed, the Accademia nevertheless seems to me to have wanted to determine its parts too precisely." His concern was that the prosciutto's small scale made determining the passage of time to such a degree of precision impossible. He looked specifically at the seventh vertical line whose 65° 26′ tangent is equivalent to 235,560 arcseconds all crammed onto a line shorter even than the 1,000 units of the equinoctial line. "I do not believe the keenest eye would distinguish the fifteenth part of a line, whence we can infer that, even if this machine was fashioned most accurately, an error of four minutes in determining the altitude of the sun or the distance from the vertex could easily occur."[57] As a consequence, Calkoen concluded the actual latitude for which the sundial was designed probably lay somewhere between 41° 34′ (41.56°) and 41° 42′ (41.7°).

The final target of Calkoen's critique was the Accademia's determination of the obliquity of the ecliptic and the date of the sundial's construction. Advances in astronomy in the thirty-five years since the Accademia's publication had resulted in greater agreement within the scholarly community about the rate of the obliquity's secular diminution. Lalande, for example, had published a table of the values for the obliquity of the ecliptic calculated from 800 BCE through 1833 CE, which had become a common reference tool.[58] Calkoen wrongly believed the Accademia had employed a secular diminution rate of 1′ 45″ obtained from de Louville's table, but in fact Carcani had settled

---

56. This equals a tangent of 0.889 or 41.63°, or 41° 37′ 48″.

57. Fifteen divided into sixty minutes is four minutes.

58. J. J. Lalande, *Tables Astronomiques: Calculées sur les observations les plus nouvelles, pour servir à la troisieme édition de l'Astronomie* (Paris 1792) 1, Table 1: "Epoques des longitudes du soleil et des argumens qui reglent ses inégalités." This was a separate volume intended to supplement his three-volume work *Astronomie* (Paris 1792); see Chapter 2.f.

Enlightenment 193

on 1'. "By common agreement of astronomers," Calkoen noted, the rate did not exceed 50", as Lalande had argued in the third edition of his handbook on *Astronomie*, which was the rate Calkoen concluded "seems to fit best with the observations and calculations."[59] For the value in 1762 Calkoen agreed with Lalande's 23° 28' 17". This differed from Carcani's by only 1". It was Carcani's value of 23° 46' 30" for the obliquity at the time the sundial was constructed that Calkoen disputed. He noted that this would have dated it to around 440 BCE, not 28 CE.[60] The true value in the Accademia's proposed year of production, he stated, was 23° 42' 25", a reduction of 4' 5" from Carcani's, which would account for about 500 years of diminution.[61] This would constitute a diminution of 14' 8" (0.24°) for the period 28 to 1762 CE, conforming to Calkoen's rate of secular diminution of about 50".[62]

Ultimately Calkoen concluded that the three- to four-minute degrees of wiggle room in establishing the latitude for which the sundial was designed, along with its corresponding obliquity of the ecliptic, prevented dating the prosciutto with the level of precision the Accademia had claimed. Consequently, he resorted to other means of dating. First was the inscription, which provided the *terminus post quem* since the calendrical changes resulting in both the months of Quintilis and Sextilis becoming Julius and Augustus could not have taken place before 27 BCE. Similarly, the eruption of Vesuvius, which Calkoen dated to 80 CE, served as the *terminus ante quem*. He therefore concluded the prosciutto dated between 725 and 830 *ab urbe condita* and that no more secure a date than this one-hundred-year window was possible.[63]

---

59. J. J. Lalande, *Astronomie* (Paris 1792) 3.56–60, a review of the scholarship on the value, with 50" per century given as the rate. Calkoen refers to this text in a footnote.

60. His calculations are difficult to reconstruct. Lalande's table gives 23° 46' 12" for 500 BCE and 23° 45' 27" for 400 BCE, a difference of about 45" for the century or 4.5" per decade, adding 27" to 23° 45' 27", which yields 23° 45' 54" for the year 460 BCE. The Accademia's value was actually closer to 660 BCE; see Chapter 4.e.

61. There is another misprint here in that the value published was 23° 22' 25", which would date it well beyond 2050; it is assumed he meant 42' rather than 22'.

62. Lalande, however, had given 23° 42' 22" for the value at the turn of the millennia, which would push Calkoen's value closer to 5 BCE. The actual value for 28 CE is 23° 41' 29". For 28 to 1762 CE, this gives a total rate of diminution of 13' 12" (0.22°), or 45" per century. Carcani had found a rate of 18' 12".

63. The traditional foundation date for Rome was 753 BCE, and the Romans dated subsequent historical events from this year, *ab urbe condita* "from the city's founding." Thus, 28 BCE was 725 years after the foundation, and the eruption in 80 CE was 833 years after. The eruption is currently dated to 79 CE; Calkoen was using a date common in his era.

3.c) "Nice Invention, for Not Forgetting Dinner": Montucla's Histoire des Mathématiques

Another French mathematician and astronomer treated the prosciutto soon after Calkoen. Jean-Étienne Montucla (1725–1799) was trained at the Jesuit college in Lyon by the same tutor as Lalande, with whom he formed a lifelong friendship (Figure 3.15). Later, in Paris, he moved in the same circle as Lalande, Diderot, and d'Alembert. Like Lalande, he was elected to the Berlin Akademie der Wissenschaften, but he never became a member of the French Académie Royale des Sciences. Montucla's *Histoire des Mathématiques*, a two-volume work first published in 1758, is considered not just the first comprehensive history of mathematics but "a history of science written from a mathematical angle."[64] Reflecting the influence of his intellectual contacts, his

FIGURE 3.15 Portrait of Jean-Etienne Montucla (1725–1799). Engraving by P. Viel. Smithsonian Libraries Image Gallery (SIL-SIL14-m005-11). Creative Commons Public Domain CC0. https://library.si.edu/image-gallery/73687.

---

64. G. Sarton, "Montucla (1725–1799): His Life and Works," *Osiris* 1 (January 1936) 519–67. The full title of Montucla's work is *Histoire des Mathématiques dans Laquelle on Rend Compte*

Enlightenment 195

book was "a brilliant contribution of the Enlightenment closely allied in spirit to the contemporary *Encyclopédie*," though it had the peculiar trait, for a mathematical treatise, of being completely bereft of mathematical symbolism in the text, Montucla, like Lalande, preferring to rely on verbal explication rather than symbolic renderings.[65] The first edition of the first volume had treated mathematics from antiquity through the seventeenth century, but it did not include a section on gnomonics. With Lalande's encouragement, Montucla produced a revised and enlarged second edition late in his life—he died four months after its publication—which included a "supplement" on the history of sundials.[66]

Montucla's supplement, subtitled the "Histoire de la Gnomonique Ancienne et Moderne," followed what had become the routine approach to such surveys of the literary and archaeological evidence, with slight variations that gave his discussion its own character. He retailed, for example, Horapollo's legend of the Egyptian *cynocephali* clocking time by urinating on the hour, not uncritically, as had the Accademia, but with the observation, "It is unfortunate that this is not confirmed by our naturalists, for they know no animal endowed with this singular property."[67] While the philosopher with his *clepsydra* in Bato's comedy and similar anecdotes are missing, Montucla enlivened his account by highlighting others. "A sundial was shown to Epicurus as an ingenious mathematical innovation which he affected to despise. 'Nice invention,' he said, 'for not forgetting dinner.'"[68] This had served as a prelude

de leurs Progrès Depuis leur Origine jusqu'à nos Jours: où l'on Expose le Tableau et le Développement des Principales Découvertes dans Toutes les Parties des Mathématiques, les Contestations qui se Sont Élevées Entre les Mathématiciens, et les Principaux Traits de la vie des Plus Célèbres (Paris 1799–1802).

65. I. Grattan-Guinness, "Tailpiece: The History of Mathematics and Its Own History," in *Companion Encyclopedia of the History and Philosophy of the Mathematical Sciences* (New York 1994) 2.1666: "Of course many basic aspects of the history of mathematics were thus left out." Unlike his acquaintances, Montucla was not exclusively an astronomer or mathematician, having held a number of government appointments until the Revolution in 1789.

66. As it happened, Lalande assumed the task of completing the third and fourth volumes of the second edition.

67. Montucla 1799, 716. The "Supplement" appears at the end of the first volume of this second edition, though it supplements the fourth volume.

68. Montucla 1799, 718. Montucla erred in ascribing this quotation to Epicurus since the original source is Diogenes Laertius (6.9.104), who credits it to Menedemus the Cynic: ὁ γοῦν Διογένης πρὸς τὸν ἐπιδεικνύντα αὐτῷ ὡροσκοπεῖον, "χρήσιμον," ἔφη, "τὸ ἔργον πρὸς τὸ μὴ ὑστερῆσαι δείπνου." Montucla's source was Martini 1777, 77–78 note 137, where he cites Diogenes Laertius and notes that by *horoscopion* was doubtless meant a *horologion*. Montucla acknowledged the works of Martini and Zuzzeri as the principal sources for his historical material.

to the lamentations of Plautus' parasite about sundials fixing the dinner hour, along with a citation from the *Greek Anthology* that had also become a regular fixture of historical surveys on ancient sundials.

> The *Greek Anthology* has preserved for us an inscription attached to a sundial which seems to relate to the use of the day by a certain order of citizens. The sense is: "Six hours of the day are dedicated to work, the following four tell mortals: Live!" These four hours indeed were marked with the Greek letters Z. H. Θ. I. and this word *ZHΘI* means "Live!" (*GA* 10.43).[69]

The sundials from the archaeological record follow, to illustrate in part the Vitruvian types, though Zuzzeri's Tusculum example now is definitively ascribed to Cicero, and the 1762 marble sundial of Pompeii is correctly described as marked with both the horary lines and the equator but not the Tropics.

Not even Montucla could escape the curse of the prosciutto that had stalked every French astronomer from d'Alembert through Lalande. Like them, he based his description on the available secondary sources and made no attempt to analyze its underlying geometry. The prosciutto is the most

---

69. D. L. Page, *Further Greek Epigrams: Epigrams before AD 50 from the Greek Anthology and Other Sources* (New York 1991) #91: Ἐξ ὧραι μόχθοις ἱκανώταται· αἱ δε μετ αὐτας/γράμμασι δεικνύμεναι ΖΗΘΙ λέγουσι βροτοῖς (Anonymous). The Greek numerals were associated with letters, so hour 7 = Z, 8 = H, etc. Montucla's literary reference, which he took from Zuzzeri 1746, 80–81, subsequently metamorphosed into an actual sundial, ascribed to Herculaneum and now "missing." In musing upon the ancients' views of the afterlife, the anonymous contributor to the May, 1823, edition of the *London Literary Gazette* 328 (May 3, 1823) 283–84, described a "dial found at Herculaneum," which he illustrated by a "plain and correct diagram" showing the semicircle of an ornate vertical sundial divided into twelve equal parts, the initial five hours labelled with the first five letters of the Greek alphabet, the central two left blank, and the last five bearing the letters "ς, Z, H, Θ, I." The author explains the use of the "ς," or digamma, as being equivalent to "6" in the Greek numerical system, though in this reconstruction of the dial it falls in the eighth hour, throwing the remaining letters out of sync as well. He quoted this epigram as its likely inspiration and remarked, "the sixth and seventh hours are left blank, on account of the heat of noon usually incapacitating the ancients for business or pleasure." Working all of this out, he noted, "must have been pleasing to the Herculaneum dilettanti." This reference, in turn, was noted, complete with errors, in 1824 in the *Bulletin des Sciences Historiques, Antiquités, Philologie* 1 (1824) 230, and it ultimately entered the annals of actual, if no-longer-extant, sundials from Herculaneum, both as a formal entry in the *Inscriptiones Graecae* (IG XIV 713: "ΑΒΓΔΕ ςΖΗΘΙ"), and in catalogs of sundials, e.g., J. Bonnin, *La Mesure du Temps dans l'Antiquité* (Paris 2015) 391 (A_155): "cadran plan vertical decouvert à Herculaneum et disparu."

curious example of a portable sundial, began Montucla. Recovered in the excavations at Portici in 1755, the Neapolitan academicians had published it in their third volume. "Its shape is that of a ham suspended by a ring attached to the foot, and the end of the tail which has been preserved takes the place of the style. The hours are inscribed on the roughly flat part of the ham." Montucla then described the seven intersecting lines of the dial and the abbreviated calendrical notation below before referring the reader to his illustrative figure, which "will supplement a longer explication." "People a little versed in gnomonics will easily recognize how one used this dial," Montucla continued, explaining how the prosciutto was held toward the sun so the shadow fell on the desired vertical, such as that representing the equinoxes.

> The shadow of the style's tip fell on the middle line which corresponded to the entry of the sun in the Ram [Aries] and in the Balance [Libra], the hour was then marked by the nearest transverse. When the sun occupied the middle of Cancer or the Twins [Gemini], one needed only to drop the end of the shadow in the middle of the interval between the first and the second line. The Neapolitan savants found this dial of great accuracy.

The accompanying engraving, however, is not a copy of the Accademia's accurate rendering but a revised version of Maréchal and David's flipped prosciutto, the differences being that the skin around the hock is more pronounced and the inscription is correctly positioned beneath the grid, even if this inverted form changed entirely the *boustrophedon* ordering (Figure 3.16).[70] No doubt the Accademicians would have relished seeing yet another Frenchman "tricked" by "false reports" from scholars with "more enthusiasm than wisdom and travail."[71]

---

70. Montucla's work was cited as the principal reference for the prosciutto sundial in C. Daremberg and E. Saglio's influential reference work the *Dictionnaire des Antiquites Grecques et Romaines d'apres les Textes et les Monuments*, vol. 3.1 (Paris 1899) 260 n. 7 (s.v. *Horologium*), but only in a footnote and without elaboration.

71. By this time, all the members of the Accademia responsible for the third volume were deceased.

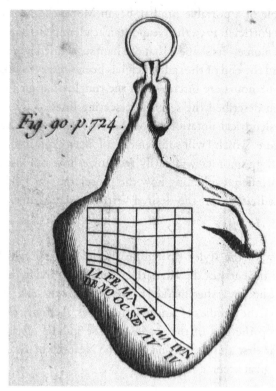

FIGURE 3.16 Montucla's prosciutto. Plate 12, figure 90 from the "Histoire de la Gnomonique Ancienne et Moderne," the supplement to the first volume of his *Histoire des Mathématiques* (Paris 1799–1802). The entire prosciutto has been reproduced in reverse, placing the gnomon/tail on the right instead of the left and with the inscription flipped as well.

3.d) Ingénieuse et assez simple: *Delambre's Trigonometric Analysis*

Jean-Baptiste Joseph Delambre (1749–1822), an important French astronomer of the early nineteenth century, evaded the prosciutto's curse by simply not publishing an image of it or mentioning the inscription (Figure 3.17). Despite a bout with smallpox as a child that had diminished his eyesight, Delambre was a keen observer of details and a voracious reader with an excellent memory and a knowledge of Greek, Latin, Italian, German, and English. After attending a Jesuit school in Amiens, he studied under Lalande at the Collège de France in Paris before becoming his assistant in 1783. He successfully documented a transit of Mercury across the sun in 1786, having detected an error in Lalande's calculations for the timing that had thrown off other astronomers. He subsequently went on to revise the existing astronomical

FIGURE 3.17 Portrait of Jean-Baptiste Joseph Delambre (1749–1822). Engraving by B. Holl after L. L. Boilly. Smithsonian Libraries Image Gallery (SIL-SIL14-D2-16). Creative Commons Public Domain CC0. https://library.si.edu/image-gallery/73387.

tables, many of them the work of Lalande.[72] In 1792 he took part in the endeavor to measure the length of the meridian arc between Dunkirk and Rodez, France, the data from which was used by the Académie des Sciences to define a new unit of measurement, the meter.[73] Napoleon named him permanent secretary of the Académie, now renamed the Institut National, in 1803, a position he retained until his death. A member of the Royal Swedish Academy of Sciences, the Royal Society of London, and the American

---

72. For example, the *Tables Astronomiques* (Paris 1806) prepared by Delambre's Bureau des Longitudes was over 370 pages long with some 160 pages of introductory explication and equations relating to the tables; Lalande's 1792 edition of the *Tables Astronomiques* was 281 pages, without explication. The value of the latitude of Naples was refined to 40° 50′ 15″; that of Rome to 41° 53′ 54″; the obliquity of the ecliptic in 1700 was 23° 28′ 49″ and that of 100 BCE 23° 42′ 26″.

73. Delambre stepped in for Jean-Dominique de Cassini, who refused to participate in this initiative of the revolutionary government. Pierre Méchain took the measurements from Rodez to Barcelona; the third member of the mission, Adrien-Marie Legendre, had also declined to participate. The project was completed in 1799.

THE PROSCIUTTO SUNDIAL

Academy of Arts and Sciences, he served as director of the Paris observatory and Professor of Astronomy at the Collège de France, becoming Chair after the death of Lalande in 1807.[74] Delambre's six-volume *Histoire de l'Astronomie*, produced in the last years of his life and still referenced today, included two volumes on ancient astronomy, the second of which, devoted primarily to Greek astronomy, included a chapter on ancient sundials.[75]

Unlike Lalande and Montucla before him, Delambre was fond of astronomical equations, and he used these extensively in his historical work to illustrate the theoretical concepts and explain the practical aspects of this evolving science. His chapter on ancient sundials is no exception. His primary focus was on the principles underpinning the design of extant sundials from antiquity. He began with a critical analysis of a quadruple vertical sundial from Greece, a sundial shaped like two open books that stood upright with a different dial inscribed on each cover. This bore the inscription *Phaidros, son of Zoilos, from Paiania, has made [it]*, in Greek. Delambre was particularly enthusiastic about this, first because he believed that Phaidros was the actual mathematician who had designed it and, second, because he had already analyzed it and could offer a more detailed assessment here.[76] His discussion of the hemicyclium type illustrates his thesis that most ancient dials, including *clepsydras*, were designed empirically through observation rather than on the basis of complex mathematical theory:

> We have spoken often of the *hemicyclium* or rather the hemisphere of Berosus. This dial is the simplest and most natural of all; it must have

---

74. *Biographical Encyclopedia of Astronomers* (New York 2014) 542–43, s.v. *Delambre, Jean-Baptiste-Joseph* (A. Ten). Royal Society of London, *Archives* EC/1791/14 (May 5, 1791); Lalande wrote in support of his nomination. His election to the Swedish Academy was in 1788 and to the American Academy in 1822, the year of his death. His name appears on the Eiffel Tower and a crater on the moon.

75. J-B. J. Delambre, *Histoire de l'Astronomie Ancienne* (Paris 1817) 2.504–19: Book IV, Chapter XVII: *Cadrans d'Athènes de Phaedre, et Cadrans Diverse*. Delambre's vast knowledge on the topic is reflected in the length of just these two volumes, the first being 556 pages and dedicated primarily to literary sources and the second volume 640 pages.

76. Now in the British Museum 1816,0610.186 (Athens, circa 300 CE; 0.50 m H × 0.99 m W). J. Spon, *Voyage d'Italie, de Dalmatie, de Grece et du Levant* (The Hague 1724) 2.127, had seen it in the courtyard of the church of Panagia Gorgoepikoos, now called the Little Metropolis in Athens. E. Visconti, who dated it to the time of Hadrian on the basis of the letter forms, had asked Delambre to analyze it for his *Mémoires sur des ouvrages de sculpture du Parthénon, et de quelques édifices de l'Acropole a Athènes* (Paris 1818) 76–87, by which time Lord Elgin had already transported it to England and, ultimately, the British Museum. Delambre's treatment of it for Visconti was largely descriptive. See also Jones 2016, 77–78 #9, fig. III-12, where it is dated 300 CE and deemed "not precisely executed…"

Enlightenment

preceded all the others. It was the most widespread, but it could never be more than of a very small dimension and was capable only of very mediocre accuracy. It could hardly be put on public display: it was a sundial that the curious placed in their country houses or in their cabinets. What sets it apart is that it does not assume any mathematical theory: to imagine and describe it, it was enough to have a clear idea of the spherical movement of the sky.[77]

This brought him to the extant ancient examples of hemicyclium sundials from Tusculum, Castelnuovo, Rignano, and, finally, the 1762 marble sundial from Pompeii.

It differs from the preceding ones in that the Tropics are not expressly marked there; we only see the equator. G.M. Martini...concludes that this fourth [Pompeian dial] must be older than the other three and that it is probably the more primitive dial, as Berosus designed it. I would be much more tempted to draw the opposite conclusion. Without the Tropics, the construction of the dial becomes much more difficult. To mark the temporary hours [hours varying according to the time of year] there, we would need a theory which only arose in the Greeks, in the time of Hipparchus, before whom we do not find in the annals of any nation not the slightest idea, not the slightest indication of any trigonometric calculation. But if the winter Tropic [of Capricorn] is missing from the Pompeii dial, the same cannot be said of the summer Tropic [of Cancer], which forms the lower edge of the preserved section of the hemisphere. Now, it suffices on the basis of this Tropic [of Cancer] and the equator to describe all the hourly lines which one will have extended unnecessarily beyond the winter Tropic: thus, the reasoning of Martini falls on its own.[78]

He then explained the process for carrying out such empirical construction:

Alongside the Chaldean hemisphere, place in an arbitrary but constant situation any plane with a perpendicular gnomon. Mark on this plane the progress of the [gnomon's] shade from hour to hour, on the solstitial and equinoctial days. Join by lines the corresponding time

77. Delambre 1817, 2.510.

78. Delambre 1817, 2.512.

points. You will see that the three analogous points will always be on the same line and you will thus have, without any theory, the temporary dials of every kind. You can draw all horizontal, vertical, declining or inclined dials, at will.

As an example, Delambre turned to the prosciutto sundial.[79]

We have one sundial whose construction we know perfectly. It is the one with the shape of a ham which was found in Portici in 1755 and described by the Academicians of Naples. It consists of seven vertical lines traversed by as many broken but practically horizontal lines which are the horary lines. The theory is extremely simple. The vertical lines serve for the entry of the sun in the different signs. One could multiply these lines indefinitely and draw them for all degrees of longitude, at least ten in ten. The horary points are determined by the tangents of the heights of the sun, taking as the ray the rectilinear and horizontal distance from the top of the style to the vertical of the day.

Let "a" be the height of the gnomon, whose foot falls on the horizontal which passes through all the vertices of the verticals. Let "E" be the part of the horizontal between the foot of the style and the vertical of the day.

$\tang A = \dfrac{E}{\alpha}; \dfrac{\alpha}{\cos A}$ will be the distance from the top of the style to the top of the vertical.

$\left(\dfrac{\alpha}{\cos A}\right) \tang$ height $\odot$ will be the distance from the hour point to the top of its vertical.

Sin height $\odot$ = sin H sin D + cos H cos D cos (hour angle of the temporary hour).[80]

---

79. Delambre had previously discussed the prosciutto in the course of his analysis of a marble convex sundial from Delos: "Cadran antique trouvé dans l'Ile de Délos, et par occasion de la Gnomonique des Anciens…," *Magazin Encyclopédique, ou Journal des Sciences, des Lettres et des Arts* (September 1814) 361–91. Following a brief description of its operation, he concluded, "The idea [of the dial's design] is ingenious and the construction fairly simple. It depends, like that of the dials above, on the theorems of Ptolemy and his predecessors. But the hour lines would become continuous curves instead of being formed of rectilinear parts which form unequal angles between them. It is a license that the craftsman has taken for convenience, and the disadvantage of which is not very serious."

80. The $\odot$ symbol represents the sun. Delambre used "temporary" to refer to the "seasonal" hours.

By means of these formulas, of which the Greeks had equivalents, in truth less convenient, one will determine on each vertical the lowering of the time point below the horizontal. The Portici dial gives the lines from hour to hour, from sunrise up to the sixth hour, that is to say until midday. The morning hours are sufficient because the same heights return in the evening at equal distances from the meridian. To observe the time, hang the dial vertically on a hook; we turn it towards the sun, until the shadow of the gnomon falls on the vertical of the day, the position of the shadow-point between the hour lines indicates the hours passed, and we estimate as we can the fraction of the current hour. For the intermediate days between the entries at the different signs, one should pass a curve through all the points of the same hour on the different verticals; but it seems that we were content to join these points by as many straight lines; whence it follows that no greater precision was sought there than that of a quarter of an hour.

Despite the simplicity of this theory, we see that the calculation of so many points had to be very long and tiring, especially for the Greeks, who had no use of tangents, secants, or logarithms. They could find the heights by the *analemma*, the length of the ray and that of the shadow, by the graphic solution of two triangles. We are ignorant of the date of this dial, of which Vitruvius makes no mention. It can be suspected that it was described by observation, using the sundial of Berosus. We could thus determine by experience, and without calculation, as many points as we thought fit; the work may have required six mornings a month apart. We can imagine that an amateur with little geometry may have preferred this method. A mathematician would not have had patience for this; he would have preferred the computation, despite its lengths. Today we would find relief in the [Astronomical] *Tables*, which give the heights of the sun for every ten minutes for all degrees of declination.

Delambre had acknowledged the earlier work of the Accademia and Martini but appears unaware of Calkoen's, whose mathematical analysis is more analogous to his own. Delambre advanced three basic equations of spherical trigonometry fundamental to analyzing horizontal sundials that could be applied to the analysis of the prosciutto. The first could be used to establish both the angle ("A") of the hypotenuse running from the tip of the gnomon to the top of the vertical line marking the target month and the length of the hypotenuse. This equation required knowing the distance from the first vertical month line ("Jun(ius)") along the top horizontal line—the horizon—to the desired month, in this case represented by "E." It also required knowing the height of the gnomon, represented here by "a." "E" could be measured off the Accademia's engraving, but the height of the gnomon required knowing the design latitude. Dividing the gnomon height into the length of the horizontal yields the tangent of the sun's altitude ("A"). The second equation established the distance along the vertical line for each of the horizontal hourly lines. This required knowing the sun's altitude at that specific hour, in addition to knowing the height of the gnomon ("a") and the distance along the horizon line ("E"). The third equation could be used to determine the sine of the sun's altitude, but this required knowing the design latitude ("H") along with the value of the declination of the sun ("D"). The "hour angle of the temporary hour" was the value of the angle of each of the six hours between sunrise and midday or the six from midday to sunset, with midday being "o," the fifth horizontal being "1," and so on up the vertical. Delambre did not provide the equation for working out these values here, but they were derived by dividing the desired hour by six and multiplying it by the inverse cosine of the negative value of the tangent of the design latitude ("H") times the tangent of the declination ("D").[81] Delambre nowhere states that he had accepted the Accademia's value for the design latitude or that he had taken measurements from their engraving to check their results, and he provided no data showing that he had applied any of his equations to a detailed analysis of the prosciutto.

Delambre's remark that its date is unknown is particularly curious, given the Accademia's meticulous effort to date it by determining the obliquity of the ecliptic and his own refinements to the table of its secular diminution. He could have considered this a component in the design of the sundial, the main topic of this particular section of his book, and critiqued the value the Accademia had determined. Moreover, he ignored entirely the clues provided by the inscription that offered at least a useful *terminus post quem* and seems unaware of the connection drawn by earlier scholars between the prosciutto and Vitruvius' *viatoria pensilia*.

3.e) *Woepcke's* Disquisitiones *and the Temporal Hour*

The only other treatment of the prosciutto in the nineteenth century to approach the level of analysis of Calkoen and Delambre was that of Franz

---

81. Delambre's last equation is the standard equation for determining the zenith distance (z_H) at a given hour on a horizontal planar sundial, where "τ" designates the "hour angle of the temporary hour"; see Gibbs 1976, 40–41; and Chapter 4.e, for additional discussion.

Woepcke (1826–1864), which appeared in 1842. In his relatively short life, Woepcke earned a reputation for his editions and translations of medieval Arabic manuscripts on mathematics. His life and work paralleled that of Calkoen in several ways. He had discussed the prosciutto in his doctoral thesis, *Disquisitiones archaeologico-mathematicae circa solaria veterum* (Berlin 1842), written in Latin, for which he was awarded his PhD at the University of Berlin before his twentieth birthday.[82] The hybrid nature of his research is signaled in the title of his thesis, which emphasizes archaeology and mathematics, and reflects the diverse backgrounds of his two thesis advisors. One was Johann Franz Encke (1791–1865), a professor of astronomy at the university and director of its observatory, known primarily for his work on comets.[83] The other was Ernst Heinrich Toelken (1795–1878) a physicist and archaeologist who was professor of art history and director of antiquities at the Berlin Museum.

Woepcke's thesis follows Calkoen's model. He divided his material into four chapters, the last of which was a detailed presentation of the equations and tables used in the analysis of his sundials. Like Calkoen's, the first chapter was devoted to an explanation of the division of the day into hours by the ancients, and the second chapter was a survey of Greek and Roman sundials.[84] This included the traditional methodology of illustrating the Vitruvian catalog of types with the archaeological examples from Tusculum and Rome. Baldini's bronze disk, about which the Accademia had been so dismissive, served as Woepcke's first example of the Vitruvian *viatoria pensilia*-type. He then turned to the prosciutto. Citing the studies of Martini, Calkoen, the Accademia, and Delambre on the prosciutto, he singled out the Accademia for their review of the mathematical principles of its design and Calkoen for the "great subtlety" of his critique of the Accademia's assessment.

Woepke's description of the sundial is relatively succinct. He noted the unevenness of the ham's surface; the tail taking the place of the gnomon; the seven vertical lines each representing two months in the calendar year, the

---

82. *Allgemeine Deutsche Biographie* (Leipzig 1898) 44.209–10, s.v. *Woepcke, Franz* (W. Cantor); and details provided in the *Vita* he appended at the end of his thesis. Calkoen had lived to 39, Woepcke to 38; Calkoen received his doctorate at 25, Woepcke at 21.

83. *Biographical Encyclopedia of Astronomers* (New York 2014) 661–62, s.v. *Encke, Johann Franz* (M. Meo); C. Bruhns, *Johann Franz Encke: Seine Leben und Wirke* (Leipzig 1869) 81–91 (Venus transits). He had reviewed the data from the 1761 and 1769 transits of Venus and deduced the solar parallax as 8.57116″ arcseconds, which was long accepted as the authoritative value; Carcani had employed 8″ for his analysis of the prosciutto.

84. The third chapter focused on an ancient sundial in Berlin (SK1049); Gibbs 1976, 218 #2023G: "spherical variant."

names of which the craftsman had abbreviated, although he did not state where these were inscribed; and the seven horizontal lines each representing two hours of the day, the upper line that marked the horizon being the only fully straight line in the series and the lower one indicating midday. The instrument was suspended from its ring and turned on its axis of suspension so the tip of the tail cast its shadow on the line representing the current month while concurrently marking the hour of the day. The accompanying engraving is a tracing of the Accademia's, in simple lines and only slightly distorted, including their restored gnomon, and with the orthography and position of the inscription properly reproduced to maintain the association between the calendrical abbreviations and their vertical lines (Figure 3.18). Unlike Calkoen, Woepcke offered no illustrations of the sundial's geometry, though he did so for many of the other equations with which his text is replete.

Woepcke offered only two equations for analyzing the prosciutto's dial. The first was essentially a slight variation on Delambre's equation for establishing the angles of the individual temporal hours throughout the year.[85] Woepcke also provided the specific equation for determining "τ," the temporal hour, which Delambre had not specified. This involved finding the arccos of the negative tangent of the latitude multiplied by the tangent of the declination and multiplying this sum by the fraction representing the temporal hour, from zero to six.[86]

Woepcke's principal new contributions to the analysis of the prosciutto came from his interest in relating the azimuth of its dial to that of the sun, a consideration no earlier scholar had made, as well as his reassessment of the Accademia's determination of the obliquity of the ecliptic. Concerning azimuth, he was interested in the fact that, in taking readings over the course of the day, both the tip of the gnomon and the grid of the sundial had to be the rotated around the axis of suspension while following the changing azimuth of the sun as it moved across the sky. This differed from the relationship, for example, between a horizontal sundial whose position was fixed in relation to the changing azimuth of the sun. This constituted an attempt to capture this relationship mathematically, putting a different twist on how to conceive of

---

85. Woepcke used arcsin instead of Delambre's sine to determine solar altitude, also known as the zenith distance. The resulting equation for the altitude of the sun/zenith distance = arcsin $(\sin \delta \cdot \sin \varphi + \cos \delta \cdot \cos \varphi \cdot \cos \tau)$. The equation for zenith distance is commonly expressed today as: $z_H = \cos^{-1}(\sin \delta \cdot \sin \varphi + \cos \delta \cdot \cos \varphi \cdot \cos \tau)$; see, e.g, Gibbs 1976, 40.

86. The full equation for τ, where n = 0–6 for the temporal hours (0 is meridiem, 1 is the 5th and 7th hours, etc.) = $\dfrac{n}{6}$ arccos $(-\tan \delta \cdot \tan \varphi)$. The equation is commonly expressed today as: $A_H = \dfrac{n}{6} \cdot \cos^{-1}(-\tan \delta \cdot \tan \varphi)$; see, e.g., Gibbs 1976, 40.

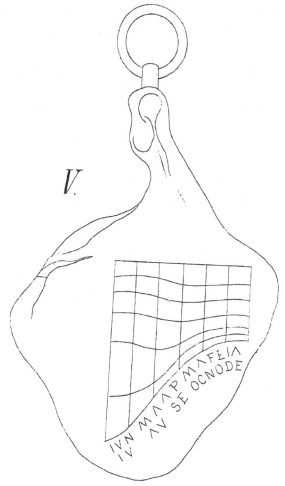

FIGURE 3.18 Franz Woepcke's (1826-1864) prosciutto. Figure 5 from his *Disquisitiones archaeologico-mathematicae circa solaria veterum* (Berlin 1842).

the prosciutto's operation in the real world, but it did not provide additional information about the design of its grid or clarify earlier scholarship. Moreover, Woepcke's presentation was theoretical in that he offered the formulae for such analysis but no data based, for example, on the Accademia's design latitude, and no subsequent scholars pursued this particular avenue of inquiry.[87]

---

87. Woepcke's equations are based on a spherical coordinate system. "Let the azimuth of the sun and of the apex 'C' differ while they are related to the axis of suspension with the arbitrary angle 'A.' Let us imagine furthermore a system of polar coordinates whose center is 'C,' whose initial plane is a plane located through 'C' and the axis of suspension; whose initial radius is a

208 THE PROSCIUTTO SUNDIAL

In his recalculation of the obliquity of the ecliptic, Woepcke first noted that the Accademia had determined the prosciutto had been designed for a latitude of 41° 39′ 45″, while Calkoen had revised this to 41° 38′ 18″. He did not state which of these he accepted, and the value is not expressed in his equations. The Accademia had lauded the remarkable subtlety with which the prosciutto had been made in using this value for the design latitude to determine that the obliquity of the ecliptic at the time of its manufacture had been 23° 46′ 30″. Woepcke observed that Calkoen had looked to other evidence for dating the prosciutto, specifically the inscription designating the months Julius and Augustus, the use of which could not have occurred prior to 27 BCE, and had calculated the obliquity as 23° 42′ 25″. Woepcke agreed with this *terminus post quem* and then factored in the fixed date for its deposition as a consequence of the eruption of Vesuvius, an event he dated to 80 CE. Settling on 27 CE as the midpoint of this period, Woepcke recalculated the obliquity of the ecliptic using Besselian years, a common means of reckoning the precise duration of a year in Woepcke's day.[88] Because a more accurate value

line in this plane from 'C' to the axis of perpendicular suspension; the hourly curves are described on the surface of the sundial with a radial vector which includes in its own plane with the initial radius the angle HCP [= the zenith distance], whose plane includes with the initial plane the arbitrary angle (180° − A). From this it follows:

$$z = r.\ \text{sine HCP} \qquad\qquad = r.\ \varphi\,(\delta)$$

$$y = r.\ \cos \text{HCP.\ sine } A \qquad = r.\ \psi\,(\delta)\ \text{sine } A$$

$$x = -\,r.\ \cos \text{HCP.\ cos } A \qquad = -\,r.\ \psi\,(\delta)\ \cos A$$

then $x^2 + y^2 + z^2 = r^2$ and

$$\delta = \varphi\left\{\frac{z}{\left(x^2 + y^2 + z^2\right)^{\frac{1}{2}}}\right\},\quad \delta = \psi,\left\{\frac{y}{\sin A\left(x^2 + y^2 + z^2\right)^{\frac{1}{2}}}\right\}\ \text{and will be}$$

$$\varphi = \left\{\frac{z}{\left(x^2 + y^2 + z^2\right)^{\frac{1}{2}}}\right\} = \psi,\left\{\frac{y}{\sin A\left(x^2 + y^2 + z^2\right)^{\frac{1}{2}}}\right\}\ \text{or even}$$

$$\varphi = \left\{\frac{z}{\left(x^2 + y^2 + z^2\right)^{\frac{1}{2}}}\right\} = \psi,\left\{\frac{-x}{\cos A\left(x^2 + y^2 + z^2\right)^{\frac{1}{2}}}\right\}.''$$

In general, altitude dials like the prosciutto depend on the sun's altitude not its azimuth.

88. Named for the German astronomer Friedrich Wilhelm Bessel (1784–1846) a Besselian Year is almost identical in length with the tropical year (defined as 365.2421988 mean solar days, the number of days between one vernal equinox and the next) but is defined as beginning at the moment at which the right ascension of the fictitious mean sun, affected by aberration and referred to the mean equinox of the date, is exactly 280°; a Besselian year, designated by the year

for the obliquity over the centuries had been established, Woepcke knew that the Accademia's value was much too great. Even Calkoen's recalculation of this on the basis of the rate of secular diminution employed in his own day remained too high.[89] Woepcke's approach allowed him to calculate a more realistic value since he could work back from the value for the obliquity established for the year 1800, as the equation for working in Besselian years required, and this was $23°\ 27'\ 54.8''$. He employed the equation $\epsilon = 23°\ 27'\ 54.8'' - \{0.457''\ (\tau - 1800)\} - \{0.00000272295''\ (\tau - 1800)^2\}$, where $\tau = 27$. As a result, Woepcke concluded that the true value for the obliquity of the ecliptic for his target year of 27 CE was $23°\ 41'\ 16.5''$.[90]

Calkoen, Delambre, and Woepcke illustrated the growing application of trigonometric equations to explain the relationship of the dial's grid to the movement of the sun through the ecliptic. At the same time, astronomers in the nineteenth century focused increasingly on the celestial realm and less on the terrestrial one, a trend reflected in the near total absence of discussions of ancient sundials in the astronomical literature of the later part of the century.[91] Other than Calkoen's and Woepcke's theses, no major works on ancient sundials appear in Houzeau and Lancaster's *Bibliographie générale de l'astronomie* (Brussels 1882); most of the publications listed under their section on *Gnomoniques* concern modern examples or the general principles of sundial design. In discussing the moveable gnomon on cylinder dials in their introductory survey, however, they refer briefly to the prosciutto, their characterization of it evidently influenced by Lalande's notion that it had just

preceded by a "B" and followed by a ".0," is one tropical year minus -0.148 seconds times the number of centuries from, e.g., B1900.0; J. D. Mulholland, "Measures of Time in Antiquity," *Proceedings of the Astronomical Society of the Pacific* 84.499 (June 1972) 358. Besselian years formed the basis of Besselian epochs, which were replaced by Julian epochs in 1984.

89. Based on the calculator provided by PHP Science Labs http://www.neoprogrammics. com/obliquity_of_the_ecliptic/, the Accademia's value would date the prosciutto to the year 665 BCE; Calkoen's would date it to 100 BCE.

90. Converting to decimal degrees, the equation is $\epsilon = 23.46522° - (-0.2250724212) - 0.00000000000179842281354576562 5 = 23.690292421198201577186454234375° = 23°\ 41'\ 25.05''$. This obliquity is closer to the year 37 CE, not 27 CE; Woepcke's value likewise is closer in date to 56 CE. The true value for 27 CE is $23°\ 41'\ 29''$ ($23.6914°$).

91. F. Arago's *Astronomie Populaire*, 4 vols. (Paris 1854) 1.44, noted that the precise date of the introduction of sundials to Rome is uncertain and that elites relied on slaves to provide them with the time of day, but he did not describe any of the extant examples. M. Hoefer in his *Histoire de l'astronomie depuis ses origins jusqu'à nos jours* (Paris 1873) also ignored ancient sundials, declaring (245) that "Vitruvius only dealt in his book *De architectura* with questions of mechanics. But what we have quoted above from him, relative to the theory of the stations and retrogrades of the planets, does not give us a high idea of his astronomical knowledge." C. Flammarion, *Astronomie Populaire* (Paris 1881), did not discuss sundials.

THE PROSCIUTTO SUNDIAL

such a mutable gnomon: "Such was the silver dial found, in the first half of the last century, among the ruins of Herculaneum."[92]

Errors and misconceptions like this continued to find their way into the literature on the prosciutto despite a half century's worth of scholarship. In the same year the first volumes of Delambre's *Histoire de l'Astronomie* appeared, William Archibald Cadell made a passing reference to the prosciutto sundial, but it appears he failed to recognize its porcine origins. Cadell (1775–1866) was a wealthy Scottish industrialist whose independent means allowed him to pursue his scientific and antiquarian interests. A member of both the Royal Society of London and the Royal Society of Edinburgh, he published a paper in 1818 in the latter society's *Transactions* entitled "On the lines that divide each semidiurnal arc into six equal parts," in which he refers to the prosciutto sundial in passing. While he clearly lacked first-hand knowledge of the sundial, he knew of the Accademia's treatment of it but perhaps did not read its Italian text since he does not mention either its distinctive form or that the tail functioned as the gnomon, the latter detail seemingly essential to his observation about the mutability of its azimuth: "In the description of the antiquities of Herculaneum there is represented and explained a dial whose azimuth is changeable, to suit the hour and the different declination of the sun. It is drawn on an irregularly curved surface of bronze, and the declinations are marked with the Roman names of the months."[93]

Among the more obscure treatments of the prosciutto appeared at roughly this same time in Domenico Romanelli's (1756–1819) *Isola di Capri: Manoscritti inediti* (Naples 1816). The principal manuscript for which Romanelli offered a text and commentary here was the *Descrizione dell' Isola di Capri* penned by Count Gastone della Torre Rezzonico (1740–1795) after a visit to the island in 1794, accompanied by Romanelli.[94] An outing to the

---

92. J.-C. Houzeau and A. Lancaster, *Bibliographie générale de l' astronomie* (Brussels 1882) 164–65, 1272–310 (*Gnomonique*).

93. W. A. Cadell, "On the Lines that Divide Each Semidiurnal Arc into Six Equal Parts," *Transactions of the Royal Society of Edinburgh* 8 (1818) 61–81, 77 (prosciutto). His travels through Italy in 1817–1818 evidently did not extend as far south as Naples, so he would not have had first-hand knowledge of the prosciutto: W. A. Cadell, *A Journey in Carniola, Italy and France in the Years 1817, 1818*, 2 vols. (Edinburgh 1820). He also read a paper entitled "Hour Lines of the Ancients" to the Royal Society of London in July 1830 (*Archives* AP/10/4), a precis of which was given in the *Abstracts of the Papers Printed in the Philosophical Transactions of the Royal Society of London* 3 (London 1837) 18–19.

94. A. Chalmers, *The General Biographical Dictionary*, vol. 26 (London 1816) 166–67, under the entry for his father, Anthony Joseph Rezzonico. Rezzonico was considered one of the best poets of his age, appointed by Ferdinando, the Bourbon Duke of Parma, as president of the Accademia di Belle Arti, but he had been obliged to leave Parma after undisclosed indiscretions. He settled in Rome and traveled frequently to Naples.

Enlightenment

Grotta di Matromania, a cave on the east side of the island said to have served as a shrine of Mithras, had triggered a detailed explication of Mithraic imagery on a marble relief in Rome.[95] In the course of this Rezzonico referred to a sundial fashioned in a cave on the island of Syros by the sixth-century BCE cosmographer Pherecydes. According to tradition, the cave's opening had been oriented toward the south, and Pherecydes had used the sun's rays entering the cave to mark the solstices.[96] Rezzonico noted that, while this was a simple and early method for marking time, there were earlier examples, known from the Babylonians and, especially, the Greeks who had adopted the gnomon as a means of tracking the sun's shadows, as Pollux had explained in his *Onomasticon*. Romanelli, an abbot and prefect of what later became the Biblioteca Nazionale in Naples, with a passion for Neapolitan antiquities, jumped at the opportunity to expand on the topic of ancient sundials with a detailed explanatory note featuring a famous example with a local pedigree: the prosciutto.[97] His presentation is merely a condensation of the Accademia's, however, with no revisions to reflect the subsequent scholarship and no new observations of his own. His principal contribution was an engraving, doubtless derived from the Accademia's, that showed a distorted prosciutto, its thigh excessively wide and bulging; the leg too short; the gnomon too long and curving, but neither broken off nor fully reconstructed; the grid overly rectilinear and insufficiently undulating; and the lettering of the inscription straightened out rather than following the dial's curve, with the abbreviations positioned incorrectly, and with Aprilis abbreviated "AD" (Figure 3.19).

---

95. Rezzonico's principal source was B. Montfaucon's *L'Antiquité expliquée et représentée en figures*, 5 vols. (Paris 1719) 1.2.373–84, plate 215, fig. 4.

96. The main reference is Diogenes 1.119 ("His sundial is also preserved in the island of Syros"), but other details about his sundial appear to be later embellishments, if not entirely apocryphal. Among the links drawn between Pherecydes' sundial and Syros is one taken from Homer's *Odyssey* 15.403–4, where Syros is said to mark "the turnings of the sun"; see M. Munn, *The Mother of the Gods, Athens, and the Tyranny of Asia: A Study of Sovereignty in Ancient Religion* (Berkeley 2006) 47–51. Rezzonico had in fact referenced the Homeric lines in his discussion, arguing that Pherecydes must have lived earlier because Homer knew of this cave.

97. Little is known about Romanelli; most biographical details are derived from his own publications. His best-known works include *Antica topografia istorica del regno di Napoli* (Naples 1815) and *Viaggio a Pompei, a Pesto e di Ritorno ad Ercolano ed a Pozzuoli* (Naples 1817). For his activity as a librarian, see V. Trombetta, "La biblioteca napoletana della 'Croce di Palazzo' nel Piano di Organizzazione dell' abate Romanelli," *La Specola: Annuario di Bibliologia e Bibliofilia*, anno II (1992–93) 169–98, and V. Trombetta, *L'Editoria a Napoli nel Decennio francese: Produzione libraria e stampa periodica tra Stato e imprenditoria privata (1806–1815)* (Milan 2011) 26–27.

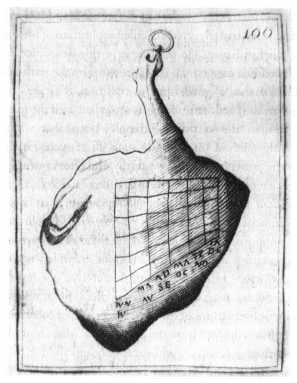

FIGURE 3.19 Domenico Romanelli's (1756–1819) prosciutto. Figure 100 from his *Isola di Capri: Manoscritti inediti* (Naples 1816).

3.f) The Dawn of "Dialling"

As the prosciutto disappeared from the pages of astronomical histories, it continued to feature in encyclopedias where the transition of ancient sundials into the domains of mathematics and horology was noted early in the nineteenth century in the context of the study of "dialling." Dialling, a science dating from at least the seventeenth century, was defined broadly as "the method of constructing sundials...a branch of mixed mathematics which depends partly on the principles of geometry and partly on those of astronomy."[98]

---

98. *Edinburgh Encylopedia* (Edinburgh 1830) 7.691, s.v. *Dialling* (W. Wallace); William Wallace (1768–1843) was professor of mathematics at Edinburgh University and an astronomer, elected in 1804 to the Royal Society of Edinburgh. A popular British manual on dialling was W. Leybourn's *The Art of Dialling* (London 1669).

Innumerable treatises have been written on [sundials] both scientific
and practical; many of the former are distinguished for the elegant
application of geometry; and the subject on the whole has been treated
so variously and with such details as to now be completely exhausted.
But it is still an object of interest to the mathematician, and not without utility to the artist, as the dial is still employed to regulate the
clock.[99]

As with the differing editions of the French *Encyclopédie*, the prosciutto's
presence in these encyclopedias depended on the edition and the training and
interests of the entry's author, but by the late nineteenth century ancient sundials in particular were increasingly viewed as quaint relics about which little
more could be said.

In the eighteenth century clocks and watches began to supersede sundials and these have gradually fallen into disuse except as an additional
ornament to a garden or in remote country districts where the old dial
on the church tower still serves as an occasional check on the modern
clock by its side. The art of constructing dials may now be looked upon
as little more than a mathematical recreation.[100]

Nevertheless, the prosciutto appeared in the 1830 edition of the *Edinburgh
Encyclopedia* at the conclusion of an historical survey, clearly modeled on
those found in earlier astronomical treatises, that included the usual anec-
dotal references to the parasites in Bato and Plautus, the Vitruvian catalog,
and the inventory of extant ancient sundials from Italy.

A very curious portable dial was dug out of the ruins of Portici in 1755
and described in the preface to the third volume of the description of
representations found in the ruins [*Le Antichità*]. Its shape is that of a
bacon ham; and it is suspended to a ring fastened to the leg. The end of

---

99. *Encylopedia Edinensis, or Dictionary of Arts Sciences and Literature*, 6 vols. (Edinburgh
1827) 2.769, s.v. *Dialling*.

100. *Encyclopedia Britannica: A Dictionary of Arts, Sciences and General Literature*, 9th ed.
(New York 1878) 7.154, s.v. *Dialling* (the American edition of the version originally printed in
Edinburgh). This entry was penned by Hugh Godfray (1821–1877), astronomer and fellow at
St. John's College, Cambridge. This edition did not mention the prosciutto, and it is missing as
well from the earlier 1810 and 1823 editions. No mention of it is made in the *Encyclopedia
Londinensis* (London 1811) 10.287–369, s.v. *Horology*.

the tail, which has been preserved, serves as a stile. Its figure is shewn in Plate CCXXVIII. Fig. 2. The hours are marked on that part of the surface which is nearly plane. There appears to be seven vertical lines intersected by as many others, and below these intervals the names of the months are written. Such as understand dialing will readily see how this dial is to be used. It must be suspended by the ring, and turned slowly round, until the shadow of the top of the stile fall [*sic*] upon the line of the month, the hour will then be indicated by the nearest transversal line.[101]

Two clues reveal that the author had relied on a single source for this woefully abbreviated treatment of the prosciutto. The phrase "Such as understand dialing" echoes "People a little versed in gnomonics will easily recognize how one used this dial," while the figure referenced in the text is to an engraving showing a reverse-image of the prosciutto; both are derived from Montucla's 1799 *Histoire* (Figure 3.20). In this case, however, the inscription compounds the error in having IAnuarius inexplicably abbreviated as "IV," while the figure's caption gives the provenance as "Portico." This version of the prosciutto would have seen substantially more popular circulation than the one in Montucla's more scholarly tome. The 1842 edition of the *Encyclopedia Britannica*, on the other hand, relied on Delambre's 1817 *Histoire* for its brief dismissal of the prosciutto's complex design:

> To the ancient dials which have been noticed here we may add a singular one found at Portici in 1755 and described by the academicians of Naples. It has the figure of a bacon ham and, like all the others, shows temporary hours. The theory of this dial is simple, but, considering the imperfect trigonometry of the Greeks, its construction by calculation might be laborious; probably it was made by the simple process, already described, from the Chaldean dial [Berosus' hemicyclium dial].[102]

In transitioning to the community of dialers it was appropriate that the prosciutto caught the attention of the British Horological Institute. Their

---

101. *Edinburgh Encylopedia* (Edinburgh 1830) 7.693, s.v. *Dialling* (W. Wallace).

102. *Encyclopedia Britannica: A Dictionary of Arts, Sciences and General Literature*, 7th ed. (Edinburgh 1842) 7.758, s.v. *Dialling* (H. Meickle). Henry Meickle, Esquire, was a civil engineer.

Enlightenment

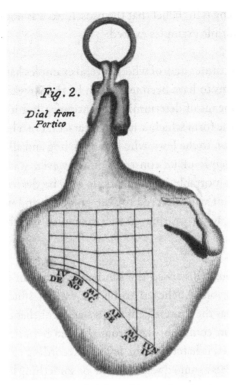

FIGURE 3.20 The prosciutto sundial from "Portico" as it appeared under the entry "Dialling" in the seventh volume of the *Edinburgh Encyclopedia* (Edinburgh 1830).

official organ, the *Horological Journal*, commenced publication in 1858, the same year in which the Institute was founded. The prosciutto was featured in the first issue of September 1858 in a running column entitled "What is Horology?" written by William Hislop (1820–1876).[103] In his previous column, Hislop had begun to catalog the examples from Tusculum, Rome, and the quadruple-faced sundial of Phaidros now in the British museum. Here, under the subtitle "The earlier sundials, and the principles of dialing," Hislop resumed his discussion with a summary of Martini's description of the 1762 Pompeian sundial, and this brought him to the prosciutto. His treatment is notable for the degree to which he bungled crucial details while injecting his own errors and resurrecting misconceptions that had been put to rest decades

---

103. *The Horological Journal* 1.1 (September 1858) 28. Hislop was a clockmaker and the Institute's honorary secretary. Excerpts continued to appear through the January 1, 1860, issue of the second volume. The *Journal* claims to be "the world's oldest monthly technical journal published continuously since" this first issue of 1858.

THE PROSCIUTTO SUNDIAL

earlier. Most striking is his belief that the prosciutto was not unique but was a type of which multiple examples existed.

> There is a very curious dial, of which several examples have been found, and which seems to have been used both by the Greeks and Romans. There are no means of determining its antiquity, but it is clear that it could not, in the form which it usually bears, that of a ham (see fig. 5), have been known to the Jews, who hated the hog and all that belonged to it. The principle of its construction, however, was applicable to other forms, as portable instruments. It may be described simply as a ham, the tail of which served for the gnomon, and which was furnished with a hook or ring at the extremity for the purpose of suspension.

Hislop had previously discussed the sundial of Achaz in the *Book of Kings* (2 Kings 20:8–11), often believed to be the earliest sundial referenced in literature.[104] This was the context behind his statement that the type of sundial under consideration could not date from that era because its porcine form would have rendered it hateful to the Jews.[105] According to Hislop, while the usual form these dials assumed was the ham, or something looking like a ham, the elements constituting a portable sundial theoretically could be applied to any shape. He then provided a reasonably succinct and clear explanation of how the gnomon and dial worked together:

> The dial is delineated on the back of the ham, on which are described six vertical lines, under which are abbreviated the names of the twelve months, beginning with January, retrograding to June, and again returning to December. Six horizontal lines traverse the vertical ones, and by their intersection show the extension of the shadow thrown by the gnomon on the sun's entering each sign of the zodiac, and consequently at every point of his path through the ecliptic. This also points out the hours of the day, the shadow descending with the rising and again ascending with the setting sun, the square compartments being

---

104. The literature on the "miracle" of Achaz' (or "Ahaz") sundial is enormous. For an early scientific explanation, see J. Morrison, "The sundial of Ahaz," *Popular Astronomy* 6 (1898) 537–49, and the detailed discussion in Rohr 1970, 8–9.

105. The remark recalls Mazzocchi's refusal to accept that such sophisticated scientific knowledge could be applied to so "vile" a creature as a pig, but it is otherwise entirely Hislop's own notion.

marked with the hours. It seems, that when it was in use it was suspended by the hook or ring, the side being presented to the sun, and that when the extremity of the shadow of the gnomon reached the extremity of the line marked with the name of the actual month, the horizontal intersection showed the hour. There are several points of obscurity as to details, which could probably only be cleared up by the inventor. This dial was found at Herculaneum in 1754, and a similar one was found at Portici in 1755.

The accompanying illustration clarifies this last, curious observation, that there were at least two prosciutto sundials, found at different times at different sites, which had led Hislop to believe it was a type of which multiple copies were reproduced in different shapes. Hislop earlier had referenced Martini's 1777 treatise on sundials and so should have known Martini's discussion of the prosciutto and seen his reproduction of the Accademia's engraving, but he clearly also had obtained a copy of Paciaudi's engraving from the *Monumenta Peloponnesia* with its "carafe" said to have been found in Herculaneum in 1754. It was the latter he chose not only to reproduce but even to enhance by emphasizing what Mazzocchi had identified as "magical or Coptic characters" inside the grid and Hislop evidently believed were symbols marking the hours (Figure 3.21). In looking to the inventor to explain his design, Hislop reveals that, despite seeming to understand the principles of its design, he was clearly as perplexed as Paciaudi had been in perceiving how the upright gnomon could cast its shadow across the entire dial. In the end, he had not even tried to match his explanation to his illustration, leaving it to his readers to puzzle it out.

3.g) *The Prosciutto's Peregrinations and the Horologists of the Nineteenth Century*

By now the prosciutto had been transferred from Portici to the Real Museo Borbonico in Naples. This new royal museum had been conceived originally in 1777 but was fully realized only between 1817 and 1819, after the return of King Ferdinand IV from Palermo, where he had fled during the French occupation of his kingdom.[106] As one of the finds deemed among *i pezzi piu*

---

106. A. Milanese, "Il Museo Reale di Napoli al tempo di Giuseppe Bonaparte e di Gioacchino Murat: La prima sistemazione del 'museo delle statue' e delle altre raccolte (1806–1815)," *Rivista dell' Istituto Nazionale d'Archeologia e Storia dell'Arte*, S. 3, XIX–XX (1996–1997) 345–405.

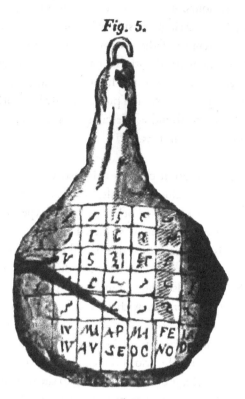

FIGURE 3.21 William Hislop's prosciutto sundial from his "What is horology" article in *The Horological Journal* 1.1 (September 1858).

singolari (the most unique pieces), the prosciutto had accompanied the king in his flight in its first departure from the foot of Vesuvius in seventeen centuries. The antiquities had been packed into sixty crates, the artifacts from each room being boxed together systematically. Only the paintings had been left behind in Portici.[107] So it was that the prosciutto landed in crate 41, together with the other precious finds displayed beside it in cabinets 36 and 37 of room 10 in the Herculanense Museum. Inside the crate, two smaller, custom-designed boxes of walnut cradled the cameos, gold rings, bracelets and *bullae*, silver cups and spoons, the famous silver mirror in bas relief depicting the

---

107. R. Paolini's *Memorie sui monumenti di antichità e di belle arti ch'esistono in Miseno, in Baoli, in Baja, in Cuma, in Pozzuoli, in Napoli, in Capua antica, in Ercolano, in Pompei, ed in Pesto* (Naples 1812) 221, 227–29. Like Cicero's imagined tourist guides in Sicily after Verres' plundering of its antiquities, Paolini had to resort to explaining the objects that "were" in the rooms in the museum at Portici prior to their departure for Palermo. He describes the finds in the ninth room in detail but does not mention the prosciutto.

"death of Cleopatra" (MANN 25940), and the gold coin of Augustus and Diana (MANN 3692) featured in the preface to the second volume of *Le Antichità*. Room 10's centerpiece, the drunken satyr reclining on a wineskin from the Villa dei Papiri (MANN 5628), traveled alone in crate 42. An inventory correlating the individual crates with their room numbers dates to December 30, 1798—only days after the king sailed off on stormy seas.[108] Precisely when the crates were transported to Palermo and whether the antiquities were unpacked once they reached their destination remains unclear. Another inventory originating in Palermo and dating to March 8, 1808, catalogs the contents of each crate, and these numbers correspond precisely with the inventory of 1798. The entry for the prosciutto reads, "a small sundial of bronze, sheathed in silver, in the form of a prosciutto." This inventory may have served as the bill of lading for the return of the crates to Naples, which must have occurred by the time the king returned to Naples in 1815.[109] Nevertheless, the prosciutto would not see the light of day again for some time as the process of organizing the exhibitions in the new museum, begun

---

108. The harrowing, deadly journey is retailed by H. Acton, *The Bourbons of Naples, 1734–1825* (London 1956) 314–20.

109. ASN, *Archivio Borbone, Inventari diversi*, Busta 304: "Inventario di tutto ciò che esiste in Palermo dei Reali Musei, Ercolanense, Capodimonte, Studi Vecchi ed altro di Napoli ordinato da farsi con R(eale) Dispaccio, 1807." The 1787 inventory is entitled "Nota delle Casse contenenti i pezzi piu singolari di questo Real Museo Ercolanese, e caratteristiche colle quali restono contrasegnate, e queste corispondono all'Inventario col quale furono consegnati i monumenti di questo Real Museo al Sig.r Colonn.o D. Francesco la Vega mio fratello Direttore del med.o ed in assenza di esso Sig.r Colonn.o si fa da me sotto consegna di n.o cinquantanove ben condizionate al Sig.r D. Pirro Paderni, p[ri]mo Ajutante di questo Real Museo e dette Casse si estraggono da questo Real Museo in vigore di dispaccio dell'Ecc.mo M.se Demarco con data de 30. dello scorso dicembre 1798, comunicato a S. E. Il Sig.r Brig.re Cav.re Macedonio, e da questi al Sig.e Colonn.o mio fratello," signed but not dated by Pirro Paderni, Camillo's son. The 1808 inventory is entitled "Inventario generale di tutto ciò, ch'esiste in Palermo dei Reali Musei, cioè Ercolanese, di Capodimonte, dei Regj Studi Vecchj, Vasi Etruschi, Quadri di Francavilla, e di Capodimonte, e Codici Manoscritti; ordinato farsi da S. M. /D. P./ con R.l Dispaccio dei 25. Luglio del passato Anno 1807. per via della R.l Segreteria di Stato, e Casa R.le, e fattane la consegna dall'Ill.mo Sig.r Girolamo Ruffo, Controloro della R.l Casa, al Custode del Museo Ercolanese D. Pirro Paderni in esecuzione del riferito veneratissimo R.l Comando della Maestà Sua." The precise chronology of the departure and return of the antiquities is unclear. An addendum to the 1808 inventory attests that the younger Paderni took an additional twelve crates to Palermo in 1806, while a letter signed by Pietro La Vega, son of Francesco, indicates that all the antiquities still remaining at Portici were headed to Naples by April 11, 1806 (ASN, *Casa Reale Antica*, Primo Inventario 1529, 31/fol. 13). For this period in the history of the museum, see also Allroggen-Bedel and Kammerer-Grothaus 1983, 91–92. See also A. Filangieri di Candida, "Monumenti ed oggetti d'arte trasportati da Napoli a Palermo nel 1806," *Napoli Nobilissima* 10 (1901) 13–15, deals only with the inventory of the twelve crates transferred in 1806, which is deemed "poco preciso" and of no importance for the identification of the objects.

already under French rule, unfolded over the course of many years; the last paintings were transferred from Portici to Naples only in 1827.

The museum in Naples was intended to bring the antiquities of the Herculanense Museum together with the sculpture and painting collection of the Museo Farnesiano di Capodimonte under a single roof, the seventeenth-century Palazzo degli Studi in Naples. Housed here as well were the royal library, the Accademia di Disegno, and the royal academies with the goal of creating a "palace of culture."[110] One of the four principal galleries on the second floor of the new museum was the "Gabinetto degli Oggetti Preziosi," a substantial amplification of the room of precious objects in the Herculanense Museum. Collected here were all the gold and silver objects, the cameos, and the carbonized food from the excavated cities. The prosciutto was displayed in the first case along with 208 other objects in silver.[111] The official inventory of the museum dating to 1848 ascribes it to Herculaneum and describes it as "A sundial in bronze covered in silver, five *once* high by three and three-fifths *once*. It has the shape of a prosciutto missing the posterior extremity (*la estremità posteriore*) that served as the gnomon. The lines indicating the hours are drawn on the rind side, under which the months of the year are read."[112]

Among the more comprehensive guidebooks to the new museum from this period is that of Stanislao d'Aloé from 1854. D'Aloé was secretary to the director of the museum and a corresponding member of the revamped

---

110. A. Milanese, "Exhibition and Experiment: A History of the Real Museo Borbonico," in E. Risser and D. Saunders eds., *The Restoration of Ancient Bronzes: Naples and Beyond* (Los Angeles 2013) 13–29. The French had advanced the project to create the new museum, which constitutes the present Museo Archeologico Nazionale di Napoli. For a brief history of the museum with an eye toward the finds from the Villa dei Papiri, see also Mattusch and Lie 2005, 75–90.

111. Among the earliest detailed guidebooks to the new museum to list the sundial was that of Edward Gerhard and Theodore Panofka, *Neapels antike Bildwerke* (Stuttgart 1828) 438: "Sonnenuhr von Bronze."

112. Museo Archeologico Nazionale di Napoli, *Inventorio Sangiorgio* (1848), Oggetti Preziosi #126: "Orologio solare di bronzo foderato in argento, alto once cinque per once tre e quattro quinti. Ha la forma di presciutto, mancante della estremità posteriore che ne forma il gnomone; dalla parte della cotenna sono tirate le linee indicanti le ore, sotto le quali si leggono i mesi dell' anno." A contemporary guide, F. Alvino's *Guide du Musee Royal Bourbon* (Naples 1840) 86, reduced it to "among all these [objects] one should note the sundial in the shape of a ham; it is of bronze covered with a thin layer of silver and has a small ring which served to suspend it."

Accademia Ercolanese d'Archeologia.[113] His description of the sundial is the most detailed available in a popular guide from this period.[114]

> Among the bronzes from Herculaneum, this sundial caught the attention of the archaeologists. On the basis of the visible traces which still remain, one judges it was of silver. The Academicians of Herculaneum who have discussed it so sagely have given it the name prosciutto for its shape and for the inequalities of its surface. The tip of the tail cast toward the front served as the gnomon to mark by its shadow the days and the months traced by its lines of which seven were vertical and parallel.

Despite the intervening century of scholarship, d'Aloé continued by offering only a recitation of the Accademia's analysis, right down to the calculations of Carcani and the date of 28 CE. He concluded by observing that, "Of all the sundials published so far this one is the most singular for its vertical form and for its lines which indicate the course of the sun in the twelve signs of the zodiac and the twelve hours of the day with the greatest accuracy."[115]

By this time the prosciutto jostled for attention with 231 other objects in the display case, all unlabeled apart from the occasional inventory number, many of them larger and shinier, and with their imagery and function more readily comprehensible to the average visitor.[116] Yet it naturally continued to generate curiosity. When a marble hemicyclium sundial inscribed in Oscan was recovered in the Stabian Baths in Pompeii in 1854, Bernardo Quaranta (1796–1867), the recently appointed perpetual secretary of the Accademia Ercolanese, attributed its design to Berosus, the first of the ancient astronomers named in Vitruvius' catalog (Vitr. 9.8.1). He may have hoped this would

---

113. Castaldi 1840, 45–48.

114. The director of the museum and president of the Accademia Ercolanese in 1854 was the numismatist Domenico Spinelli (1788–1863), and he and d'Aloé also oversaw the royal excavations; see *Almanacco Reale del Regno delle Due Sicilie per l'Anno 1854* (Naples 1854) 78.

115. S. D'Aloé, *Nouveau guide du Musée royal bourbon* (Naples 1854) 99–102. G. B. Finati, *Le Musée Royal-Bourbon* (Naples 1843) 31, offered only a brief description of the prosciutto but in the context of the remarkable collection of antiquities surrounding it. Cf. L. Giustiniani, *Guida per lo Real Museo Borbonico*[2] (Naples 1824) 96–97, in Italian with facing pages in English: "[in the third cupboard with 159 objects] A Sundial of bronze, plated with silver."

116. The prosciutto's current inventory number (MANN 25494), painted in red on the instrument's reverse, was assigned during Giuseppe Fiorelli's tenure as director of the museum (1863–1875).

serve to distinguish it from the other sundials in the Naples museum in marble and, especially, bronze, "like that one with the bizarre shape of a prosciutto that claims more attention of viewers than the others."[117]

The establishment of the Real Museo Borbonico constituted a 180-degree change in attitude toward the royal collection of antiquities. The museum in Portici had restricted access to the public, prohibited drawing and taking notes, and maintained strict control, through the agency of Paderni and his successor, over which antiquities were displayed and what details about them were divulged to visitors; there were no guidebooks. The new museum encouraged broad public access to a substantially larger number of the finds and encouraged artists to learn from, sketch, and be inspired by the collections on display. The proliferation of guidebooks like d'Aloé's was one byproduct of this new approach, as was its correlate the greatly enhanced familiarity with the antiquities by a larger audience.

The prosciutto was among the beneficiaries of this improved accessibility, as Quaranta's observation demonstrated. In 1872, Margaret Gatty, aka Mrs. Alfred Gatty (1809–1873), a British author of popular children's books, published the first edition of her *The Book of Sun-Dials* (London 1872). This was a compendium of inscribed mottos, of a primarily moralizing character, read on sundials in Britain, the continent, and farther afield, that Gatty had begun to compile in 1835.[118] This first edition did not mention the prosciutto; it did not even rank among those featured in an addendum of "Further Notes on Remarkable Sundials." In most cases Gatty explained the context of these sundials, but she did not treat their design and operation, pronouncing in the first sentence of this first edition that, "If anyone should open these pages expecting to find in them an astronomically scientific account of sundials...he will be disappointed." The prosciutto first appeared in the expanded "Further Notes" of the second "new and enlarged" edition of 1889, edited by

---

117. B. Quaranta, *L'Orologio a sole di Beroso scoperto in Pompei addì XXIII di settembre MDCCCLIV* (Naples 1854) 8. The marble sundial is MANN 2541. Its Oscan dedication helps date it to the first century BCE, making it the oldest known sundial from Pompeii. See Gibbs 1976, 291; Jones 2016, 77, fig. III-13; Pagliano 2018, 88–113. The attribution to Berosus is not universally accepted.

118. Because she did not travel, Gatty had relied on contributions sent to her by acquaintances. *Dictionary of National Biography* (London 1908) 7.946–48, s.v. *Gatty, Margaret* (G. C. Boase): her most popular book was *Aunt Judy's Tale* (1858), but she was also a renowned algologist and published a book on British seaweeds, for which see J. A. Bryant et al., "Life and Work of Margaret Gatty (1809–1873) with Particular Reference to British Seaweeds (1863)," *Archives of Natural History* 43.1 (April 2016) 131–47. Her book on sundials, published just before her death, is largely ignored in her biographies and bibliographies, despite its popularity.

Gatty's daughter Horatia K. F. Eden together with Eleanor Lloyd after Gatty's death. Under their editorial stewardship, the book's original aim was expanded to include more details on extant sundials and their design. Among the "interesting specimens of ancient dials" in the museum at Naples there was one that "has often been noticed."

> It was found at Herculaneum in 1754 [sic] and is of bronze faced with silver, shaped like a ham, and engraved on one side with seven lines enclosing spaces under which the Latin names of the months appear abbreviated. These are again divided into six irregular spaces by cross-lines for the hours, and a small gnomon, shaped like a tail, projects over one side; the dial hangs from a small ring. It was for some time thought to be the only existing specimen of a portable Roman dial, but this is not the case; for another was brought to light not long ago at Aquileia, near the Gulf of Trieste, and belongs to Dr. Gregorutti of Bapieriano.[119]

The popularity of Gatty's book was such that two revised editions followed. The third, "enlarged" edition, published in 1890, contained an "Introduction to Addenda" in which Eden explained that, "Before passing on to the dials with mottoes we will introduce a drawing of the portable dial found at Herculaneum, which Miss Lloyd lately had the opportunity of making from the original; it is two-thirds of the actual size. The dial is preserved in the Naples museum." Lloyd was an artist credited with providing many of the illustrations in the book, some made during her travels. Her sketch shows a somewhat distorted prosciutto, narrower and slightly elongated, the tail extending out from the left side and pointing up like a raised forearm, the undulating lines of the grid exaggerated, but with the inscription accurately transcribed and positioned (Figure 3.22). Lloyd's drawing appears to be the first rendering of the prosciutto sundial based on first-hand inspection rather

---

119. A. Gatty, *The Book of Sun-Dials*[2] (London 1889) 397. The Aquileia dial was first published by F. Kenner in 1880 in his "Römische Sonnenuhren aus Aquileia," *Mittheilungen der k.k. Central-Commission zur Erforschung und Erhaltung der Kunst- und historischen Denkmale* 6 (1880) 18–20. Kenner described this disk-shaped portable dial and provided a sketch of its obverse and reverse, albeit at a very small scale. Because the obverse is inscribed RO(ma) and the reverse RA(venna), Kenner concluded the two fan-shaped grids, radiating down from near the top of the disk, were designed for those latitudes, which he calculated at 41° 25′ and 44° 25′, respectively. The months on the obverse are abbreviated with their initial two letters, except for three for July, while those on the reverse are single letters; both inscriptions are *boustrophedon*. Long given up for lost, this sundial is in the Civici Musei di Trieste, Italy; see Talbert 2019, 971–81. See also Talbert 2017, 35–36, fig. 2.6.

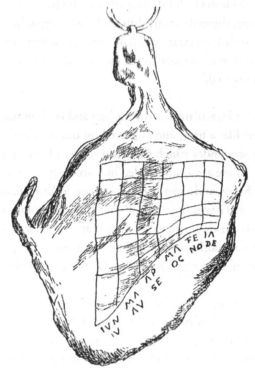

DIAL FROM HERCULANEUM (SCALE ⅔).

FIGURE 3.22 Margaret Gatty's (aka Mrs. Alfred Gatty) (1809–1873) prosciutto, as drawn by Eleanor Lloyd. Gatty, *The Book of Sun-Dials*² (London 1889).

than being derived from an earlier engraving like the Accademia's or word-of-mouth like Paciaudi's. It is also the only known published drawing of the prosciutto by a woman.

The "now enlarged and re-edited" fourth edition of 1900 took the treatment of the prosciutto one step further. While a feature of the second edition had been a new "Appendix on the construction of sun-dials" by J. Wigham Richardson, this last edition included a chapter on "Portable Sundials" written by Lewis Evans, fellow of the Society of Antiquaries and the Royal Astronomical Society. Richardson (1837–1908) was a shipbuilder while Evans (1853–1930) had taken over his father's paper manufacturing business. Both were of independent means and passionate about sundials.[120] Evans'

---

120. Evans' elder brother was Sir Arthur Evans, the archaeologist and excavator of Knossos, Crete, and curator of the Ashmolean Museum at Oxford. His excavations were funded in part by profits from the family business; see *Dictionary of National Biography, 1941–1950* (Oxford 1959) 240–43, s.v. *Evans, Sir Arthur* (J. L. Myres). Their father, Sir John, who founded the

Enlightenment

FIGURE 3.23 The "Sundial Man," Lewis Evans (1853–1930), holding a diptych sundial from his collection. Evans penned the description of the prosciutto in the second edition of Gatty's *The Book of Sun-Dials*[2] (London 1889). Painting in the Frogmore Paper Mill, Hemel Hempstead UK. Image reproduced courtesy of the Apsley Paper Trail.

fascination with sundials, and especially astrolabes, eventually led to his being given the sobriquet "Sundial Man" (Figure 3.23). He founded the History of Science Museum at Oxford University, where the extensive collection of sundials and books he donated now constitute the Lewis Evans Collection in the museum library.[121]

paper manufacturing business, also was an archaeologist with a specialty in numismatics; see *Dictionary of National Biography, Second Supplement* (London 1912) 1.634–37, s.v. *Evans, Sir John* (W. Wroth).

121. Little is known of Evans' biography beyond that provided on the website of the History of Science Museum (https://www.hsm.ox.ac.uk/lewis-evans-biographical-account); also, P. de Clercq, "Lewis Evans and the White City Exhibitions," *Sphaera* 11 (Spring 2000), the museum's online newsletter (https://www.mhs.ox.ac.uk/sphaera/index.htm). For his and his family's involvement with the Frogmore Paper Mill, see https://frogmorepapermill.org.uk/paper-valley-people/.

THE PROSCIUTTO SUNDIAL

Evans opened his contribution by remarking on the variety and "protean" form of portable dials, their widespread use over the ages, and how a "vast amount of time and thought was expended on their design and construction."

There is no doubt that fixed dials preceded portable ones by many ages, and that the length of his own shadow long continued to be the only visible timekeeper that a man carried about with him, and one that was in recognized use in classical times, to which period the earliest known specimen of a portable dial must also be ascribed. This dial, which is made in the shape of a ham, was found in excavations at Herculaneum in 1754 [sic], and is now in the Naples museum, where Miss Lloyd made the drawing from which the illustration is taken.

Its material is bronze, and on its flat side are vertical lines enclosing six spaces, below which are engraved the shortened names of the months, with the winter months under the shortest space and the summer under the longest; while across the upright lines are curved ones dividing the spaces each into six sections to represent six hours from sunrise to noon, and from noon to sunset, in accordance with the plan adopted in other Roman dials, which gave to the day twelve long hours in summer and twelve short ones in winter. The tail-piece on the left must originally have been much longer so as to come round in front of the hour lines in such a way that its shadow would fall on the proper month space and show the hour when the dial was suspended by the rind and turned towards the sun. The age of this dial is fixed within narrow limits by the fact that it must have been made after B.C. 28 when the month Sextilis was changed to Augustus in honour of the emperor, and before A.D. 79 when the great eruption of Vesuvius buried Herculaneum, while it seems probable that it was made after A.D. 63 when the town was greatly injured by an earthquake.

An instrument of this kind could only be used in one latitude, and that the later Romans knew and felt the disadvantage of this fact is shown by another dial on the same principle found a few years ago at Aquileia and described by Dr. Kenner.[122]

---

122. Gatty 1900, 185–86.

Nothing in his treatment indicates what earlier source or sources Evans had consulted, though the erroneous date of 1754 exposes the earlier edition of Gatty's book as one. He certainly had examined Lloyd's drawing, correcting the gnomon's position and elongating it to ensure its shadow fell on the dial. That he did not mention its silver coating or that it was designed according to the zodiacal calendar suggests he did not do much research on it; that the dial adhered to the seasonal hours common to the ancients and worked at only a single latitude were traits shared by the other portable dials he was describing.[123] Evans' principal contribution was his proposal that the date of the prosciutto could be refined to the sixteen-year period between the earthquake that struck the Vesuvian region in 63 CE and the eruption of 79 CE. This was entirely conjectural, however, there being no evidence for confining its construction and use to this relatively narrow post-seismic period.[124]

The popularity of Richardson's manual on sundial design may have inspired Abel Souchon to publish his *La Construction des Cadrans Solaires* (Paris 1905), a short guide on the principles of designing various kinds of sundials. Souchon (1843–1906) appears to be the last in the long line of nineteenth-century French astronomers to write about the prosciutto. Like Lalande and Delambre before him, Souchon was a member of the French Bureau des Longitudes, which still oversaw the Paris Observatory as well as the preparation of the *Connaisance des Temps*, the annual astronomical tables and the successor to Lalande's *Connaissance de Mouvemens Célestes pour l'Annee*. Several years earlier, Souchon had published a lengthy treatise entitled *Traité d'astronomie théorique contenant l'exposition du calcul des perturbations planétaires et lunaires et son application à l'explication et à la formation des tables astronomiques avec une introduction historique et de nombreux exemples numériques* (Paris 1891), so like Montucla he was an astronomer serving in a governmental post. Like several of his compatriots, Souchon could not evade the curse of the prosciutto whose principal symptom was a seriously flawed treatment of it. His book on sundials should have been thoroughly informed by the work of his predecessors, yet he appears singularly oblivious even to the

---

123. Copies of the works of Martini, Calkoen, and Woepcke are in the library of the Lewis Evans Collection in the History of Science Museum, suggesting they originally could have belonged to Evans.

124. The currently accepted date for the earthquake that struck the region is 62 CE, but Evans probably had relied on the date provided in the most popular available discussion of the buried cities, A. Mau's *Pompeii: Its Life and Art* (New York 1899), recently translated into English by F. Kelsey. Seneca (*Quaest.Nat.* 6.1.2) names the consuls for 63 CE; Tacitus (*Ann.* 15.22.2) places the event in 62 CE, but there likely was more than one temblor prior to the eruption.

228   THE PROSCIUTTO SUNDIAL

tradition of generic sundial studies that came before him. "Our little Book," he pronounced in the preface, "is preceded by a History of Gnomonics which we have tried to make as complete as possible. We think that this subject, which has never been treated, could be of great interest to gnomonists and especially to astronomers."[125] The concision of his historical presentation is notable.[126] He managed to arrive at Vitruvius on the third short page and on the fourth began his treatment of *cadrans portatifs*, portable sundials, which he evidently defined as any ancient sundial not permanently affixed to a structure and of a small enough size to be moved. Zuzzeri's sundial from Tusculum, designed according to Berosus the Chaldean, was his first example, and the prosciutto was his other. His description of it is an odd hybrid of the *Encyclopédie's* serrated gnomon and Paciaudi's carafe, the dial consisting of twelve lines with hollow squares but with no mention at all of its prosciutto shape. His sense of obligation to include this "curious" dial in his history evidently was greater than his desire to understand how it was constructed, though like Paciaudi he was clearly perplexed by the design he had before him, which he believed required unspecified "precautions" to make work.

> Shortly after [the 1746 Tusculum dial], in 1755, a small portable dial, all in gilded copper, of bi-convex shape, fairly flattened and thinning upwards, was discovered at Portici, near the ruins of Herculaneum. On one of its faces, crisscrossed with sinuosities, was placed a serrated style whose length was roughly equal to a quarter of the diameter of this dial. The upper surface, at least the one that we can look at as such, was coated with a layer of silver and this is where the style was placed. This dial was divided by twelve parallel lines which formed as many small, somewhat hollow square spaces. The last six squares contained the initial letters of the name for each month. The arrangement of these letters was quite remarkable, in that they went alternately from right to left and from left to right. The different hours of the day were indicated by curved lines which cut the perpendiculars, much like in cylindrical dials. To use this dial, it was necessary to take special precautions which

---

125. A. Souchon, *La Construction des Cadrans Solaires: Ses principes, sa pratique précédé d'une histoire de la gnomonique* (Paris 1905) viii.

126. Turner 1989, 316 n. 13, deemed Souchon's historical introduction "hardly better than that of his seventeenth- and eighteenth-century predecessors, being a farrago of unverified stories and confused chronology."

Enlightenment

would make its use quite inconvenient. This dial is one of the most curious we have ever seen.[127]

The prosciutto's subsequent appearance in two unusual venues underscores the benefits of its enhanced visibility in the Naples museum and how this fostered its familiarity to a wider audience. The August 1909 issue of the journal *Popular Mechanics*, whose tagline was "Written so you can understand it," featured an article entitled "Time and Its Measurement." The author was James Arthur, a skilled machinist trained in Scotland under A. S. Herschel, the grandson of the astronomer Sir William Herschel (1738–1822), who had helped discover Uranus (Figure 3.24). After moving to New York in 1871, he established Arthur Machine Works, a manufacturer of specialized machinery. Arthur (1841–1930), like the industrialists Richardson and Evans, had a passion for clocks and timepieces, and, like Evans, he had collected a substantial number of them.[128] He supplemented his physical collection with extensive travels to see first-hand many other devices for keeping time. His travels had landed him in the Naples museum where he had encountered the prosciutto, which was suspended appropriately from the top shelf in a glass case along with other silver objects, examined it in person, and appears to have either discussed its basic features with local staff or learned about them from a guidebook. He described it in his article immediately after comparing the sundial of Achaz to another in India that employed steps rather than hour lines, though the connections between these and the prosciutto are not explained.

Figure 2 shows a pocket or portable sundial taken from the ruins of Herculaneum and now in the Museo National, Naples. It is bronze, was silver plated and is in the form of a ham suspended from the hock

---

127. Souchon 1905, 5.

128. J. Arthur, "Time and Its Measurement," *Popular Mechanics* (August 1909) 154; reprinted separately as J. Arthur, *Time and its Measurement* (Chicago 1909) 16 (prosciutto), where the frontispiece describes Arthur as "an enthusiastic scientist, a successful inventor and extensive traveler, who has for years been making a study of clocks, watches, and time measuring devices." His collection of over 1,500 timepieces was considered one of the largest in the country at the time of his death. In 1926 he donated this to New York University along with funding to support a foundation, but after 1964 the entire collection was broken up with some pieces and archival material transferred on permanent loan to the Smithsonian to become the James Arthur Clock and Watch Collection, 1743–1967 (NMAH.AC.0130), and what remains now in the National Clock and Watch Museum in Columbia, PA. See also R. G. Aitken, "The James Arthur Foundation," *Astronomical Society of the Pacific Leaflets* 5.238 (January 1949) 313–20; his obituary in the *New York Times* April 28, 1930, "James Arthur Dies in His 89th Year."

FIGURE 3.24 Portrait of the industrialist James Arthur (1841–1930) from his *Time and Its Measurement* (Chicago 1909).

joint. From the tail, evidently bent from its original position, which forms the gnomon, lines radiate and across these wavy lines are traced. It is about 5 in. long and 3 in. wide. Being in the corner of a glass case I was unable to get small details, but museum authorities state that names of months are engraved on it, so it would be a good guess that these wavy lines had something to do with the long and short days.

This final remark, especially, suggests that Arthur did not fully comprehend how the prosciutto functioned and the reader, too, would have been hard-pressed to "understand it" much beyond its unusual shape. The accompanying illustration only complicates matters further (Figure 3.25). He must have drawn it from memory, for while it bears a resemblance to its familiar grocery store form, its finer details are completely botched. Rather than the usual quadrant-shaped grid, the seven lines representing the zodiacal signs radiate out from the upper left-hand corner where a large, sinuous gnomon extends

Enlightenment 231

FIGURE 3.25 James Arthur's prosciutto from his book *Time and Its Measurement* (Chicago 1909).

out to fall on seven undulating horary lines that are entirely inverted compared to the actual grid.[129] The dilettante nature of his knowledge of ancient sundials is further illustrated by his observations on a sundial he encountered in a visit to Pompeii.

> In a restored flower garden, within one of the large houses in the ruins of Pompeii, may be seen a sundial of the Armillary type, presumably in its original position. I could not get close to it, as the restored garden is

---

129. He may have relied on the description provided by D. Monaco in his popular *A Complete Guide to the Small Bronzes and Gems in the Naples Museum according to the New Arrangement* (Naples 1889) 33: "25496: Bronze sun-dial faced with silver, in the shape of a ham. The hours are indicated by radiating lines, across which run irregular horizontal lines. Below these are the names of the months. The tail served as a gnomon (Herculaneum)." Alternatively, he may have conflated it with the dial on small circular portable sundials such as that found in Mainz-Linsenverg now in the Landesmuseum (2321) at Mayence. This has twelve small sockets aligned to the left of a dial that radiates down on the right. A moveable gnomon would be inserted in the socket corresponding with the month. For a photograph, see Bonnin 2015, 141 fig. 37, 395 (A_316).

232 THE PROSCIUTTO SUNDIAL

railed in, but it looks as if the plane of the equator and the position of the earth's axis must have been known to the maker.[130]

In addition to failing to recognize what must have been a traditional hemicyclium sundial, he appears unaware that its ancient fabricator would have been fully versed in the basic principles of astronomy underlying its design.

A second notice appeared at this time in the *Deutsche Tieraerztliche Wochenshrift*, the German veterinary weekly. The short article was entitled "Ham from Herculaneum: A Small Contribution to the History of Animal Foods." The author was clearly delighted to stumble upon the prosciutto during a visit to Naples and wanted to share it with the journal's readers as an example of representations of ham in Roman art. This is the only known firsthand assessment of the prosciutto by a doctor of veterinary medicine. As such he could apply the proper terminology, identifying this specifically as a "left ham" (*linken Hinterschinken*), meaning it was not the lower quality pork shoulder (*Vorderschinken*) but a proper ham made from the left rear leg.

We know that ham was a well-known and popular dish for the Greeks and Romans. This is also clear from the fact that the Romans gave the shape of ham to certain items of daily use. I saw a small bronze representation of a ham from the late first century AD this summer in Naples. In the collection from Herculaneum in the museum there is a bronze sundial in the shape of a left ham, silver-plated on top. The inscription

IV MA AP MA FA IA
IV AV SE OC NO DE

shows the names of the months abbreviated to two letters each, decreasing from June to January on the first line and in correct order on the second. The part of the rind side of the ham not covered with letters is covered by lines crossing into fields. The remnant of the tail and the ham leg can clearly be seen.[131]

---

130. Arthur 1909, 154: "Both these dials were in use about the beginning of our era and were covered by the great eruption of Vesuvius in 79 A.D., which destroyed Pompeii and Herculaneum." Armillary sundials have a complex design in that they represent both the terrestrial and celestial spheres through a set of interlocking metal rings forming a spherical framework around an inner gnomon. They date as early as the fourth century BCE, but none have been found at Pompeii.

131. R. Froehner, "Schinken aus Herkulanum: Ein kleiner Beitrag zur Geschichte der animalischen Nahrungsmittel," *Deutsche Tieraerztliche Wochenshrift* 16 (January 4, 1908) 12. While the byline credits a "Richard" Froehner, it seems more likely to have been authored by

Enlightenment

Froehner had done his research, citing the recent publication of the inscription in the *Corpus Inscriptionum Latinarum* (*CIL*), the authoritative catalog of Latin inscriptions edited by the renowned German epigrapher Theodor Mommsen in 1883. Mommsen, for his part, ascribed the reading of the inscription he published (*CIL* 10.8071, 23) to his student, Heinrich Dressel, a German archaeologist, numismatist, and epigrapher, who had examined it in the Naples museum.[132] Mommsen also cited the earlier versions of the text provided in the *Memorie*, Paciaudi, *Le Antichità*, and even the popular British book *Pompeiana* by William Gell.

Gell had presented the prosciutto in much the same manner as the Accademia had the 1762 marble sundial from Pompeii: as a decorative vignette featured at the end of the explication for one of his plates with a more detailed treatment allotted to an addendum at the end of the book. His engraving is virtually identical to the Accademia's, with the same careful stippling and shadowed details, the gnomon the same shape and position, and the inscription accurately copied and located; only the grid is slightly narrower (Figure 3.26). Although Gell wrongly attributed the Accademia's analysis to the fourth volume of *Le Antichità*, he successfully summarized their conclusions, even if he was not entirely satisfied with their explication. "It is added," he remarked, "that it had been observed to act nearly correctly through the whole day: but it is not explained whether the instrument was made to turn with the revolving sun, without which it is evident that it could not have acted at all; and if so, it would appear to have been intended for momentary use, and to have required adjustment whenever made use of." Gell must mean that the Accademia had not made it clear that in order to function its orientation needed to change to follow the course of the sun through the ecliptic because its position was not fixed as in other sundials, such as a vertical dial attached to the side of a building. By "momentary use" he must mean that because readings were taken when desired, the sundial needed to be reoriented each time for the shadow to fall at the necessary point.[133]

Reinhard Froehner (1868–1955), a well-known veterinary historian who published a three-volume *Cultural History of Veterinary Medicine* (*Kulturgeschichte der Tierheilkunde* (Konstanz 1952–1968). His other main example from Roman art was a "butcher's shield" (*Metzgerschild*) from Rome decorated with five prosciuttos, perhaps a pastry mold.

132. The entry for the prosciutto was included in the section cataloging *Suppellex aenea, argentea, aenea reperta Pompeiis et Herculanei* (gold, silver, and bronze utensils from Pompeii and Herculaneum) credited to Dressel and Giulio De Petra.

133. W. Gell and J. Gandy, *Pompeiana: The Topography, Edifices, and Ornaments* (London 1817–1819) 222, 269–70. Sir William Gell (1777–1836) was a topographer and artist, a Fellow of the Royal Society, the Society of Antiquaries, and the Society of Dilettanti, and the principal

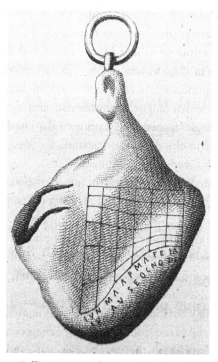

FIGURE 3.26 William Gell's prosciutto, from W. Gell and J. P. Gandy's *Pompeiana: The Topography, Edifices, and Ornaments* (London 1821).

Even though Gell had clearly transcribed the inscription separately from the engraving, Mommsen still managed to publish an erroneous version of it in his authoritative text, and Froehner, who probably had not bothered to transcribe it in situ, had unwittingly reproduced it. Junius is abbreviated "IV" rather than "IVN," and Mommsen even stated that seven lines separated the six columns of text, both errors seen in the *Memorie*'s original printed version but not followed by Froehner. If the open form of the "P" in Aprilis or the late nomenclature applied to the months had piqued his epigrapher's curiosity, Mommsen did not so indicate, though in general this was not the venue for this type of critical analysis. In providing the context for the inscription, he had simply stated that it was found "on a bronze sundial (*horologio solari*) covered with a sheet of silver which has the shape of a ham (*perna*) with a ring by which it is suspended."

author of the text. His co-author, John Peter Gandy (1787–1850), was also a member of the Society of Dilettanti but an architect and so not likely involved in dealing with the prosciutto. Several different editions were printed and so pagination may differ.

3.h) Comparetti and De Petra and the Villa dei Papiri

Mommsen's entry in the *CIL* appeared in the same year as Domenico Comparetti and Giulio De Petra's magisterial study of the extant archival documents and the finds from the Bourbon era excavations in the Villa dei Papiri that had returned the prosciutto to its proper provenance. Their publication included the text of the original excavation daybooks describing the discovery of a *pernil di metal* (bronze ham) with *algunas rayas y señales* (some stripes and symbols), De Petra's proposal that it had been found in the rooms situated to the east of the villa's atrium, and a detailed catalog entry.[134] The latter included a reproduction of the inscription that attempted to capture the sans-serif style and the lack of middle bars in the "A"s but with the "P" of Aprilis given a closed rather than an open form. They highlighted the *boustrophedon* arrangement and described the tail as *frammentata* (fragmented).[135]

Comparetti and De Petra's other significant contribution was to publish the first black-and-white photograph of the prosciutto. This was produced through the process known as *fototipia*, or phototype, invented in 1869 by Michele Danesi, a printer in Rome with an established reputation for high-quality reproductions of works of art. His new process rendered a significantly better tonal range to published photographs and therefore was in high demand at this time for fine-art publications. Comparetti singled out Danesi as the printer of choice, and each of the plates bears the stamp "Fototipia Danesi Roma."[136] Although Comparetti acknowledged that the scale of all the photographs had to be altered for publication purposes and reference therefore would need to be made to the dimensions provided in the catalog entries, the prosciutto is shown at a sufficiently large size to reveal all its details. Only the front of the prosciutto with the dial was reproduced, but for the first time it was possible to see clearly the actual position of the broken gnomon tail, the irregularities of the surface, the shallow grooves making up the grid, and

---

134. Comparetti and De Petra's study was followed by A. Ruggiero's published compilation of all the excavation records from Herculaneum, where the full text of the entry also appeared: *StErc* 175 (June 15, 1755).

135. Comparetti and De Petra 1883, 177 (excavation record), 233 (provenance), 286 #110 (catalog entry).

136. *Dizionario Biografico degli Italiani* (Rome 1986) 32.564–67, s.v. *Danesi* (M. Miraglia). The reproductions of inscriptions in the book were lithographs printed by Giulio Steiger in Florence.

FIGURE 3.27 The first published photograph of the prosciutto. From Domenico Comparetti and Giulio De Petra's *La Villa Ercolanese dei Pisoni* (Turin 1883).

the precise relationship between the placement of the calendrical abbreviations and the vertical month lines (Figure 3.27).[137]

The specialized nature of Comparetti and De Petra's work, however, meant that both their successful clinching of the prosciutto's provenance in the Villa dei Papiri and their detailed photograph largely escaped notice. This was especially the case with horologists for whom it primarily remained the "ham of Portici." One exception to this was Ernst von Bassermann-Jordan's 1914 modest study on sundials, which reproduced Comparetti and De Petra's photograph but nowhere described the prosciutto.[138] Even Sir Charles

---

137. Comparetti and De Petra 1883, vi, Plate XVII #4.

138. E. von Bassermann-Jordan, *Uhren: Ein Handbuch für Sammler und Liebhaber* (Berlin 1914) 50, Fig. 39: "Altrömische Reise-Sonnenuhr (*viatorium pensile*) in Schinkenform. Aus Herkulanum." *The Book of Old Clocks and Watches* (New York 1964), an English translation of the enlarged text of his *Uhren*, did not mention or illustrate the prosciutto. A professor of art history and archaeology at the University of Munich, Bassermann-Jordan was involved in

Waldstein (1856–1927), the Cambridge professor who had launched an international effort to resume open-air excavations at Herculaneum, had failed to associate the prosciutto with the Villa. He had sought to catalog all the finds from Herculaneum in the Naples museum and had gathered these in an appendix to his book, *Herculaneum: Past, Present and Future* (London 1908), which had made this remarkable collection accessible to an English audience. By Waldstein's day the prosciutto had migrated to the west wing on the ground floor, in the fifth room featuring bronzes from Herculaneum, and was displayed in a glass case to the left of the room's window. "Bronze sun-dial, faced with silver, shaped like a ham, H[ei]g[h]t mill[imeters] 118," was its catalog entry, penned by Antonio Sogliano (1854–1942), a student of De Petra's and the director of the Naples museum and the excavations at Pompeii at that time. Waldstein had marked all the finds from the Villa dei Papiri with an asterisk, but the prosciutto was not so designated. This may have been a simple oversight, since Sogliano also had been responsible for the entry that appeared at this same time in the *Guida Illustrata del Museo Nazionale di Napoli*, as the museum was now called, the principal popular, but official, guidebook in Italian published in 1908. One of the guide's principal aims, in addition to concision, had been to secure the provenance for as many antiquities as possible, and so the prosciutto is identified as coming from "Ercolano (dalla villa suburbana)," one of the names given to the Villa dei Papiri at that time.[139]

Mommsen's successor as secretary of the Berlin Akademie der Wissenschaften had been Hermann Alexander Diels (1848–1922), a prolific scholar of Classical texts, known primarily for his editions of the fragments of the Presocratic philosophers and Lucretius (Figure 3.28). He was one of the founders of the *Thesaurus Linguae Latinae*, a project to create the definitive dictionary of Latin that remains in production today. Although he devoted considerable efforts to producing readings of the texts of the charred scrolls from the Villa dei Papiri, his philological proclivities make him an otherwise unlikely candidate as a source on archaeological artifacts like the prosciutto. Yet he had long been fascinated with the connection between ancient technology and science and had given a series of lectures in 1912 at the University

---

excavations at the Greek sites of Orchomenos and Aegina, funded in part by his father, Emil, a vintner and financier. He published several studies of timepieces and at the time of his death had a collection of over 300 watches and clocks; *Neue Deutsche Biographie*, vol. 1 (Berlin 1953) 623, s.v. *Bassermann-Jordan, Ernst von* (E. Zinner); see also Chapter 3.i.

139. *Guida Illustrata del Museo Nazionale di Napoli*, A. Ruesch, ed. (Naples 1908) 223 #898. De Petra was among the contributors to the guide.

FIGURE 3.28 The Classical philologist Hermann Alexander Diels (1848–1922). From O. Kern, *Hermann Diels und Carl Robert: Ein biographischer versuch* (Leipzig 1927).

of Salzburg on the topic, which he published together in his *Antike Technik* (Leipzig 1914). In the expanded second edition of 1920 he included a chapter on ancient sundials, a field of study illustrating amply how "the craft is closely related to the science."[140] His was a mostly descriptive treatment demonstrating how the examples from the archaeological record could shed light on the literary sources; there is not a mathematical equation to be found, though he

---

140. H. Diels, *Antike Technik*² (Leipzig 1920) 155. For Diel's remarkably productive scholarly life and achievements, see E. E. Schütrumpf, "Hermann Diels," in W. W. Briggs and W. C. M. Calder III, eds., *Classical Scholarship: A Biographical Encyclopedia* (New York 1990) 52–60. W. Calder III, "What Sort of Fellow Was Diels?," in W. Calder III and J. Mansfeld, *Hermann Diels (1848–1922) et la science de l' antiquité* (Geneva 1999) 1–28, appears unaware of Diels' work on ancient technology, and none of the contributions in this volume touch on the topic either. Diels had earlier proposed that the owner of the Villa dei Papiri had been one Marcus Octavius whose name appeared at the end of two scrolls recovered in the Villa (P.Herc. 993/1149 and 336/1150), an identification that has not found much traction; see H. Diels, "Stichometrisches," *Hermes* 17 (1882) 383–84; M. Capasso, "Gli studi ercolanesi di Hermann Usener nel suo carteggio inedito con Hermann Diels," in M. Capasso et al., *Momenti della storia degli studi classici fra Ottocento e Novecento* (Naples 1987) 116–17; and see Chapter 4.a.

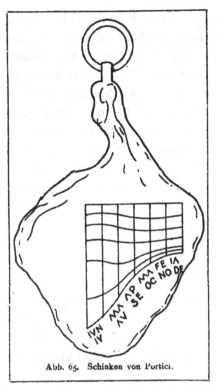

FIGURE 3.29 Diels's prosciutto. Figure 65 from his *Antike Technik*[2] (Leipzig 1920).

occasionally referenced degrees of latitude. The prosciutto appeared at the end of Diel's discussion of *viatoria pensilia*, of which he provided only one other example, and he gave its provenance as Portici, revealing that he was unaware of the most recent scholarship assigning it to the Villa dei Papiri. He credited the work of the Accademia and Woepcke as sources for his brief description and offered a reproduction of Woepcke's rendering, allowing the image to serve as a substitute for a detailed explanation of the inscription and other salient features (Figure 3.29).

Diels was one of the only scholars to highlight the humorous nature of its form, picking up on the Accademia's reference to it as an "artist's joke" while suggesting that the element of humor extended to the inherent inaccuracy of its design. Like others before him, however, he does not appear to have fully comprehended the complexity of the design, and because Woepcke's drawing did not reconstruct the gnomon tail, Diel's readers likely were at a loss to understand how the shadow was cast on the dial and how the hours were read.

240 THE PROSCIUTTO SUNDIAL

Finally, I mention the famous ham from Portici (found in 1755), which can also be counted among the *Horologia pensilia*. This bronze piece is jokingly shaped as a ham. Seven vertical lines intersect on the surface with seven horizontal curves. To determine the hour, hang the ham on the top of the ring and turn it until the sunbeam falls on the shadow of the tip of the gnomon formed by the animal's tail to the month in question, which is given below. The hours of the morning and afternoon were then read from the vertical lines in the usual way. Of course, this prank piece (*Scherzartikel*) only offered time in a very approximate way, but, as I said, the ancients did not yet know the saying: [in English] "time is money," but sadly others: *Eile mit Weile* [make haste slowly] (σπεῦδε βραδέως).[141]

3.i) Drecker's "Little Clock"

The German contributions to the study of the prosciutto continued through the early 1920s as a consequence of the work of the art historian and horologist Bassermann-Jordan, who earlier had published a photograph but not a description of the prosciutto. He had spearheaded a multi-volume work entitled *Geschichte der Zeitmessung und der Uhren* with each volume written by specialists in their fields, but the scope of the project and the intended number of volumes is unknown since only three in the series appeared. The first two focused on Egyptian and Arabic timekeeping while the third was a general survey on sundial theory by Joseph Drecker.[142] Drecker (1856–1931),

---

141. Diels 1920, 191–20.

142. That the three extant fascicules were labeled B, E, and F of volume 1 shows that more contributions to the series had been planned. L. Borchard, *Die altägyptische Zeitmessung*, Band 1, Lieferung B (E. von Bassermann-Jordan, ed., *Geschichte der Zeitmessung und der Uhren*, Berlin 1920). Borchardt (1863–1938), a German Egyptologist, had founded and then served as director of the Deutsches Archäologische Institut in Cairo from 1907 to 1928 and had directed the excavations at Heliopolis. He is best known for finding the bust of Nefertiti at Amarna now in the Berlin Neues Museum (ÄM 21300). In confirmation of Horapollo's observation that cynocephali clocked the hours by urinating, Borchardt illustrated a *clepsydra* from Edfu, now in Cairo, which he dated to ca. 100 CE, that bears a seated cynocephalus (which he refers to as a baboon) in relief near the vessel's base with a perforation through its penis (Borchardt 1920, 22 n. 3, Plate 9). In the introduction to a new edition of Borchardt's book in 2013, the editors note that Borchardt's contribution "fell into oblivion" and consequently "his significant results were not duly taken into account by either science historians or Egyptologists" (L. Borchard, *Die altägyptische Zeitmessung*, D. Wuensch and K.P. Sommer, eds. (Göttingen 2013). The second volume was Carl Schoy's, *Die Gnomonik der Araber*, Band 1, Lieferung F (E. von Bassermann-Jordan, ed., *Geschichte der Zeitmessung und der Uhren*, Berlin 1923). Schoy was an astronomer, mathematician, and historian of mathematics specializing in Arabic studies at the

FIGURE 3.30 Joseph Drecker. Photograph courtesy the *Dorsten Lexicon*. http://www.dorsten-lexikon.de/drecker-joseph/.

a German mathematician and scientist with a passion for horology, had reputedly amassed the second-largest collection of sundials in the world by the time of his death (Figure 3.30).[143] Of his horological publications his *Die Theorie der Sonnenuhren*, the third volume of the *Geschichte*, which appeared in 1925, made the most significant contribution to the field and, specifically, to the analysis of the prosciutto.[144] He devoted his twelfth chapter to ancient

University of Frankfurt am Main whose skills at combining these diverse fields were compared to those of Woepcke: D. Smith, "The Early Contributions of Carl Schoy," *The American Mathematical Monthly* 33.1 (January 1926) 28–31.

143. Drecker was born and died in Dorsten, Germany, and his biography appears in the *Dorsten Lexicon* (http://www.dorsten-lexikon.de/drecker-joseph/), which states that he sold his collection to a Dutchman with its whereabouts unknown. In fact, fifty-six of his sundials were acquired from Frederik Casparus Wieder (1874–1943) of Noordwijk in 1959 and ended up in the Collection of Historical Scientific Instruments at Harvard (http://waywiser.fas.harvard.edu/people/8771/joseph-drecker;jsessionid=984F9B14401B8A6CCFC776F7A818CF4B/objects).

144. J. Drecker, *Die Theorie der Sonnenuhren*, Band 1, Lieferung E (E. von Bassermann-Jordan, ed., *Geschichte der Zeitmessung und der Uhren*, Berlin 1925), 58–60.

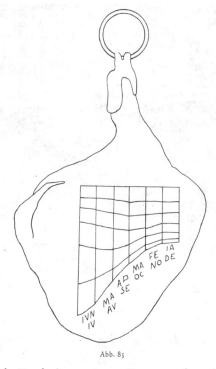

Abb. 85

FIGURE 3.31 Joseph Drecker's prosciutto. Figure 85 from his *Die Theorie der Sonnenuhren*, Band 1, Lieferung E (E. von Bassermann-Jordan, ed., *Geschichte der Zeitmessung und der Uhren*, Berlin 1925).

portable sundials, Vitruvius' *viatoria pensilia*, the oldest example of which he identified as the prosciutto, dating it to the period between 27 BCE and 79 CE on the basis of the abbreviation AV(gustus) and the eruption of Vesuvius. While acknowledging that several previous studies had been devoted to the prosciutto, Drecker singled out only the work of the Accademia and Calkoen, whose results he criticized as unsatisfactory because Calkoen had made "the unjustified assumption that the tip of the gnomon must have been perpendicular to the zero point of the first vertical" along the horizon line. His figure of the sundial, which more closely resembles that of Woepcke than Calkoen, illustrates that he believed the gnomon tip lay to the left of the dial, as the Accademia had shown it in their engraving and as Drecker's own calculations had indicated (Figure 3.31).[145] He was as dependent as his predecessors on the

---

145. Drecker's figure 85 (Plate 9) is a simple line drawing of the prosciutto with middle bars added to each "A" in the inscription and no attempt to accurately reproduce the relationship of the abbreviations to their verticals (see Figure 3.31).

numbers provided by the Accademia, but he was somewhat more critical of the Accademia's methodology. First, echoing Calkoen's critique, he highlighted the exaggerated precision resulting from their conversion of the equinox line into 1,000 units, pointing out that this would equate with each unit being only 0.00003 m. The Accademia's inclusion of the last two zeroes on their published value of "0.88100" for the tangent of the latitude amplified the exaggeration. He also rejected the excessive pedantry of factoring refraction and the solar parallax into the design latitude. Finally, he ridiculed their attempt to calculate the obliquity of the ecliptic from the midday shadow on the winter solstice and their conclusion that the prosciutto was made in 28 CE: they just as easily could have calculated "a few thousand years earlier or later with exactly the same justification and certainty."

After providing an overview of the design of the grid on the prosciutto's "not entirely flat" surface and the relationship between the horizontal and vertical lines to the temporal hours and the signs of the zodiac, Drecker explained how the instrument was suspended from its ring and the shadow of the (missing) gnomon tail read along the vertical corresponding to the date. Only around the culmination of the sun at midday could the reading be uncertain, he observed, until it could be determined if the shadow was ascending or descending along the vertical. He then explained the basic design of the grid, illustrated by a labeled diagram. The first dimensions that needed to be determined were the distances from the gnomon tip to the start of the vertical lines along the horizon line. These verticals, he noted, could be spaced any distance from one another. He labeled the lines from the gnomon tip to the vertical month lines g_1 to g_7. The elevation angles of the sun were then plotted for each hour of the day at the declination of the sun for which the vertical had been designed. These he obtained using the simple equation $s = g \cdot \tan(h)$, where "s" was the length of the vertical month line, "g" the distance from the gnomon to the vertical along the horizon line, and "h" the angle of the sun, which he asserted the ancients would have determined through an *analemma*, the design of which he explained in detail (Figure 3.32).

Drecker sought to establish the position of the tip of the gnomon in relation to the dial, a crucial detail the Accademia had left unspecified, as well as to verify the overall accuracy of the prosciutto's design. To do so, he retained key elements of the Accademia's results for his two-pronged analysis. These included the distances in units from the gnomon tip to the central equinox line and the winter solstice line along the horizon line, determined by the Accademia to be 881 and 1482 units respectively, and the lengths in units of

244   THE PROSCIUTTO SUNDIAL

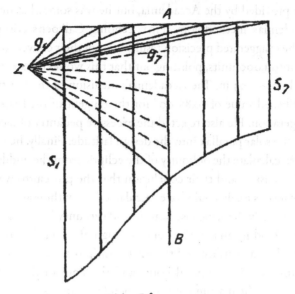

Abb. 86

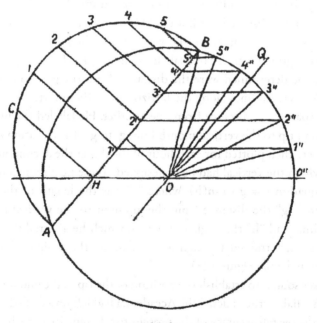

Abb. 87

FIGURE 3.32 Drecker's illustration of the prosciutto's geometry (Figure 86) and his rendering of an *analemma* (Figure 87), from his *Die Theorie der Sonnenuhren* (Berlin 1925).

Enlightenment

the vertical lines on the grid. His first step amounted to a check on the Accademia's calculations. For this he used the base latitude they had determined, stripped of solar parallax and refraction. At 41° 28′ (= 41.4667°) this yielded the necessary tangent of 0.881 corresponding to the length of the gnomon's shadow from its tip to the equinox line along the horizon line. Because the Accademia had failed to specify the length of the horizon line, he defined it as 1,320 units. This he divided equally into six units of 220 separating the vertical lines.[146] The intersection of circles measuring 881 and 1,482 units in diameter taken from the intersection of the equinox and winter solstice lines along this horizon line located the position of the gnomon tip and from this intersection he could determine the distances to the five remaining month lines. This yielded lengths along the horizon line of 500, 570, 710, 881, 1070, 1280, and 1482. To the Accademia's latitude he then added or subtracted the values of the angle of solar declination at each of the vertical lines representing the zodiacal months, using ± 11° 38′ (= 11.63°), ± 20° 26′ (= 20.43°), and ± 23° 37′ (= 23.61°).[147] He then calculated the tangents for each of these angles and divided the lengths of his lines by these tangent values to produce the required lengths for each of the vertical month lines.[148] These compared favorably with the lengths the Accademia had provided for the month lines, with the exception of the first two lines, which were markedly shorter. Though Drecker did not explicitly state so himself, this process had located the gnomon tip 2.83 mm to the left of the dial and 15.60 mm above the prosciutto's surface.

Drecker then ran the numbers again but rounded the latitude down to 41°, the tangent of which is 0.869. Again he found the angle of the declination of the sun for each of the verticals and their associated tangents and then used the Accademia's stated lengths for the verticals to establish the length of the shadow from the gnomon tip to the start of the verticals along the horizon line: 523, 579, 700, 870, 1053, 1269 and 1445. He then singled out the length

146. Drecker believed the equinox line's 1,000 units was equal to 30 mm rather than its actual 31.23 mm. This means his horizon line equaled 41.22 mm rather than 41.35 mm and his distance between the verticals was 7.33 mm rather than their actual average of 6.89 mm. The resulting slight differences between his measurements and the actual prosciutto run through his analysis. Calkoen's reconstructed horizon line was 1,260 units with the verticals spaced 210 units apart.

147. This produced angles of 53.01°, 61.81°, 64.99°, 41.48°, 29.75°, 20.95°, and 17.77° for the seven month lines, from left to right. Drecker's values for declination differed slightly from those commonly employed today.

148. The different declinations Drecker employed resulted in different tangents, and this, in turn, affected the resulting vertical line lengths: 1575, 1489, 1242, 1000, 806, 687, and 687, compared to the Accademia's lengths of 1686, 1543, 1244, 1000, 804, 691, and 687.

he had established in this manner for the central equinox line, 870, drew intersecting circles from the points of intersection of the vertical lines along his reconstructed horizon line, and determined the actual distances, which varied only slightly from the calculated ones: 520, 580, 705, 870, 1060, 1255 and 1455. By dividing these by each line's tangent, he obtained the length for each of the vertical month lines independently of those provided by the Accademia and concluded that the two were close: they differed by only nine units at the most.[149] "And now the surprising result," concluded Drecker, "that this little clock shows the noon hours, one can say, correctly without any errors because nine units are only 0.27 mm. The remaining hours of the day are, as far as I have calculated them, correct to the same extent."[150] He did not, however, offer any detailed calculations to support this claim.

3.j) De Solla Price's "Mass of Bronze"

Despite the availability of detailed engravings of both sides and a black-and-white close-up photograph of its front, doubts continued to linger that this instrument had been designed to represent a prosciutto. One notable skeptic was the physicist and historian of science Derek de Solla Price (1922–1983) (Figure 3.33). De Solla Price is best known for his early contributions to the decipherment and analysis of the so-called Antikythera Mechanism, the earliest known analog computer from the Hellenistic period consisting of thirty meshed bronze disks that could be used to predict astronomical events.[151] In 1969, he published an article on ancient portable sundials, a study sparked

---

149. Drecker's vertical lengths were 1,677, 1,546, 1,253, 1,001, 810, 683 and 686, which differed +9, -3, -9, -1, -6, +8 and 0 units from the Accademia's lengths.

150. Drecker linked his study of the prosciutto to Schoy's book on Arabic time keeping in this same series by comparing the prosciutto's design to a later Islamic sundial known as the *Sakhe-al-Jeradah*, or "grasshopper tail," described in a thirteenth-century manuscript by Aboul Hhassan Ali of Morocco. These combine aspects of a cylinder dial and a flag dial in consisting of a rectangular plate held vertically and inscribed with a grid of the zodiacal months similar in shape to the prosciutto's but with only a horizontal line and a curving meridiem line: the temporal hours are not indicated. One form of this sundial has a moveable gnomon—and so possibly another source of Lalande's notion about a "moveable gnomon" on the prosciutto—while the other had a fixed gnomon in the upper left-hand corner of the dial. Both were designed to function in only a single latitude. Schoy 1923, 55, and, Drecker's other source, J. J. Sedillot, *Traité des Instruments Astronomiques des Arabes*, 2 vols. (Paris 1835) 2.440–44, figs. 73–74.

151. D. de S. Price, "Gears from the Greeks: The Antikythera Mechanism, a Calendar Computer from c. 80 BCE," *Transactions of the American Philosophical Society* 64.7 (1974) 1–70. See also A. Jones, *A Portable Cosmos: Revealing the Antikythera Mechanism, Scientific Wonder of the Ancient World* (Oxford 2017).

Enlightenment

FIGURE 3.33 The physicist and historian of science Derek J. de Solla Price (1922–1983) with a model of the Antikythera Mechanism. Family photo in public domain. https://commons.wikimedia.org/wiki/File:DerekdeSollaPrice.jpg.

by the discovery six years earlier of the bronze disk from a Byzantine-era portable dial during excavations at the Roman site of Aphrodisias, Turkey.[152] He cataloged eleven portable dials and placed each into three different categories according to Vitruvius' terminology (Vitr. 9.8.1): those designed for general observations (*pros ta historumena*), those designed for all latitudes (*pros pan klima*), and those intended for use within only a single latitude (*viatoria pensilia*). The Aphrodisias dial, marked with the abbreviated names of cities ranging in latitude from 23° to 39°, fit his definition for a *pros pan klima* type. The prosciutto was one of only two examples that were properly portable but useable in only one latitude.[153]

De Solla Price earlier had included the prosciutto in a chapter he contributed to Singer's *History of Technology* entitled "Precision Instruments: to 1500." In a section on portable sundials, he underscored the difference between them and common fixed sundials, specifically the hemicyclium type, which was "sufficiently accurate for daily purposes and even for the timing of

---

152. D. de Solla Price, "Portable Sundials in Antiquity, Including an Account of a New Example from Aphrodisias," *Centaurus* 14.1 (1969) 242–66. For the sundial from Aphrodisias, see also Talbert 2017, 60–63.

153. The other was an ivory disk found near Mainz inscribed with the twelve months of the year abbreviated to three letters and placed horizontally opposite each other with sockets into which a moveable gnomon could be inserted to cast a shadow on a grid radiating out from the disk's center; see De Solla Price 1969, 246–47, fig. 2. Arnaldi and Schaldach 1997, 109: the hour grid on the Mainz sundial "is so crude it is not possible to calculate the underlying latitude." The radiant grid recalls Arthur's botched drawing of the prosciutto (see Figure 3.25).

eclipses." "More important technically," he continued, "are the numerous and often ingenious portable sundials. Their small size demanded a certain accuracy in construction, and their portability led to problems that do not arise with an instrument rigidly fixed with respect to the meridian and horizontal planes."[154] The earlier Egyptian shadow clock was a type of altitude sundial with a vertical gnomon that cast a shadow along a horizontal graduated rod, the graduations having been determined empirically.

> A Roman "ham" dial from Herculaneum shows a marked improvement on the Egyptian forms, in that it consists of a flat plate which may be suspended from a ring so that it automatically takes up a vertical position and does not need to be levelled by independent means. In the "ham" dial the gnomon protrudes from the surface of the plate, which is oriented until the tip of the shadow falls on the graduations of the appropriate column containing divisions for the month of the year.[155]

From his repeated reference to it as a "ham" dial and his avoidance of applying any porcine terminology identifying its prosciutto form or its pig's tail, it is clear that de Solla Price believed he was describing a type of Roman portable sundial that happened in this case to have the shape of a ham, rather than singling it out as a unique example. The prosciutto, in other words, was a variant of a type that consisted of a flat plane suspended vertically with a single fixed gnomon projecting from that plane.[156] According to de Solla Price, even this improvement on the Egyptians had its limitations, however. "In practice it is found that the use of a single fixed gnomon leads to inconvenience in graduating the instrument," he observed, but, in language echoing Lalande's solution for the perceived limitations of the prosciutto, this was a problem resolved subsequently by having a moveable gnomon, as in the proper shepherd's or

---

154. D. de Solla Price, "Precision Instruments: to 1500," in C. Singer et al., eds., *History of Technology* (Oxford 1957) 3.594–96.

155. De Solla Price 1954, 596.

156. The notion that the prosciutto was a "type" rather than unique was reiterated in a review essay by D. King, "Towards a History from Antiquity to the Renaissance of Sundials and Other Instruments for Reckoning Time by the Sun and the Stars," *Annals of Science* 61 (2004) 376: in discussing portable sundials he included "the Graeco-Roman 'ham dials,' several of which survive and have been studied in Price." It is true, as Piaggio had observed, that dials similar to that on the prosciutto can be projected on any flat, vertically-oriented plane given an appropriately calibrated gnomon and grid.

Enlightenment

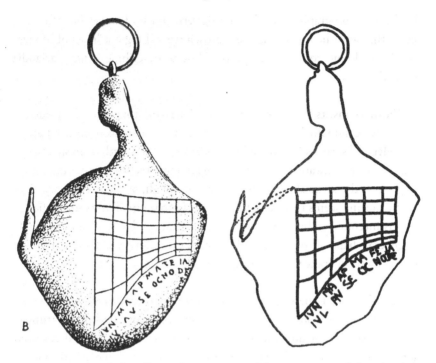

FIGURE 3.34 Derek de Solla Price's two versions of the prosciutto. *Left*, from his chapter on "Precision Instruments: to 1500," in C. Singer et al., eds., *History of Technology* (Oxford 1957) figure 351b. Courtesy Oxford University Press. *Right*, from his "Portable Sundials in Antiquity, Including an Account of a New Example from Aphrodisias," *Centaurus* 14.1 (1969) figure 1. Courtesy John Wiley and Sons—Books.

cylinder dial. He illustrated his remarks with a black-and-white line drawing of the prosciutto based on that of the Accademia (Figure 3.34, left). This was carefully stippled to give it shape and volume, its grid a reasonably close copy, and the inscription properly transcribed and positioned, but with the broken gnomon-tail turned out and extending up, perpendicular to the grid's horizon line, making it difficult to comprehend exactly how it could be oriented so its shadow "falls on the graduations" of the dial.

De Solla Price offered a revised version of his illustration for his later study of portable sundials. He appears to have based this not on his earlier drawing but on Gatty's, reduced to a simple line drawing in a jagged outline, as if intentionally trying to mask its actual form, and with the stub of the gnomon projecting upright on the left (Figure 3.34, right). July is now abbreviated as IVL instead of IV, and a reconstruction of the gnomon shown in dotted line extends over to end at the upper left-hand corner of the dial. He cited the Accademia's preface, so he knew their engraving, and he acknowledged the "good photograph" published by Bassermann-Jordan, yet he refused to accept that what he was seeing was a prosciutto, calling it a "mass of bronze" and, in so doing, coming dangerously close to describing it like Paciaudi's "carafe."

> The dial consists of an irregularly shaped mass of bronze which appears to have originally been silver-plated, and to which has been fixed an eyelet and suspension ring of brass or bronze. The length from suspension ring downwards is 116 mm, the maximum breadth 80 mm, and the greatest thickness about 17 mm. Some, if not all, of the irregularity of the outline seems to be due to damage, so that one doubts if the dial was intended to be shaped like a ham; more probably it follows the general shape of the engraved lines with the addition of a neck above and a protuberance for the long conical gnomon.

Earlier scholars had described the surface of the prosciutto as irregular, with "concavities" and "convexities," but no source had suggested the instrument had been damaged in any way. The shank had never been described as a neck, not even in the *Memorie*'s characterization of it as a flask where this portion "narrows," or the tail as merely a protuberance. As a result, readers not previously acquainted with this sundial were unaware that the "long conical gnomon" was actually the animal's tail or even were aware that the gnomon tail was no longer present, especially given his subsequent statement.

> This gnomon is now bent far from its natural position. It can readily be shown that the tip of the gnomon should in fact lie directly above the top left hand corner of the engraved grid of hour and month lines, at a height above equal to about one-third of the equinoctial shadow (shown by dotted line in Fig. 1). One may readily determine that the dial is indeed, within the accuracy of measurement, made for the latitude of Herculaneum; it is not adaptable for other latitudes.

The "dotted line" in de Solla Price's figure shows the orientation of the gnomon—"the true position of the gnomon" in the figure's caption—not the equinoctial shadow, the length of which he did not provide so the reader could not determine the height at which he believed the tip of the gnomon stood above the grid. Given the 31.23 mm length of the central equinoctial line—a figure he does not provide—de Solla Price's calculation that the height of the gnomon tip above the grid was one-third the length of the

Enlightenment 251

equinoctial shadow would yield 10.4 mm. Such a 1:3 ratio does not, however, correspond with any known equations for establishing the height of a gnomon from the length of its equinoctial shadow. The confidence with which he assigns its latitude to Herculaneum (40° 48′ 26″ = 40.80°) ignores the Accademia's conclusion that it was made to work farther north toward Rome (41° 39′ 45″ = 41.66°), though it is true that the differences in hourly readings would be negligible. He concluded with a clear description of the dial's design that underscores its unique character and the benefits of its specific design, but he did not explain how a reading from it was taken.

> In its technical gnomonics the dial is unique though strongly related to a well-known class of later dials (the Shepherd's Dial or Chilindre being the most common) where a movable gnomon juts out from the top of a set of vertical scales giving the shadow lengths of the gnomon for each of the months of the year. Each scale serves a pair of months with equal solar declination, and each is divided by a set of seven hour lines so that the time of day can be read from the altitude of the sun. The hours used here are the seasonal hour scheme in which the intervals from sunrise to sunset and from sunset to sunrise are each divided into twelve equal parts. The horizontal line on top of the grid corresponds to both sunrise and sunset and the lower edge of the grid corresponds to the length of the noon shadow throughout the year. The peculiarity of this particular dial is that the scale and gnomon are placed so that the vertical scale and therefore the shadow for the winter sun is much further from the gnomon tip than those of the summer. This ingeniously reduced the amount of variation in shadow length and gives a scale which is easier to read in winter.[157]

---

157. De Solla Price 1969, 244–45, fig. 1. E. Winter, *Zeitzeichen: Zur Entwicklung und Verwendung antiker Zeitmesser* (De Gruyter Berlin 2013) 2.372–73, relied exclusively on De Solla Price's scholarship for her catalog entry on the prosciutto, accepting his calculations, his conclusion that it was designed for Herculaneum's latitude, and even his belief that it was not designed as a prosciutto but that its odd form was due to "the state of preservation of the piece: the flat, originally disc-shaped metal instrument is so corroded in some places that only a small part of the original outer edge has been preserved along its upper edge with a small eyelet for hanging." R. Hannah, "Sundials," in G. Irby, ed., *A Companion to Science, Technology, and Medicine in Ancient Greece and Rome* (London 2016) 1103–4, appears similarly skeptical that it was actually modeled on a prosciutto: "Known as the 'ham dial' because its distorted bronze plate looks just like a small leg of ham, it consists of a spike on one side, which threw a shadow on to a series of crisscrossing lines on the plate, from which one could read the hour of the day."

He concluded by noting that the use of the abbreviations for July and August required that the prosciutto date after the year 27 BCE and before the eruption of Vesuvius, in so doing rejecting the Accademia's date of 28 CE based on their determination of the diminution in the obliquity of the ecliptic.

3.k) Nothing New Under the Sun: Prosciutto Studies in the Late Twentieth Century

The effort to identify the provenance of the antiquities in the museum initiated under Spinelli and continued by his successor, Giuseppe Fiorelli, coupled with Comparetti and Dé Petra's thorough research, meant that the finds from the Villa dei Papiri were increasingly displayed in close proximity with one another in the museum.[158] Amedeo Maiuri's further reorganization of the museum through the 1960s resulted in the Villa's marble sculptures being gathered together on the ground floor while the prosciutto was displayed in a room on the upper floor alongside small busts and statuettes from the Villa.[159] In 1973 a gallery of four rooms dedicated exclusively to the most spectacular finds from the Villa was inaugurated, a significant break from the long-standing practice of displaying the artifacts by genre or period.[160] The prosciutto was featured in the first room (Sala 117), in a glass case that also held small bronze fountain figures from the atrium and stood across from Weber's detailed plan of the Villa.[161] It has remained in this gallery to this day, apart from its most recent forays out of Italy for international exhibitions.

Relatively few recent studies have touched upon the prosciutto, and these tend to be of either an archaeological or a horological nature. Some largely

---

158. Giuseppe Fiorelli (1823–1896) was director of the Museo Archeologico from 1863 to 1875, director of excavations at Pompeii and Herculaneum from 1860 to 1875, and a professor of archaeology at the University of Naples Federico II. He is credited with beginning the practice of taking plaster casts of the cavities left by human victims of the eruption at Pompeii and produced a number of important publications, including the *Pompeianarum Antiquitatum Historia* (1860–1864), his edition of the original excavation manuscripts from Pompeii.

159. Amedeo Maiuri was Soprintendente alle Antichità della Campania e del Molise from 1924 to 1961, including director of the Museo Archeologico di Napoli, professor and chair of Antichità Pompeiane ed Ercolanesi at the University of Naples, professor of ancient history at the Istituto "Suor Orsola Benincasa" in Naples, and one of the most prolific scholars on the antiquities of Pompeii and Herculaneum. He excavated substantial portions of both Herculaneum and Pompeii.

160. Rooms 114 to 117 on the first floor in the museum are dedicated to the principal finds from the Villa dei Papiri.

161. Mattusch and Lie 2005, 89.

Enlightenment

reiterate past discussions, while others seek to interrogate the sundial in new ways. In 1986, the prosciutto was cataloged as part of Maria Rita Wojcik's comprehensive review of the Villa's finds. She was primarily interested in the Villa's decorative program and so only provided limited bibliography and a cursory description that highlighted the *boustrophedon* arrangement of the inscription and the broken tail. Wojcik assigned it a *terminus post quem* of 8 BCE exclusively on the basis of the abbreviation of AVgustus, accepting the traditional date for the Senate's granting of that honorific title. She ignored entirely its astronomical aspects. Her most significant contribution was a new black-and-white photograph, with no scale provided but showing it greater than life size and highlighting the inscription and the grid in the process.[162]

Among the more popular treatments are those published by Nicola Severino, an Italian independent scholar and historian of ancient and medieval gnomonics. Severino had included a section entitled "The True Story of the Prosciutto of Portici" in his self-published *Antologia di Storia della Gnomonica* from 1995. The "true story," however, is merely a synopsis of the Accademia's preface but with the additional observation that Rene Rohr had dated it to the first century CE.[163] Severino also described the prosciutto in a contribution to *The Compendium*, the journal of the North American Sundial Society. This again was essentially a translation into English of the Accademia's principal findings but included a concise explanation of how to operate it. In addition to providing copies of Price's illustrations and the Accademia's engravings, which otherwise might not have been accessible to his audience, Severino included one of his own making, a remarkably crude rendering of Price's "mass of bronze" with the hour lines labeled in Arabic numerals but lacking the calendrical inscription, alongside two smaller rectilinear renditions of the front and right side showing how the sun's beam was intended to strike the gnomon (Figure 3.35).[164]

---

162. Wojcik 1986, #O2, plate 129. Wojcik cataloged it as a "sporadic" find and rejected Comparetti and De Petra's conclusion, remarking only that it "comes from our villa, even if it is impossible to determine the original provenance."

163. N. Severino *Antologia di Storia della Gnomonica* (Roccasecca 1995) 19–20; a second edition (Roccasecca 2001) 51–54, offers the same text with reproductions of the Accademia and Woepcke/Diels, along with a third line drawing of unknown origin showing it with an enlarged gnomon/tail angled far to the left of the grid, its tip ending before reaching the prosciutto's body.

164. N. Severino, "The Portici Ham," *The Compendium: Journal of the North American Sundial Society* 4.2 (June 1997) 23–25. In his *Storia della Gnomonica* of 2011, a survey of many of the standard texts, material remains, and discussions of sundials referenced in the present study, he recycles his "The True Story of the Prosciutto of Portici," adding now that Rohr had published

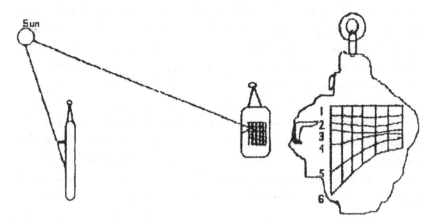

FIGURE 3.35 Nicola Severino's prosciutto from his article "The Portici Ham" in *The Compendium: Journal of the North American Sundial Society* 4.2 (June 1997). Courtesy of the North American Sundial Society.

Other recent studies of the prosciutto include those by Karlheinz Schaldach, a German historian of gnomonics and astronomy and an independent researcher with numerous publications on ancient and medieval sundials. In 1997 Schaldach had published an article together with Mario Arnaldi, another Italian historian of gnomonics, revealing that what had long been considered a simple cylindrical lidded vial of ivory in the museum at Este, Italy, was actually a cylindrical sundial with two gnomons of different lengths tucked into the lid, each gnomon designed for use at a different time of the year within the latitude for which the dial was designed. Typical for this type, the grid consisted of twelve month lines wrapping around the cylinder. Three-letter abbreviations of the month names label each line, but those of July and August are missing, so the authors proposed that the change from Quintilis and Sextilis had occurred recently enough that they were not in customary use (Figure 3.36). They therefore dated the dial to the early Augustan period, though it had been recovered in a first-century CE tomb. Their study had included a survey of known portable sundials, based on Price's earlier categorization according to types but now numbering some nineteen examples. They limited their comments on the prosciutto to, "Because of its strange

---

a "brutissima figura" (the ugliest drawing) of the prosciutto in the 1986 Italian translation of his book on *Sundials*, but this edition was not available for the present study and there is no drawing of it in the English edition of 1970.

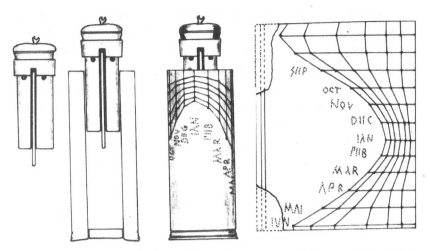

FIGURE 3.36 The Este Cylinder Dial from the first century CE, designed for the latitude of 45° (Museo Nazionale Atestino di Este 15397). The gnomon could be retracted into the cylinder for travel. The grid of twelve vertical month lines wraps around the cylinder but is missing labels for July and August. Adapted from Arnaldi and Schaldach 1997, figures 2 and 3.

ham-like appearance it is completely different from the other instruments, all of which are circular in form."[165]

Schaldach went on to explore portable sundials in greater detail and in the following year issued "A Plea for a New Look at Roman Portable Dials" in the *Bulletin of the British Sundial Society* in which he summarized some of his observations about this type of sundial. He concluded, for example, that all were of bronze or ivory and most were not well made, "a calculated error of more than thirty minutes not being unusual." The user could easily control for such inaccuracies, however, by consulting the various fixed dials available in most ancient communities that, the author presumes, offered more precise readings. He proposed that their inherent inaccuracies were due to the fact that they were made by craftsmen, not astronomers or those familiar with gnomonics, and cited Vitruvius' statement that "Many have also left instructions for making *viatoria pensilia*" (Vitr. 9.8.1): like other artisans, the makers

---

165. Arnaldi and Schaldach 1997, 107–17. The dial was found in the Tomb of the Physician, so called because, along with other finds, it contained a set of surgical instruments. Museo Nazionale Atestino di Este, MNAE: IG MNA 15397; 6.2 cm high, 2.5 cm diameter. The gnomons measured 27 mm and 21 mm and the dial appears to have been designed for Este's latitude (45°). Odium for Julio-Claudians seems a more likely explanation for the exclusion of these honorific month names. For the Este dial, see also H. Higton, *Sundials: An illustrated history of portable dials* (Wappinger Falls NY 2001) 11–18, who, in a chapter on ancient portable dials, describes this dial in detail but seems unaware of the prosciutto's existence.

256   THE PROSCIUTTO SUNDIAL

of portable dials had relied on extant models and copybook drawings. This process of replication caused additional errors to creep into these dials' grids over time. Moreover, Schaldach believed that earlier portable dials would have had their month lines properly marked to reflect the entrance of the sun into each zodiacal sign but that, also as a consequence of replication, the distinction between the sign and the name of the month gradually was lost. As a result, a single month name came to be associated with each of the six segments created by the seven month lines. In the Este cylinder dial, for example, the month names marked the beginning of each segment while in the prosciutto each segment bears the name of two months (see Figure 3.36). This was not evidence that the prosciutto is more recent than the Este dial but was the result of errors in replication. Schaldach is not entirely accurate in this, however, because the month names on the prosciutto are staggered, not paired, so that each month's abbreviation is associated with its prevailing zodiacal sign and precedes the vertical line it labels, in precisely the manner he proposed was earlier. He believed that all portable dials had used the value of 24° for the obliquity of the ecliptic and that this was the case with the prosciutto as well.[166] As a consequence, the declinations associated with each of its seven month lines were, from left to right, 24°, 20.6°, 11.7°, 0°, −11.7°, −20.6°, and −24°. He rejected Price's assertion that the prosciutto was designed for Herculaneum's latitude and, though he did not explicitly say so, rejected as well the Accademia's 41° 39′ 45″, arguing instead for a generic latitude associated with a "climate" rather than a city, and settled on 38°.[167] Finally, he stated that the gnomon tip should have rested 14 cm [sic] above the intersection of the first vertical month line and the horizon line. He compared the resulting hour curves with those on the prosciutto's actual grid in an accompanying graphic representation, the first of its kind in studies of the prosciutto. This reveals a relatively close correspondence between the curves on the right that encompass the period from the autumnal to the vernal equinox and the greatest variations occurring in the fourth- and fifth-hour curves in the period from the vernal equinox to the autumnal equinox. He did not explain, however, how he had derived the measurements used for establishing these hour

---

166. K. Scaldach, "A Plea for a New Look at Roman Portable Dials," *Bulletin of the British Sundial Society* 98.1 (1998) 48–50. This is the value employed in antiquity; the value today is 23.45°; see Chapter 4.e.

167. Athens, Greece, and Ephesus, Turkey (Roman Asia Minor) are roughly at the 38° latitude.

Enlightenment 257

curves.[168] He subsequently concluded that, "Compared with the rest of the portable dials the errors one encounters on [the Ham dial and the Este dial] are surprisingly small. This either means that the lines of the grid have been constructed or that they were conscientiously copied very closely to the original, and this again would mean that these grids were relatively recent inventions at the time when these instruments were constructed."

Schaldach's proposals concerning the prosciutto in the *Bulletin* went largely unnoticed because of the journal's limited circulation. This was less the case for his more popular book *Römische Sonnenuhren: Eine Einführung in die antike Gnomonik* (Frankfurt 1998). In a section devoted to portable sundials, he reiterated his 38° latitude and 14 cm [*sic*] height for the gnomon tip. He also critically assessed the earlier treatments of Calkoen and Drecker. Drecker, he noted, believed the prosciutto was designed for optimal operation at noon for its latitude and therefore had shifted his proposed location for the tip of the gnomon to the left of the upper left-hand corner of the dial. In a change from his earlier conclusion that craftsmen had relied on copybooks, Schaldach here argued that the dial was not based on an established template but derived from empirical evidence, as Delambre had earlier proposed.

> If one assumes, however, that the grid on the ham is the result of observations rather than being based on a template, then a location for the gnomon tip is more likely to be vertical above this point of intersection, since this could be determined more easily. Beeck [Calkoen] had already made calculations for this, but Drecker's results were better. In fact, the results can be improved by choosing a lower local latitude [38°] and a distance of 14 cm [*sic*] above the plate.

Schaldach concluded his discussion of the prosciutto by briefly explaining how it functioned. While he correctly stressed the need to know the current month, taking an accurate reading required knowledge of the dates when the sun entered and exited each of the zodiacal signs since the solstice lines—the first and seventh vertical lines—mark the entry of the sun into Cancer, being operable for the dates from roughly May 22 to June 24, and Capricorn, from November 22 to December 22, respectively, not, as Schaldach wrote, the "beginning" of July and January.

---

168. Schaldach 1998, 49–50, fig. 4. In both his text and the figure's caption Schaldach gave "14 cm" for the value of the height above the grid, but this should have been expressed in millimeters. The illustration, presumably measured from the Accademia's engraving, appears accurate in the length of the month lines but is slightly too narrow as published.

To use it, it is necessary to know the respective month. The summer solstice line marks the beginning of July. The winter solstice line on the far right represents the beginning of January. If one turns the clock so that the curved tip of the curved gnomon falls exactly on the respective vertical of the date, the hour is read. The horizontal [horizon line] stands for sunrise or sunset, the curved [lines] below for the start of the second or twelfth hours, etc., the lowest line is midday.[169]

In 2016 the prosciutto traveled for the first time to the United States in an exhibition organized by the Institute for the Study of the Ancient World at New York University entitled "Time and Cosmos in Greco-Roman Antiquity." Schaldach contributed a chapter on Greek and Roman time-pieces for the exhibition catalog. He singled out the prosciutto as the "most famous and oldest known example of a *viatorium pensilium*" [*sic*], noting as he had earlier that the ham shape distinguished it from other portable dials that tended to be circular. The presentation is brief and highlights the aspects that were most likely to catch the unacquainted but inquisitive exhibition viewer's eye: the inscription and the design of the grid. He pointed out that the months were paired in columns, rather than that they were staggered, and, as others had done before him, abbreviated IV(lius) as IVL(ius). He explained how the tip of the gnomon would have stood slightly above the surface of the ham, at the same level as the horizon line, but did not specify how high. Now, however, he stated that it was designed to function at the latitude of Herculaneum rather than 38°, as he had earlier proposed.[170]

An important contribution to the long history of scholarship on the prosciutto was published in 2008 by Gianni Ferrari (1938–2019), an Italian electrical engineer and technical software designer who applied his skills to his interest in astronomy. Like Severino and Arnaldi, Ferrari was a member of the Italian community of sundial aficionados and a frequent contributor to popular journals and conferences. At the Fifteenth Italian National Seminar on Gnomonica, Ferrari offered a paper on the prosciutto entitled "L'Orologio

---

169. K. Schaldach, *Römische Sonnenuhren: Eine Einführung in die antike Gnomonik* (Frankfurt 1998) 41–47. Schaldach incorrectly identified the second horizontal line as marking the twelfth hour; it marks the start of the eleventh.

170. K. Schaldach, "Measuring the Hours: Sundials, Water Clocks, and Portable Sundials," in A. Jones, ed., *Time and Cosmos in Greco-Roman Antiquity* (Princeton 2016) 83, 186 #15, fig. III-16; dated to end of first century.

romano conosciuto come 'Prosciutto di Portici.'"[171] Ferrari's study is important for his efforts to interrogate the prosciutto from new angles and offer a critical assessment of problems in its design. His diachronic analysis of it influenced the present study in several ways. He had been struck by the contradictory and inaccurate information provided by previous scholars, but while he did not pursue a detailed investigation of those earlier sources, he did collect depictions of the prosciutto and place them side-by-side to highlight the recurring errors and inadequate representations over time. Despite recognizing its shortcomings, he accepted the Accademia's analysis. "The description by the Neapolitan academics is not only the earliest but also, in my opinion, the best… it is the first and perhaps only quantitative study of the object." He acknowledged that Delambre's discussion was also good but does not cite the work of Calkoen, Woepcke or Drecker. Noting that its discovery halfway between Portici and Herculaneum had led to some confusion about where it was found, he looked to the Bourbon excavation records and concluded it came from the Villa dei Papiri, but he appears unaware of Comparetti and De Petra's identification of the rooms east of the atrium as its likely provenance. Ferrari emphasized that the grid's vertical lines correspond to the start of each sign of the zodiac or the twenty-first day of each month, that the hours are seasonal, and that its readings were valid for only a single latitude. He lauded the calendrical revisions of Caesar and Augustus for changing the calendar from a lunar year to a solar one in which the sun's celestial coordinates repeat on almost the same day in subsequent years: this recurring cycle ensured sundials like the prosciutto retained their accuracy from year to year.[172]

---

171. The XV Seminario Nazionale di Gnomonica was organized by the Sezione Quadranti Solari della Unione Astrofili Italiani (UAI) and held in Monclassico, Trento, Italy, from May 30 to June 1, 2008. Ferrari's Italian contribution is available at http://www.elsolieltemps.com/pdf/gnomonica/158.pdf. In 2019, an English translation was published in the journal of the North American Sundial Society with the title "The Roman Sundial Known as the Ham of Portici," *The Compendium* 26.2 (2019) 19–32. This improved on the Italian version with the addition of a color photograph and slight corrections in citations but some of the explanations behind his calculations are not as detailed. Like Lalande, Ferrari gave his name to a celestial object, in this case an asteroid orbiting between Mars and Jupiter (Asteroid 315046), discovered by Italian astronomers, named "Gianniferrari" in 2017; see https://ssd.jpl.nasa.gov/sbdb.cgi?sstr=2007+CG51&orb=1.

172. R. Hannah, *Greek and Roman Calendars: Constructions of Time in the Classical World* (London 2005) 112–38. Caesar had introduced the solar calendar as part of his reforms of 46 BCE. According to Pliny the Elder (*Nat. Hist.* 18.211), he had received advice from Egyptian astronomers, in particular one Greek-Egyptian named Sosigenes, during his time with Cleopatra. Macrobius (*Saturnalia* 1.14.2), writing in the fifth century CE, states that a scribe named Marcus Flavius had drawn up a table that fixed the position of each day. Augustus eliminated the practice of inserting intercalary months and fixed the number of days in the months to set the calendar correct.

Ferrari's principal thesis was that the Accademia's calculations were flawed. They had failed, however, to provide all the necessary dimensions to reconstruct their analysis and determine the source of their errors. They had given the lengths for all the month lines and the distance from their reconstructed gnomon to the lines representing the equinoxes and the winter solstice but not the width of the grid nor the exact location of the gnomon's tip. They had expressed these not in actual Neapolitan *oncie* but in what Ferrari termed "arbitrary units" and so Ferrari, like others before him, had converted his own metrical numbers into these units as well. He had measured the grid in the Accademia's engraving and an enlargement of a black-and-white photograph and superimposed the resulting grids to illustrate their close agreement.[173] While the dimensions he derived using this methodology differ slightly from the actual prosciutto, and therefore in his conversion of the units into millimeters, the resulting errors are relatively insignificant given the dial's small scale.[174]

Using the lengths that Carcani had given for the distance from his reconstructed tip to the equinoctial and winter solstice lines, Ferrari concluded that the Accademia's proposed values for both the latitude and solar declination probably were off by ± 6 units, for an average latitude of 41.38° ± 0.36° (= 41.74° to 41.02°; 41° 44′ 24″ to 41° 1′ 12″) and an average declination of 23.74° ± 0.65° (= 24.39° to 23.09°; 24° 23′ 24″ to 23° 5′ 24″). He suspected that the Accademia may have realized their calculations were inaccurate, and so had offered only this limited data, or that they had erred in their haste to complete their investigations aimed at discrediting the authors of the *Encyclopédie*. In particular, he questioned the precision of the dimensions locating the tip's distance from the equinoctial and winter solstice lines and concluded that their method for determining the date of the prosciutto's construction was wrong. Noting that their 21′ rate for the diminution of the ecliptic would date it to around 600 BCE he proposed the date of 8 BCE

---

173. Ferrari did not cite his source for the photograph but, if it had not been supplied by an acquaintance, the lime deposits on some of the inscription's characters are similar to those seen in Wojcik's plate, though he did not list Wojcik in his bibliography.

174. Ferrari had measured the equinoctial line as 33.5 mm, equivalent to the Accademia's 1,000 arbitrary units, so each unit was 0.0335 mm, though several of his calculations appear to use around 0.0303 mm. The actual length of the equinoctial line is 31.23 mm. His dimensions also differed from those of the Accademia and are not consistent: his length for the first line representing the summer solstice was 1690 units (54.11 mm) rather than the Accademia's 1686, but he states that these units are equivalent to 56.6 mm, which would be 1,896 units. The actual length is 52.48 mm. Ferrari used the Accademia's unit values for most of his calculations, however.

Enlightenment 261

based not on astronomical calculations but on the use of the honorific "Augustus" in the calendrical inscription.

Ferrari then used the average of his values for latitude and declination to determine a gnomon that was situated 17.7 mm above the upper left-hand corner of the dial, the intersection of the lines representing the summer solstice and the horizon. This resulted in a horizon line that was longer than that on the prosciutto. In questioning whether the Accademia had reconstructed the gnomon in the wrong position, he based his subsequent inquiries only on dimensions he could derive directly from the engraving and the photograph. First, he found that a latitude of 44.5° and a declination of 20° generated hour curves reasonably close to those on the prosciutto. Such a design seems unlikely, however, as it employs implausible coordinates and reconstructs the gnomon tip an unwieldy 23 mm from the surface; Ferrari had rejected it. He then tried the latitude for Rome (41.63°) with the value used by Vitruvius for the declination (24°) and applied these to two different gnomons positioned left of the grid, one 15.6° left with a gnomon height of 15.9 mm and the other 32.8° left with a gnomon height of 14.8 mm. These, too, generated hour curves reasonably close to those on the prosciutto, but he doubted that the ancient craftsman would have designed it in this way since virtually every sundial from antiquity aligned the gnomon tip directly above the grid. Employing these same values, he assumed the length of the horizon line was wrong. With a gnomon height of 17.7 mm he found that the horizon line needed to be 44.2 mm instead of the 40.8 mm he obtained from the engraving. While the hour curves thus generated agreed favorably with the prosciutto in the summer months, they extended well past the grid in the winter months and therefore would not fit on the surface of the prosciutto. He hypothesized that the craftsman had begun drawing the grid too far to the right and, to compensate, had shortened the distance between the month lines. The craftsman may have thought that he could reduce the distance between the month lines in the same way they can be contracted with no ill effects on cylinder dials.[175] By "squeezing" the grid with the wider horizon line to fit the scale on the prosciutto, Ferrari was able to closely match the hour curves; at only some four mm wider than the existing grid, the correct one would have fit if the craftsman had begun farther to the left. This, however, resulted in substantial inaccuracies of over an hour in the winter months because the plane, or azimuth,

---

175. Because the gnomon is rotated to stand perpendicular to, and cast its shadow down, each month line, rather than casting its shadow across the grid, the month lines on an altitude dial can be spaced as close or far apart as the diameter of the cylinder on which they are inscribed.

of the dial needed to be rotated from its proper axis to align the shadow to the month line. His investigations suggested that the dial had been designed graphically, drawing on tables of values for the height of the sun at various hours and on the days when the sun entered each sign obtained using Vitruvius' method for drawing an *analemma*. Ultimately Ferrari marveled that so many previous scholars had accepted the Accademia's conclusion that the prosciutto was accurate to within two to three minutes: not only would the miniscule distance that averaged 6.7 mm between the third and fourth hour lines on the equinoxes require the user to distinguish between 0.2 and 0.3 mm, but the shadow produced by the tip alone would have been about 0.3 mm in diameter, making such a precise reading nearly impossible.

The Prosciutto Sundial: Casting Light on an Ancient Roman Timepiece from the Villa dei Papiri in Herculaneum.
Christopher Parslow, Oxford University Press. © Oxford University Press 2024. DOI: 10.1093/9780197749418.003.0003

4

Observations

THE ACCADEMIA HAD posited three possible sources of inspiration behind this "artist's joke" of inscribing a sundial onto a prosciutto. They mused that it might have been intended to represent an association between pigs and the owner's name. Or it might have been symbolic of the owner's gluttonous lifestyle. Alternatively, they had looked to the catalog of common types provided by Vitruvius and hypothesized that it had derived its name from its shape: the "Perna" type. This last can be ruled out, despite de Solla Price's belief, since no other examples of the prosciutto sundial have been recovered, nor have any ancient sundials inscribed on comparable objects been found. No subsequent scholars have pursued potential links between the design of the sundial and the owner of the Villa dei Papiri in Herculaneum because they either were ignorant of its discovery in this grand, luxuriously appointed seaside residence or simply did not investigate the possible significance of its provenance there.

Modern prosopographical studies derived from the extant epigraphical record appear to rule out any direct connection between the shape and the owner's name. The Accademia had suggested the *nomen* Suillus (swine) and the *cognomina* Perna (ham) and Scrofa (sow), but there is no archaeological evidence for individuals so named in Herculaneum or Pompeii. Inscriptions, primarily graffiti, attest to a Vaccula (heifer), an Asella (little ass), a Panthera (panther), and even a Lampyris (glowworm), but the closest *cognomen* with porcine associations is Aper (boar).[1] Varro specifically notes that the *nomen* Porcius was derived from *porcus* (Varro *Rust.* 2.1.20) and while Marcus Porcius, son of Marcus, was one of the first magistrates of the Roman colony

---

1. Excavations in Regio VII.6.3 (Casa di M. Spurius Saturninus et D. Volcius) in Pompeii recovered a fragment of painted stucco with the word PORCVLVS scratched into the surface in small letters. This has been read as "piglet," the diminutive of *porcus*, with the suggestion it was a nickname or slur, but there is nothing to secure its use as a name. S. E. Chamorro and M. C. Sánchez, "A Note on a New Graffito from Pompeii: *Porculus* (Casa della Diana Arcaizzante)," *ZPE* 206 (2018) 234–35.

established at Pompeii in the 80s BCE, the Porcian clan did not otherwise establish deep roots in the area. The owner could have been a *pernarius*, a seller of hams, the sundial perhaps viewed as illustrating the passage of time required to properly cure a *perna*, though its provenance in this rich Villa makes this less likely, and there are no references to *pernarii* in the epigraphical record.[2]

4.a) *The Prosciutto and Philodemus in the Villa dei Papiri*

Establishing a link between the prosciutto and the lifestyle of its owner depends upon identifying the proprietor or proprietors of the Villa dei Papiri. In the absence of secure epigraphical evidence, scholars continue to debate this question, basing their conjectures largely on circumstantial evidence drawn from the eponymous charred scrolls. The scrolls had been recovered between 1752 and 1754 and already by 1755 Paderni could report to the Royal Society that one scroll, a treatise on music, had been unrolled and that "the name of the author, who was called Philodemus, is found written twice, at the end of the piece."[3] Another member of the Society, in observations appended to Paderni's report, further identified the author as the Greek philosopher Philodemus of Gadara (110–30 BCE), an Epicurean previously known only for a collection of his epigrams, some erotic. Philodemus was a contemporary of Cicero who features, albeit anonymously, in one of Cicero's most scathing political speeches, the *In Pisonem* of 55 BCE. The object of Cicero's attack was the former consul (58 BCE), and later censor (50 BCE), Lucius Calpurnius Piso Caesoninus (ca. 100–43 BCE), Julius Caesar's father-in-law through Piso's daughter Calpurnia and, more importantly, culpable in Cicero's eyes for his exile to Macedonia and the destruction of his house on the Palatine hill in

---

2. Castrén 1975, 210–11 (Porcius), 247–65: e.g., Aper (*CIL* 10.817, 824, 893, 1008), Vaccula (*CIL* 10.818). The number of derivatives of the Latin *porcus*, hog or pig, and *porca*, sow, is prodigious: the diminutives *porculus*, *porcellus*, and *porcula*; the adjectives *porcinus*, *porcarius*, or *porciliaris*, the latter applied specifically to a young *porcilia*, *porcella*, or *porcellinus*; *porculator*, a pig breeder and *porcinarius*, a pig seller; *porculatio*, the breeding of pigs; and *porcinarium*, a pig sty.

3. C. Paderni, "An Account of the Late Discoveries of Antiquities at Herculaneum etc. in Two Letters from Camillo Paderni…to Thomas Hollis…(June 28, 1755)," *PhilTrans* 49 (1755–1756) 503–4; this is the same letter describing the ithyphallic satyr tripod seen by La Condamine. Gray 1754, 825, was aware in 1754 that one of the papyri dealt with music and another with Epicurean philosophy. Similarly, Lalande 1769, 132, listed this "poem sur la musique" among the unrolled scrolls he had seen but did not name its author or those of the others whose contents had been identified. For a comprehensive summary of the scrolls, see D. Sider, *The Library of the Villa dei Papiri at Herculaneum* (Los Angeles 2005).

the year of Piso's consulship. Cicero states that at that time Philodemus lived with Piso, scarcely leaving his side (Cic. *In Pis.* 68), and that Piso had taken up Philodemus' brand of Epicureanism in which the pursuit of pleasure was life's principal goal. Philodemus had beguiled Piso, according to Cicero, writing poems to and for him that captured his life and lusts in verse (Cic. *In Pis.* 70–71), while Piso excelled in licentiousness yet was utterly lacking in taste, refinement, and elegance, heaping his tables not with the embossed tableware and fine seafood emblematic of his elite standing but with spoiled meat (*carne subrancida*) and common store-bought bread and wine (Cic. *In Pis.* 67). This characterization of Piso, even accounting for the rhetorical exaggeration, hardly seems compatible with the luxurious appointments found in the Villa dei Papiri. Nevertheless modern scholars have seized upon their close ties, the fragments of texts in the collection of scrolls by Epicurean authors like Lucretius, and the preponderance of scrolls attributable to Philodemus, including one dedicated to a certain Piso, to argue that the core of the collection constituted Philodemus' own library and that Philodemus had lived in this Villa, which Piso had built.[4]

Comparetti was the most ardent early advocate of Piso Caesoninus as the Villa's original owner, having set forth his arguments in 1879, but the association between Piso and the Villa had been made already earlier that century, if not earlier.[5] The Accademia, in finally publishing the *editio princeps* of the papyrus containing Philodemus' *On Music* in 1793, had made only passing reference to the connection drawn by Cicero between Philodemus and Piso but had ignored entirely the manuscript's original archaeological context and so had not associated Piso with the Villa. The link between the two already had begun to take root in 1810, however, when it was stated confidently that the scrolls had been found "in the remains of a house, supposed to have

---

4. For a thorough and balanced survey of the history of scholarship on the ownership question, see M. Capasso, "Who lived in the Villa dei Papiri," in M. Zarmakoupi, ed., *The Villa of the Papyri at Herculaneum: Archaeology, Reception, and Digital Reconstruction* (Sozomena, Studies in the Recovery of Ancient Texts 1, Berlin 2010) 89–113. A fervent advocate of the identification is M. Gigante, *Philodemus in Italy* (Ann Arbor 1995) 1–13, while a more measured assessment of the evidence is provided by D. Blank, "Philodemus," in E. Zalta, ed., *The Stanford Encyclopedia of Philosophy* (Spring 2019 edition): https://plato.stanford.edu/archives/spr2019/entries/philodemus/. For a critical overview of the most recent scholarship on the decorative program, date, and ownership of the Villa, see A. Angeli, "La Villa dei Papiri e gli scavi sub divo fra archeologia, filologia e papirologia," *Studi di egittologia e di papirologia* 16 (2019) 9–70.

5. Comparetti 1879, 1–52, a more detailed presentation of the evidence than D. Comparetti, "La Villa dei Pisoni in Ercolano e la sua biblioteca," in *Pompei e la regione sotterrata dal Vesuvio nell' anno 79* (Naples 1879) 159–76.

266   THE PROSCIUTTO SUNDIAL

belonged to L. Piso."[6] This evidently was based exclusively on the cohabitation of Philodemus and Piso alluded to by Cicero. Comparetti had built his case in support of a scenario in which Piso had constructed the Villa and Philodemus had lived there, revising his own manuscripts in the company of the few other Epicurean texts contained in the library. Comparetti argued that the apparent lack of furniture in the Villa was consistent with Cicero's characterization of Piso and that Piso had acquired many of the sculptures displayed in the Villa during his proconsulship in Macedonia (57–55 BCE), and hence after Cicero's attack. Among these was a bronze bust commonly identified as Seneca, which Comparetti argued neatly fit Cicero's depiction of Piso.[7]

Mommsen had rejected Comparetti's proposal on the grounds that there was no inscriptional evidence for the *gens Calpurnia* in Herculaneum, as remains the case today.[8] This opened the debate up for those defending the Pisonian connection with evidence that bolstered the case both for his presence, at least, in the Bay of Naples and for his philhellenism as a means of countering Cicero's negative image and justifying the obvious richness of the Villa.[9] On the other hand it also opened the door for hypotheses advocating

---

6. The comment appears in W. Drummond and R. Walpole, *Herculanesia: Or Archaeological and Philological Dissertations Containing a Manuscript Found among the Ruins of Herculaneum* (London 1810) ix. Robert Walpole (1781–1856) was a classical scholar while William Drummond (1769–1828), fellow of the Royal Society, had twice served as Envoy Extraordinary to Naples (1801–1803 and 1806–1808). Drummond must have learned of the presumed connection between Piso and the Villa in the course of acquiring material for this publication which Drummond and Walpole had dedicated to the Prince of Wales, the future King George IV. During this period, the prince had negotiated with King Ferdinand to allow his representative, John Hayter, to unroll and read some of the scrolls, a process the Bourbon court had by now largely abandoned; see J. Hayter, *A Report upon the Herculaneum Manuscripts in a Second Letter Addressed, by Permission, to his Royal Highness, the Prince Regent* (London 1811) 31, where he claims in a note that Cicero could see Piso's villa from his own villa at Puteoli across the Bay.

7. The portrait of Seneca or "pseudo-Seneca" (MANN 5616) was the one seen by Gray in 1754 and mentioned in his letter to the Royal Society (see Chapter 1.c).

8. T. Mommsen, "Inschriftbüsten I: Aus Herculaneum," *Antike Zeit* 38 (1880) 32–36. He also reinterpreted Comparetti's reading of a partial inscription naming Piso to "[Th]espis." Mommsen did not take into consideration the fact that there similarly is no trace of Cicero in Pompeii, despite frequent references to his *Pompeianum* in his letters (e.g., *ad Att.* 2.4, 4.9, 10.15, 14.17)

9. See especially H. Bloch, "L. Calpurnius Piso Caesoninus in Samothrace and Herculaneum," *AJA* 44.4 (1940) 485–93; J. D'Arms, *Romans on the Bay of Naples* (Cambridge 1970) 173–74, dates the Villa firmly mid-first century BCE and follows Bloch in accepting that it probably remained in the Calpurnii family through the eruption. D'Arms accepts L. R. Taylor's suggestion (*Voting Districts of the Roman Republic*, Rome 1960, 200, 311) that the citizens of Herculaneum were enrolled in the same Menenian tribe as Piso precisely because of his connections there: this is viewed as countering Mommsen's objections to the lack of epigraphical evidence.

alternative proprietors whose likely ownership of the Villa could be supported by other types of evidence. Diels, who was Mommsen's successor at the Berlin Akademie and would later characterize the prosciutto as a "prank piece," had speculated that one Marcus Octavius, whose name, like Philodemus', appears at the end of two scrolls, may have been the Villa's owner. He did not, however, later go on to consider the prosciutto in light of this proposal.[10]

Subsequent scholars drew on the epigraphical record of Herculaneum to advance alternative proprietors, particularly those who had made significant contributions to the town, had been honored by its citizens, or whose status fit the Villa's grandeur. The list constitutes a who's who of Herculaneum elite society featuring the three most prominent families: the Appii Claudii, the Mammii, and the Balbi.[11] Documented evidence for any of these figures in the Villa's archaeological record is as nonexistent as it is for Piso Caesoninus, however, and no individual or family has been widely embraced. Scholars currently consider strong evidence for the link between Piso and the Villa to be the invocation "O Piso" read near the end of a charred scroll containing Philodemus' *On the Good King according to Homer* (P.Herc. 1507), but this goes further toward confirming Cicero's observation concerning the volume of Philodemus' writings addressed to Piso than establishing Piso's ownership of the Villa.[12]

---

10. Diels 1882, 383–84. Capasso 2010, 96, believed Octavius was a previous owner of the scrolls that became part of the library near the end of the first century BCE.

11. Wojcik 1986, 276–84, argued the first owner was Appius Claudius Pulcher, followed by the Mammii. Pulcher, consul of 54 BCE, was credited with building the Villa while his homonymous nephew, consul of 38 BCE, was honored as *patronus* of Herculaneum with a *bisellium* in the theater. Lucius Annius Mammianus built the theater while his freedman, the *Augustalis* Lucius Mammius Maximus, was honored with a statue on the theater's *summa cavea*. G. Guadagno, "Note prosopografiche ercolanesi: I Mammii e L. Mammius Maximus," *CronErc* 14 (1984) 155 n. 63, and L. A. Scatozza Horicht and F. Longo Auricchio, "Dopo il Comparetti-De Petra," *CronErc* (1987) 157–67, believed the owner was Marcus Nonius Balbus, praetor and proconsul of Crete and Cyrene and another *patronus* of Herculaneum, who built the basilica, walls, and gates, and was honored along with his son with several statues erected throughout the town.

12. The text of P.Herc. 1507, Philodemus' "On the Good King," was first edited in 1844 (S. Cirillo, *Herculanensium Voluminum quae supersunt: Collectio prior*, Napoli 1844), but the reading of Piso's name is credited to the emendation proposed by S. Sudhaus, "Philodemum," *Rheinisches Museum für Philology* 64 (1909) 475–76, to the 1909 text of A. Olivieri, *Philodemi peri tou kath Omeron agathou basileos libellus* (Leipzig 1909) 66, a full century after Walpole and Drummond's association of Piso and the Villa.

4.b) Hoc eius erat matutinum: *Piso Pontifex and the Augustan Calendrical Reforms*

Comparetti's identification of Piso Caesoninus has been accepted in a preponderance of the modern scholarship, especially that of a papyrological nature, because it accorded as well with the assumption that the construction of the Villa dated from the early to mid-first century BCE. When excavations in the 1990s together with detailed studies of the painted décor and the sculpture shifted that date closer to the middle of the century, and therefore closer to the deaths of Piso and Philodemus, Comparetti's proposal for the Villa's second owner received increased attention as the individual most responsible for the Villa as it subsequently evolved.[13] In order to account for scrolls dating after the deaths of Piso and Philodemus, including a Latin poem (*P.Herc.* 817) on the Battle at Actium (31 BCE), and to help bridge the chronological gap between the death of Piso Caesoninus in 43 BCE and the 79 CE eruption of Vesuvius, Comparetti had suggested that the Villa had passed into the hands of Piso's son, Lucius Calpurnius Piso Pontifex (48 BCE–32 CE).

Piso Pontifex, so named because of his appointment to the pontifical college, was *consul ordinarius* in 15 BCE, *proconsul* in Syria and Pamphylia (13 BCE), and served as *praefectus urbi* in Rome for twenty years, all benefits of the close ties he maintained with the emperors Augustus and Tiberius.[14] Just as his father had been a patron of Philodemus and a beneficiary of his writings, so too Piso Pontifex was a patron of the poet Antipater of Thessalonica, who dedicated many epigrams to him.[15] The literary sources paint him as a triumphant military leader, an effective magistrate, and an able administrator whom the Senate awarded a public funeral (*Tac. Ann.* 6.10–11),

---

13. Comparetti and De Petra 1883, 10–11. For example, M. Guidobaldi, "L'Impronta epicurea nella Villa dei Papiri alla luce delle recenti indagini archeologiche," in M. Beretta, F. Citti, and A. Iannucci, eds., *Il Culto di Epicuro: Testi, Iconografia e Paesaggio* (Florence 2014) 160, concluded on the basis of archaeological excavations that the multi-leveled Villa was constructed at one time in the third quarter of the first century BCE with the belvedere added in the Augustan/Tiberian period.

14. The bronze bust of a male found in Herculaneum, often ascribed to the Villa dei Papiri, has been identified as a portrait of Piso Pontifex (MANN 5601); see K. Lapatin, ed., *Buried by Vesuvius: The Villa dei Papiri at Herculaneum* (Los Angeles 2019) 146–47.

15. Many of these are published in the *Greek Anthology*, including, e.g., in the edition of A. S. F. Gow and D. L. Page, *The Greek Anthology: The Garland of Philip, and Some Contemporary Epigrams* (London 1968) #31, 40–44. Some of the epigrams refer specifically to Piso's evident role as *proconsul* in Macedonia, where he may have met Antipater. Another (Gow-Page #30), the dedication on a statue of Dionysus, connects Piso and his house with the god of wine: "I, Dionysus, the fellow-soldier of the Italian Piso, I am set here to guard his house and bring him good fortune. A worthy house have you entered, Dionysus. Fitting is the house for Bacchus, and Bacchus for the house."

Observations 269

while at the same time they highlight his love of leisure (Vell Pat. 2.98.3) and wine. Both Pliny the Elder and Suetonius credit his position as urban prefect as a reward for two days of binge drinking with Tiberius (Pliny *HN* 14.22.145; Suet. *Tib.* 42). Seneca the Younger called him a drunk who spent most of the night at dinner parties and slept until the sixth hour: "such was his morning" (*hoc eius erat matutinum*), he quipped (Seneca *Ep. Mor.* 83).[16] In this scenario, after inheriting the Villa, Piso Pontifex would have maintained the collection of Philodemus' texts in the library, adding additional scrolls, some of an Epicurean nature, and expanding its sculptural decoration in a manner that further enhanced its Epicurean evocations.[17]

The *pontifices* were one of the three priestly colleges of the Imperial era. By tradition, they oversaw the most important aspects of public and private religion, performing sacrifices and religious rites; handling adoptions, wills, and funerary rites; expiating prodigies; and weighing religious matters referred to them by the Senate. While they had lost any influence they held during the Republic in the political sphere, they remained highly visible through their participation in Imperial processions and state sacrifices and banquets, and the position retained its high level of prestige. Both Augustus and Tiberius had been elected *pontifex maximus* by the people, who in turn also elected the *pontifices*, the candidate's names having been put forth to them by the members of the college in the traditional manner. Piso Pontifex would have been one of a relatively small group of primarily patrician elites to hold this lifetime position during the reigns of these two first emperors, his election possibly occurring as early as 12 BCE.[18]

---

16. Seneca *Ep. Mor.* 83: *Maiorem noctis partem in convivio exigebat; usque in horam fere sextam dormiebat: hoc eius erat matutinum.*

17. Pandermalis 1983, 39–40, who saw the dual natures of Piso Pontifex in the Villa's sculptural display and argued for its Epicurean qualities. E. Leach, "[Review of] M.R. Wojcik, *La Villa dei Papiri ad Ercolano,*" *AJA* 92 (1988) 145–46, favors Piso Pontifex as the owner and notes that his death in 32 CE serves as evidence for the year of first transference of the property's ownership. J. Fish, "Lucius Calpurnius Piso Caesoninus: A Philosophical Statesman of the Late Roman Republic," in K. Lapatin, ed., *Buried by Vesuvius: The Villa dei Papiri in Herculaneum* (Los Angeles 2019) 5–9, summarizes the evidence for both Pisones but underscores the Epicurean sentiments of Velleius Paterculus' (*Histories* 2.98) description of Piso Pontifex. The identification of one of the scrolls as Seneca the Elder's history of Rome from the Civil Wars to 40 CE (*PHerc* 1067; V. Piano, "Il *PHerc* 1067 latino: Il rotolo, il testo, l'autore," *CronErc* 47 [2017] 163–250) is evidence that accessions to the library continued after the death of both Pisones. The Villa's owner at the time of the eruption is unknown and could have been one of the proposed alternatives documented epigraphically elsewhere in Herculaneum. In this case, of course, the prosciutto would have belonged to someone else.

18. M. W. Lewis, *The Official Priests of Rome under the Julio-Claudians: A Study of the Nobility from 44 BC to 68 AD* (Rome 1955) 7–18, 29. The date of Piso Pontifex' election is uncertain; Lewis suggests pre-12 BCE, the year Augustus assumed the title of *pontifex maximus*, and

270 THE PROSCIUTTO SUNDIAL

One of the crucial functions of the *pontifices* in the Republic had been to oversee the state calendar: to determine the dates of the equinoxes and solstices; to set the dates for annual festivals, state sacrifices, and political elections; to declare on which days public business could be conducted; and to establish when intercalary days were required to maintain the correspondence between the celestial cycle and the state calendar. They were essentially the astronomers of the Roman state, though doubtless themselves advised by actual ancient astronomers.[19] The calendrical modifications instituted by Julius Caesar and Augustus had largely obviated the need for the *pontifices* on calendrical matters, but there had been one significant event in which Piso Pontifex may have played a key role: consultation with the Senate on the formulation of the *Lex Pacuvia de mense sextilii* of 8 BCE that renamed the ninth month in honor of the emperor at the same time that earlier Augustan measures had resulted in the calendar finally corresponding to the solar year as Julius Caesar had intended.[20] In such a light, this silver-plated astronomical instrument can be seen as a powerful symbol of status and privilege, a constant reminder of Piso Pontifex' position, his close ties to the Imperial family, and the pontifical skills that allowed him to decipher the sundial's complexities, even if the lifestyle depicted in the literary sources meant he would have consulted it only during the last six hours of the day and especially to demonstrate to his guests that the *convivium* had begun at the invited hour. If it had been a gift from Augustus or Tiberius, so much the better. This all assumes, however, that one of his descendants had retained ownership of the Villa and its accoutrements down to the time of the eruption.

4.c) *Epicurus' Pig?*

Connecting the prosciutto to the strands of Epicureanism running through the physical evidence of the Villa is easier and does not necessitate tying it

before Piso was *quaestor* because he was the member of a patrician family. He was honored with a statue from the Claudian era in the forum of Velleia that singled out his positions of *pontifex* and *consul*: L. *Calpurnio/L. F. Pisoni/Pontif Cos* (*CIL* 11.1182). For his membership in the Arval Brethren, see *CIL* 6.2023a and 2024.

19. According to Pliny the Elder (Pliny *H.N.* 18.57), Sosigenes of Alexandria had advised Julius Caesar on his calendrical revisions, while Plutarch put his reform of the calendar in the hands of "the best philosophers and mathematicians" (*Caesar* 59.1: Καῖσαρ δὲ τοῖς ἀρίστοις τῶν φιλοσόφων καὶ μαθηματικῶν τὸ πρόβλημα προθείς).

20. For a summary of calendrical reforms in the period of Caesar and Augustus, see R. Hannah, *Greek and Roman Calendars: Constructions of Time in the Classical World* (London 2005) 112–30.

Observations 271

directly to the Pisones. Pigs had long been celebrated in the literary sources for their singular capacity to supply food. In his philosophical dialog, *de Natura Deorum (On the Nature of the Gods)*, dating from 45 BCE, Cicero recorded an observation made by the third-century BCE Stoic philosopher Chrysippus of Soli (ca. 279–206 BCE) from which a popular adage would be derived. In the course of cataloging the benefits to man of the various beasts— sheep for their wool, dogs for faithful companionship, oxen for the plow— Cicero's narrator, Balbus, muses, "In fact what does the pig produce other than food, for Chrysippus even says that its soul was given to it as salt so it would not become rotten, and nature produced no more fertile beast suitable for feeding mankind" (Cicero *Nat.D.* 2.160).[21] Salt, of course, was the principal ingredient in the process of curing hams to produce prosciutto, a recipe spelled out in detail by that most famous member of a porcine clan, Marcus Porcius Cato, in his second-century BCE work, *de Agricultura* (Cato 162.1–3).[22] The philosophical notion appears to be that the pig's soul functioned the same as salt to preserve it for man's enjoyment: prosciutto still on the hoof. Varro, in his *de Re Rustica*, dating from 37 BCE, would further underscore the association between the bounty of pigs and their value for banquets. Before praising the size and quantity of prosciutto from Gaul, Varro recast Chrysippus by remarking that, "There is a saying that the race of pigs is expressly given by nature to set forth a banquet; and that accordingly a soul was given them just as salt, to preserve the flesh" (Varro *Rust.* 2.4.10).[23] Pliny the Elder, writing in the 70s, subsequently observed, "The pig is the most irrational (*maxime brutum*) of animals and it is said not inelegantly that its soul was given to it in the place of salt" (Pliny *Nat. Hist.* 8.77.207).[24]

Pliny's characterization of pigs as "irrational" captured one trait that had made pigs an ideal symbol for Epicureans. Animals in general, and pigs

---

21. Cicero *Nat.D.* 2.160: *Sus vero quid habet praeter escam; cui quidem ne putesceret animam ipsam pro sale datam dicit esse Chrysippus; qua pecude, quod erat ad vescendum hominibus apta, nihil genuit natura fecundius.* Chrysippus, too, left his mark on the Villa's remains in fragments of at least two of his texts, his *Logical Questions* (PHerc. 307) and his *On Providence* (PHerc. 1038 and 1421), along with perhaps a third, as yet unidentified (PHerc. 1020).

22. Columella also discusses the salting process in his *Res Rustica* 12.55.1–4. For cured meats in antiquity in general, see F. Frost, "Sausage and Meat Preservation in Antiquity," *GRBS* 40 (1999) 241–52, who notes the close parallel between ancient and modern methods of curing hams and observes, "With enough salt any piece of meat can be made to last forever."

23. Varro *Rust.* 2.4.10: *Suillum pecus donatum ab natura dicunt ad epulandum; itaque iis animam datam esse proinde ac salem, quae servaret carnem.*

24. Pliny *HN* 8.77.207: *Animalium hoc maxime brutum, animamque ei pro sale datam non inlepide existimabatur.*

THE PROSCIUTTO SUNDIAL

especially, were viewed as seeking pleasure naturally, without need of reasoning: they wallowed in the muck and gorged themselves, oblivious to the threat of the butcher's knife.[25] Pigs embodied *ataraxia*, tranquility, a goal of the Epicurean lifestyle, though Epicurus had advocated emulating pigs only insofar as achieving their tranquil state: true *ataraxia* required the use of reason and the enjoyment of such pleasures as friendship and community. This pursuit of pleasure and its association with pigs was viewed by Epicurus' critics as reducing man to the level of pigs concerned only with stuffing their faces. This, in turn, had led to the common association of Epicureanism with hedonism and gluttony, and the notion of *carpe diem* (Horace *Odes* 1.11).[26]

The relationship between Epicurean hedonism and pigs was rendered in high relief on a silver cup from Boscoreale now in the Louvre. This *modiolus*, or single-handled cup, one of a pair from a *ministerium* (dinner service) of around a hundred silver pieces, is roughly contemporary with the prosciutto, likely dating to the Augustan-Tiberian period.[27] Encircling its sides are six human skeletons striking various poses, but the garland strung above their heads, together with inscriptions like "Enjoy life while you can for tomorrow is uncertain," clearly indicate they are set in the context of a banquet.[28] Similar artistic renderings combined with textual invocations such as "Rejoice while you are alive" illustrate that by this time skeletons, too, were associated with the Epicurean lifestyle. Skeletons even appear together with sundials to serve

---

25. Cicero *de Finibus* 1.11.19: "Every animal, as soon as it is born, seeks pleasure and delights in it as the highest good while it avoids pain as the highest evil and so far as possible avoids it; this it does as long as it remains unperverted, its nature itself judging honestly and uncorrupted." Diogenes Laertius 137–38: animals are "delighted by pleasure and distressed by pain by their natural instincts without need of reason."

26. J. Warren, *Epicurus and Democritean Ethics: An Archaeology of* ataraxia (Cambridge 2002) 129–49, and D. Konstan, "Epicurean Happiness: A Pig's Life," *Journal of Ancient Philosophy* 6.1 (2012) 1–24. For the relationship between food and Epicureanism in general, see P. Schade, "Food and Ancient Philosophy," in J. Wilkins and R. Nadeau, eds., *A Companion to Food in the Ancient World* (Chichester 2015) 72–75.

27. Louvre Museum, Département des Antiquités grecques, étrusques et romaines, Inventory Bj 1923, MNC 1981 (dated 25 BCE-50 CE). A. Olivier, jr., "The Changing Fashions of Roman Silver," *Record of the Art Museum Princeton University* 63 (2004) 2, who designates it as *argenta potoria*. A date around 30 CE is commonly accepted.

28. A. Héron de Villefosse, *Le trésor de Boscoreale* (Fondation Eugène Piot 5, Paris 1899) 58–68, Plate 8.1–2; and, virtually identical, A. Héron de Villefosse, *L'Argenterie et les bijoux d'or du Trésor de Boscoreale: Description des pièces conservées au Musée du Louvre* (Paris 1903) 41–48, Plate 8.1–2. Three of the skeletons are labeled as "Clotho," one of the Fates; "Sophocles the Athenian"; and "Moschion the Athenian" accompanied by a smaller skeleton. On the other cup (Plate 7.1–2) are three unidentified skeletons together with Menander and Archilochus, Euripides, Monimus the Cynic and, perhaps, Demetrius of Phalera.

Observations

as reminders, like Trimalchio's *horologium in triclinio* with its trumpeter (Petronius 26), of the rapidity with which life was wasting away.[29] What must be the principal vignette on the Boscoreale cup consists of two skeletons flanking a round tripod that bears a squat cylindrical object usually identified as a cake (Figure 4.1).[30] The skeleton on the left, carrying a satchel on his shoulder and holding a staff in his left hand, looks directly at and jabs with his right index finger as if rebuking the skeleton standing on the opposite side of the tripod. This other skeleton appears focused on the cake and reaches toward it with his right hand. Inscriptions help clarify the scene for the viewer as the conflict between two opposing philosophical schools: "Zeno the Athenian," the Stoic philosopher, is written to the right of the figure on the left, "Epicurus the Athenian" is written to the left of the figure on the right, and the Epicurean sentiment "Pleasure is the *telos* (highest goal)" is emblazoned directly above the cake.[31] For many scholars this notion is reinforced by a quadruped positioned at Epicurus' feet like a faithful dog, which they

---

29. K. M. D. Dunbabin, "*Sic erimus cuncti*...the Skeleton in Graeco-Roman Art," *Jahrbuch des Deutschen Archäologischen Instituts* 101 (1986) 185–255, esp. 224–30, figs. 37–38. Dunbabin's survey includes (206, figs. 16–19) a sarcophagus from Herakleion with reliefs of a skeleton at a banquet featuring a cake displayed on a tripod, young servants and entertainers, and a sundial. K. M. D. Dunbabin, *The Roman Banquet: Images of Conviviality* (Cambridge 2003) 132–40, discusses a mosaic from the House of the Sundial in Antioch, dating from the mid- to late third century CE, that depicts a conical sundial mounted on a tall column and an inscription in Greek reading ἐνάτη παρήλασεν ("the ninth hour is past") as a man rushes past, presumably on his way to dinner. A tripartite mosaic from a triclinium in another late antique house in Antioch underscores how these notions of death, banqueting, and the swift passage of time had become closely associated by this period. One frame depicts a reclining skeleton with a cup in one hand, two loaves of bread, and an amphora under the inscription *ΕΥΦΡΟΣΥΝΟΣ* ("enjoy"). Another scene shows a column topped with a sundial labeled with the Greek letter Θ (*theta*) to indicate the ninth hour (or perhaps for θάνατος "death"). A togate man points at the sundial while in such great haste to follow the inscribed directive τρεχέδ(ε)ιπνος ("run to dinner") that he has lost a sandal, while the word ἄκαιρος ("ill-timed") is inscribed above a bearded man to his right; see H. Pamir and N. Sezgin, "The Sundial and Convivium Scene on the Mosaic from the Rescue Excavation in a Late Antique House of Antioch" *Andalya* 19 (2016) 251–80; and K. M. D. Dunbabin et al., "A Sundial, a Raven, and a Missed Dinner Party on a Mosaic at Tarsus," *JRA* 32 (2019) 329–58. See also, J. Bonnin, "*Horologia et memento mori*...Les Hommes, la mort et le temps dans l'Antiquité gréco-romaine," *Latomus* 72.2 (June 2013) 468–91.

30. Héron de Villefosse identified it as a cake while Warren 2002, 131, calls it a cooking pot, but it appears to have several layers so a cake is more likely. P. Zanker, *The Mask of Socrates: The Image of the Intellectual in Antiquity*, trans. A. Shapiro (Berkeley 1996) 209–10, fig. 109, identifies it as a "large cake."

31. Warren 2002, 131–32, notes that the satchel and staff are more often symbols of Cynic philosophers.

FIGURE 4.1 Silver cup (*modiolus*) showing banqueting skeletons with inscribed invocations, with close-up of Zenon's (*left*) interaction with Epicurus (*right*), the "tripod with cake," the two quadrupeds and the inscription "Pleasure is the *telos* (highest goal)." From Boscoreale, dated to the Augustan/Tiberian period (25 BCE–50 CE). Louvre Museum, Département des Antiquités grecques, étrusques et romaines, Inventory Bj1923, MNC 1981. Photo: Hervé Lewandowski.

identify as a small pig, raised up on its hind hooves, its forelegs lifted, and its snout pointing toward the cake as if intent on gratifying its gluttony.[32]

---

32. Héron de Villefosse's identification of it as a "petit cochon" has been widely accepted. Zanker 1996, 209, calls it the "proverbial pig" sniffing the cake and Warren 2002, 131, too, refers to a "pot-bellied pig." Dunbabin saw two pigs, and there clearly are two quadrupeds superimposed here but with attributes that make their identifications difficult. The short snout and tail of the one in the foreground suggests a pig, but the long length and pose of its visible front and

FIGURE 4.2 Bronze statuette of leaping month-old female piglet, 40.0 cm high (MANN 4893). Found on May 17, 1756, near the northeast corner of the rectangular peristyle in the Villa dei Papiri, Herculaneum. Su concessione del Ministero della Cultura—Museo Archeologico Nazionale di Napoli.

A statue of a pig leaping in a similar manner was found in the peristyle of the Villa dei Papiri and provides a further link among pigs, Epicurean philosophy, and the Villa's owner. Pigs were "a well-known Epcurean symbol."[33] While the statue itself is small in dimensions, it is a full-scale reproduction of a month-old female piglet (Figure 4.2).[34] Only the toes of its rear hooves

hind legs more closely resemble those of a dog, as does the extended snout and almond-shaped eye of the quadruped behind which might be sitting rather than lunging at the cake; both look up intently at the cake and both appear to be collared. The shape of the quadruped in the rear recalls that of the plaster cast taken of a pig at the Villa Regina in Boscoreale (Inv. 41530; 1.00 m L × 0.62 m H), with long legs and long snout, one of the two breeds of pig most common in Roman Italy (MacKinnon 2001, 661). The heads of the two dogs that appear on the second *modiolus* resemble that of the quadruped in the rear though their jumping pose and long legs are more similar to the "pig" in the foreground here.

33. Gigante 1995, 10, following Pandermalis 1983, 25–26.

34. MANN 4893; 0.40 m high; found May 17, 1756 (at number 48 on Weber's plan) near the northeast corner of the rectangular peristyle. Comparetti 1879, 272 #65: "cinghiale"; Wojcik

276

THE PROSCIUTTO SUNDIAL

touch the ground as it elevates its forelegs, bent gently at the knees, the tip of its curling tail pointing straight up, its ears pressed back against its head to enhance the sense of forward motion. It exudes a pure sense of carefree joie de vivre. Rather than illustrating the hedonism of the Boscoreale cup, this seems to capture the "kinetic pleasures" as defined by Epicurus: pleasures of the mind like pleasing smells and tastes, or the joy of bounding through the sty, as opposed to the "static pleasures" of an absence of pain or distress in body and soul. Epicurus believed the combination of the two pleasures constituted the highest form and ultimate goal of human happiness.[35] "I'm a pig from the sty of Epicurus," remarked Horace after dispensing some Epicurean wisdom in a letter to the poet Tibullus, "fat, sleek, and my skin well cared-for" (Horace *Ep.* 1.4.15–16).[36]

Pork had played a central role on the menu of at least one gathering hosted by Philodemus. An epigram ascribed to him catalogs the contributions his circle of friends had made to this banquet, perhaps one held in the Villa dei Papiri itself. Philodemus had provided pork liver to accompany five portions of pork and the simpler fare of cabbage, salted fish, and spring onions contributed by the others. The festivities were set to kick off precisely at the tenth hour (δεκάτας), Pliny's winter dinner time (*Epistle* 3.1.9), or just as the tail's shadow had risen to fall on the prosciutto's third horizontal.[37]

Whether or not the prosciutto's actual owner had been Piso Pontifex, utilizing this sundial to summon diners at the appropriate hour would have required a near-pontifical level of calendrical and astronomical knowledge. Its design differed significantly from the more common hemicyclium types. Those were fixed in place and operated naturally by the sun's rays: the position of the gnomon's shadow on the dial needed merely be consulted, its readings consistent regardless of the time of year. Even Plautus' parasite (Aulus Gellius *NA* 3.3) could read them on a basic level and organize his day around them. Temporal accuracy never seems to have been a great concern to the Romans,

---

1986, 119, D10, who also calls it a "piccolo cinghiale"; Mattusch and Lie 2005, 327–31, figs. 5.282, 288–90, provide the proper identification and age.

35. Konstan 2012, 14–18; Lapatin 2019, 160–61.

36. Konstan 2012, 1–2; Warren 2002, 134, notes that the Latin term for sty, *grex*, was often used of philosophical schools, a pun that Horace no doubt intended. In a previous line, Horace had admonished Tibullus to "Treat every day that dawns for you as the last."

37. Gigante, 59–60, argued that Philodemus was the true author of an epigram (XI.35) in the *Anthologia Palatina* that G. Kaibel, *Philodemi Gadarensis Epigrammata* (Greifswald 1885) xxvii, had rejected as his. Gigante had mused that the event had taken place in the Villa as well as arguing that Philodemus contributed pork, rather than goose, liver.

Observations

however. The literary and historical sources make it clear that early sundials imported from different latitudes were displayed and used in Rome for centuries with seemingly no one troubled by their inaccurate readings. Even Augustus' own elaborate meridian on the Campus Martius from 10 BCE, its gnomon an Egyptian obelisk and its design "contrived by the mathematician Novius Facundus," was famously already out of alignment by the time of Pliny the Elder (Pliny *H.N.* 36.72–73).[38] Given the difficulty of determining the seasonal hours without ready access to a sundial, it is likely that events took place only more or less at the scheduled hour, and quarter and even half hours were not common points of reference. This is why sundials also were common features of public baths, where inscriptional and literary evidence indicates access by the sexes was temporally based: women from first light through the seventh hour and men from the eighth hour of the day to the second hour of night.[39]

In the prosciutto's case, while it would have been possible for anyone familiar with its basic principles to get a rough sense of the time of day, orienting it correctly to the sun's rays to obtain a precise reading necessitated a degree of human agency and technical skills that went well beyond simply casting the tail's shadow onto the dial. Taking an accurate reading required prior knowledge of four essential elements, all of which figured into how the user oriented and rotated the suspended prosciutto with respect to the sun's location in the sky.

First, the operator needed to know the time of year: whether it was before or after the summer or winter solstice or before or after the spring or fall

---

38. Called the *solarium* or *horologium Augusti* in the sources, scholars debate whether it was a full-scale sundial or simply a meridian, though the archaeological evidence and Pliny's description supports the latter. For summaries of the principal arguments, see E. Buchner, *Die Sonnenuhr des Augustus: Nachdruck aus RM 1976 und 1980 und Nachtrag über die Ausgrabung 1980/1981* (Mainz am Rhein 1982), who strongly argued for a sundial based on his own excavations; and P. Heslin, "Augustus, Domitian and the so-called Horologium Augusti," *Journal of Roman Studies* 97 (2007) 1–20, and M. Schutz, "The Horologium on the Campus Martius," *Journal of Roman Archaeology* 24 (2011) 78–86, for the meridian; additional discussions of the Augustan dial were published together with Schutz' article in this volume.

39. These are the hours stipulated in *CIL* 2.5181.20–21, from Metallum Vipascense in Roman Spain. Lucian mentions both a sundial and a *clepsydra* that bellowed like a bull in the baths of Hippias (*Hippias* 8). An inscription in Rome (*CIL* 6.1261), but likely from Tusculum, lists the names of property owners and the hours of the day during which they could draw water from a channel, perhaps the Aqua Crabra; e.g., *ab hora secunda ad horam sextam* (from the second to the sixth hour). For a summary of the times for other activities recorded in the Roman literary sources, all of which refer to specific hours, see Hannah 2009, 135–39. Talbert 2019, 981–88, presents evidence for the ways in which aspects of Roman daily life, in urban and rural settings, were regulated by reference to the seasonal hours.

equinox. In this way, the gnomon's shadow could be situated to fall on the proper section of the dial, with the knowledge that the days were getting longer or shorter. The operator also needed to know the specific zodiacal month corresponding with the time of year. This would allow the gnomon's shadow to be aligned with respect to the proper vertical on the grid. Knowing the exact day within the zodiacal month would allow the operator to position the shadow's tip either closer to the corresponding vertical, if it was earlier in the passing of the month, or closer to the subsequent vertical, if it was late in the month. Whether that subsequent vertical was to the left or right also depended on knowing the time of year. Having a general sense of the time of day would facilitate aligning the shadow to the proper horizontal. Knowing whether it was ante meridiem or post meridiem would signal in which direction the shadow would be moving on the dial: down for ante meridiem, up for post meridiem. The greatest degree of accuracy was realized when the reading corresponded with the seasonal hour, since the differing lengths of those hours over the course of the year meant different durations for quarter and half hours as well, and this would further complicate positioning the gnomon shadow at the exact point in the grid. Determining the value of a single seasonal hour at a particular time of the year was no mean feat. It required knowing the precise time of sunrise and sunset in order to ascertain the total number of minutes in that day. Dividing this total by the twelve hours of the day yielded the length of a single hour. The beginning of each seasonal hour then needed to be determined by adding the length of each hour over the course of the day, starting with the time of sunrise. These values changed each and every day. The spread ranged from forty-five-minute hours on the winter solstice to seventy-five-minute hours on the summer solstice. Finally, and painfully obvious, knowledge of all of these elements was rendered useless by the lack of sunshine, and the sharper and more direct the sun the more accurate the readings. As a consequence, while the prosciutto's firm grounding in astronomical science allowed it to provide remarkably precise readings across the zodiacal year, the average user would not have had a grasp on all these factors and would have had to settle for an approximation.

4.d) Designing and Casting the Prosciutto Sundial

In favoring form over function, the prosciutto's design posed a further challenge since its surface does not provide the flat vertical plane required for precision readings across the dial. On the other hand, as a scaled model of an actual prosciutto, it is a tour de force executed by a master craftsman (see

Figure 1.1). It is a solid-cast bronze taken from an original model made of sculpted beeswax. The artisan's careful study from nature captured a prosciutto crafted from a pig's left hind quarter in all its glorious details, from the manner in which the rind wraps around the edges of the outer surface to meet the cut on the reverse that released it from the rest of the body and exposed the head of the femur bone, to the natural bulge at the stifle joint where the femur is joined to the fibula and tibia, to the protrusion of the tarsal bones down to the narrowing of the shank at the hock. The craftsman's desire to accurately reproduce the prosciutto's shape is on full display on the reverse where the undulating surface created by the exposed meat conforming to the underlying bone and muscle structure is especially vivid (see Figure 2.12). This realistic rendering causes the greatest distortions on the front, where a slight depression running diagonally from below the stifle joint to near the tail together with an outward bulge and pronounced curvature along the upper rump alter how the gnomon's shadow falls on the dial. The resulting deformations are particularly pronounced for meridiem readings on the first through third verticals and for all readings along the seventh vertical, but, as the Accademia and later scholars noted, irregularities across the entire surface cause slight deviations in how the shadow falls on the dial. What creative spark had inspired the artisan to project the grid of a portable sundial onto a hanging prosciutto, other than the vague similarities between the shape of the dial and that of a prosciutto and the fact that both required suspension, remains a mystery.[40]

Onto this wax reproduction of a prosciutto the artisan transferred key aspects of the dial using a sharp stylus (Figure 4.3). Almost certainly he relied on a scaled drawing that had been meticulously drafted to ensure the horary lines were horizontal and intersected the vertical month lines at the correct points. The drawing, perhaps on vellum, must have been superimposed onto the wax model and the tip of the stylus used to pierce through the drawing into the beeswax at the points of intersection. The resulting circular depressions remain visible at numerous points in the grid, some deeper than others. They are especially clear along the lines representing the horizon, the meridiem, and the winter solstice. Elsewhere they likely disappeared into the grid lines. It is less likely that calipers were used to measure the grid points off the drawing and transfer them onto the wax, given the small scale and the

---

40. J. Mersman, "Moving Shadows, Moving Sun: Early Modern Sundials Restaging Miracles," *Nuncius* 30 (2015) 98, mused that the irregular shape of the grid may have inspired the "supposedly ham-shaped dial."

280 THE PROSCIUTTO SUNDIAL

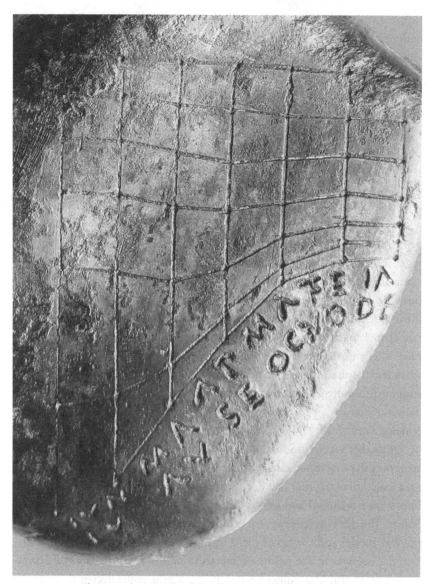

FIGURE 4.3 Close-up photograph of the prosciutto sundial's dial showing craftsman's guide points, grid lines, and calendrical inscription. Author's photo. Su concessione del Ministero della Cultura—Museo Archeologico Nazionale di Napoli.

difficulties of keeping the points properly aligned that such a technique would introduce. The deformations in the wax surface doubtless caused some points to fall slightly out of their proper position, especially along the curving edges, and this resulted in small deviations in the horary readings. For the sundial to provide the most accurate readings in the design latitude, these points on the grid also had to be positioned precisely the correct distance from the tip of the gnomon. The degree of precision required to place the grid in the proper position on the model before the gnomon was in place is strong evidence that the design called for the gnomon tip to rest perpendicular to the intersection of the horizon line and the first vertical line. Positioning the tip here would only require knowing how high it needed to be above the grid; attempting to position it accurately in space to the left of the grid would have been difficult.

Below the resulting network of points the artisan scribed the calendrical abbreviations in freehand with the stylus. The use of sans-serif characters and lack of middle bars made it easier to etch each letter into the wax at this small scale—each barely 3.00 mm high—although the tendency for the stylus to drag in the wax is evident in the shape of some of the letters. As the craftsman worked closer to the curving edge on the right, the letters became less neatly positioned. As a result, none of the abbreviations are tied securely to their corresponding vertical month line so that the upper row of abbreviations could be associated with the verticals to their left and the lower row with those to their right. This may be because the lines of the grid were carved into the bronze with a small flat chisel only after casting, when a straight edge could be used to guide the chisel in scoring the lines between the points marking the intersections. That the grid lines were engraved subsequent to casting is indicated by their remarkably consistent width and depth, and the fact that they are rectilinear channels rather than the grooves a pointed stylus would make in the beeswax, though imperfections in the bronze caused the chisel to carve coarser lines at a few points.

Maintaining the proper alignment and scale of the grid was not the only challenge in this transfer process. The craftsman also needed to ensure the dial would align correctly with the earth's horizon when the finished product was suspended. It is here where the virtuosity of the artisan and the cleverness of the prosciutto design shine through, for if this markedly irregularly shaped object was not perfectly balanced like a set of fine scales, the dial might dip down to one side or another when suspended and throw off the readings entirely. The awkward shape and angle of the shank, to which the moveable suspension ring was ultimately attached, would have complicated the calibration of the balance, especially since it is not naturally centered on the hind quarter. The wax model probably provided sufficient feedback on the efficacy of the design in trial runs, though how the bronze would swing from its suspension ring would have been difficult to emulate in wax. Whether it would translate into a workable bronze product must have remained an open question until the final cast was made. Ultimately the weight of the bronze helped transform the whole into a gravity-driven, balanced plumb bob. Its weight, at around 285 grams, helped counteract in part that other nemesis to taking accurate readings: the wind.

That the tail was put to work as the gnomon captures another aspect in which the craftsman took creative liberties in the sundial's design. The natural tail would have been long enough realistically to reach the required spot on the hind quarter, but it was one of the first parts to go in the butchering and curing process. Entirely fanciful is the tail's gravity-defying fixed curve up and out from the dial. It could not have been curled at any point because the sun's rays would transform this complex shape into a large mass of shadow that could obscure readings on the dial, especially those on the left side. The thinner the gnomon/tail the better, for at this scale a thicker gnomon casts a larger shadow, which, especially in the months with shorter seasonal hours, can block out the space of many minutes or even, in the case of the winter solstice, almost an entire seasonal hour: this is another reason why the pig and its naturally slender tail was ideal for this design. More critically, the tip needed to be positioned precisely in relation to the dial and according to the design latitude since at this scale the difference of a millimeter or more would throw off the accuracy of the entire dial. For this reason, the bronze tail most likely was soldered onto the final cast bronze rather than being worked in wax in the original model even though a seamlessly attached gnomon would have been stronger and serve as further evidence of the artisan's bronze-casting skills. This would avoid the risk of having the position and shape of the gnomon become distorted or damaged in the process of being cast. Maintaining so slender a gnomon in the astronomically correct position would have been a challenge, however. The user seeking the highest degree of accuracy would have needed to bend it back into place on a regular basis as it was bumped and jostled over time, and the resulting stress would weaken the thin bronze. This may explain why the prosciutto was found with its gnomon missing: if it was not ignominiously snapped off during the eruption, it may have broken off long before then.

To cast the final product, the beeswax model was encased in a fine clay slip, its plasticity sufficient to penetrate the grid points and lettering, with additional layers of clay added for solidity. In the course of firing the beeswax melted, allowing it to be poured out and leaving a negative impression of the sundial in the hardened clay mold. Molten bronze then was poured into the hollow mold and allowed to cool before the clay was broken away. The required destruction of the beeswax model and the clay mold ensures there was only a single version. The grid lines were likely scored at this point and any surface imperfections sanded smooth. The sundial was then silvered, perhaps through the process known as "close plating" where the object is either dipped in molten tin or wiped with an even layer of tin solder, allowed to cool, and then silver foil pressed into place. Gently heating the object melts the tin solder and binds the silver to the bronze.[41] Traces of silvering may be visible on the back side of the prosciutto's shank, inside some of the grooves of the dial, and in small patches on both sides, but elsewhere it has worn away or oxidized, a condition that had led to the early confusion about whether it was covered in silver or not. While the silvering doubtless added the desired elegant touch of prestige, it also had the practical effect of brightening the surface to make the shadows easier to discern on the dial. The last step was to mount the moveable suspension ring, which had been cast separately, by inserting it into the catch fashioned at the top of the hock and soldering its ends together.[42]

4.e) "The Craft Is Closely Related to the Science"

As extraordinary as the design of the prosciutto's dial appears, it is unlikely to have been entirely unique in its day. Vitruvius clearly stated that "Many authors have left instructions on how to fashion portable hanging sundials (*viatoria pensilia*)" (Vitr. 9.8.1), signaling that the available prototypes must have been many and varied already at the time he was writing.[43] That he could recite a catalog of dial types and provide the names of their designers is a relatively rare example from the ancient world of the enduring association between a popular consumer product and the specific individual whose scientific knowledge and creative genius underpinned its design. While the inventor behind this *perna* sundial is unknown and this singular monument of his ingenuity suggests this form had not been replicated to challenge the popular acclaim of Apollonius' *arachnen* (spider) or Dionysodorus' *conum* (conical) types, the design of the dial alone, with its gnomon rising from the left to cast

---

41. S. La Niece, "Silver Plating on Copper, Bronze and Brass," *The Antiquaries Journal* 70 (1990) 102–14. Determining the precise process employed requires chemical analysis of the prosciutto's surface. The melting point of tin is 449.5°F/231.9°C; the melting point of bronze, depending on the amount of copper included in the alloy is 1500°F/850°C, while the melting point of (pure) silver is 1762° F/961°C.

42. The suspension ring is assumed to be ancient since both Alcubierrre and Paderni specifically refer to it in their descriptions of the sundial soon after its discovery.

43. Vitr. 9.8.1: *Item ex his generibus viatoria pensilia uti fierent, plures scripta relinquerunt.*

284   THE PROSCIUTTO SUNDIAL

its shadow across the horary grid, may well have been duplicated at other scales and projected onto other shapes.[44]

Vitruvius knew that a gnomon of a given height casts shadows of differing lengths depending on its geographical position, singling out the differences among Athens, Alexandria, Rome, and Placentia and noting that the design of sundials had to conform accordingly (Vitr. 9.1.1). He described how to graph the seasonal hours for a given latitude using what he termed an *analemma*, a geometric drawing depicting the relationship between the altitude of the sun and the length of the shadows cast by a vertical gnomon at that latitude. Vitruvius considered the *analemma* a fundamental tool in the design of sundials (Vitr. 9.7.2–6, 9.8.1).[45] Drafting one helped determine the length of the meridiem shadow for each month at the design latitude, a concept commonly not expressed in degrees in Vitruvius' day but according to the ratio between the height of a gnomon and the length of its shadow. Tables of the ratios at different latitudes had been worked out long before Vitruvius, perhaps already by Eratosthenes of Cyrene (ca. 276–195 BCE) in the third century BCE or Hipparchus of Nicaea (ca. 190–120 BCE) in the second century BCE.[46] Vitruvius therefore knew that the ratio between gnomon height and equinoctial shadow length in Alexandria was 3:5; in Athens, 3:4; in Rhodes, 5:7; and in Tarentum, 9:11 (Vitr. 9.7.1).[47] In walking his reader through the

---

44. Gibbs 1976, 59–65, surveys modern attempts to associate Vitruvius' list of inventors with known sundial designs and shapes. Portions of a small rectangular box (11 × 5 × 5 cm) of ivory and bronze identified in part as a portable sundial of unique design were found in a shop in Pompeii in 1912 (Regio I.vi.3) designated the *officina* (office) of Verus the *mensor* (surveyor) from other instruments associated with surveying, including a *groma* (measuring rod), found there. The box's lid was incised with thirteen lines radiating from a hole set in the middle of one of the long sides with at least two intersecting lines. Other incised lines of various lengths adorned the two surviving sides of the box and these are believed to have served as measured scales (*tabulae rationabilium linearum*) for use in drawing plans in the field. The box is MANN 1762. Because the design corresponds with no known Roman sundial, it seems more likely to have been used for surveying. For the excavations, see *NSc* (1912) 252–56 (M. Della Corte); for the identification as a sundial, see M. Della Corte, "Groma," *Monumenti Antichi*, vol. 28 (Milan 1922) 85–88, fig. 19; and G. Sena Chiesa and M. P. Lavizzari Pedrazzini, eds., *Tesori dei Postumia* (Milan 1998) 272, inv. II.7 "Meridiana" (called a *horologius viatorum* [*sic*]).

45. Vitruvius' text in these sections appears to be corrupt and his description is incomplete.

46. C. Marx, "Analysis of the Latitudinal data of Eratosthenes and Hipparchus," *Mathematics and Mechanics of Complex Systems* 3 (2015) 309–39.

47. He may have singled out Placentia (Piacenza, Italy) in his earlier observation because the ratio of the equinoctial shadow at that latitude is 1:1. Pliny the Elder offered similar ratios for a wider number of geographical locations and included the length of the longest day (June 21) in each in terms of equinoctial hours (Pliny *HN* 6.212–20): for the latitude of Naples, where the equinoctial shadow was 6:7, the longest day was equal to fifteen equinoctial hours; for the latitude of Rome the longest day was equal to 15 1/9 or 15 1/5 equinoctial hours.

necessary steps in drawing an *analemma* he had used Rome, where the length of the equinoctial shadow was 8:9. Pliny the Elder, later in the first century, likewise stated that a nine-foot gnomon at Rome cast an eight-foot shadow (Pliny *HN* 6.217). This ratio is equivalent to the absolute value of 0.888, or a one-meter-high vertical gnomon casting a horizontal shadow 0.88 m long. This constitutes the tangent of the resulting right angle whose opposite is the gnomon and whose adjacent is the shadow, the inverse of which is 41.61°, which is equivalent to what Vitruvius and his contemporaries believed was the latitude of Rome. The shadow lengths for each seasonal hour similarly were known to be proportions of the meridiem shadow dependent on the altitude of the sun, and these values, too, would have been determined well before Vitruvius. This suggests that tabular compilations of the ratios of the seasonal hours for each zodiacal month for the latitude of Rome would have been readily available or, at least, that any competent designer of sundials working in this latitude already would have calculated them, whether via an *analemma* or through simple observation. These values would have been valid for a single gnomon of fixed height casting a shadow along each month line. What best reveals the scientific acumen of the designer of the prosciutto's dial is the fact that they knew that a single gnomon of the height required to cast the hourly shadows along the length of the long summer solstice line also would function correctly for the hours making up the shorter equinox line, as well as the even shorter winter solstice line, by simply rotating the face of the prosciutto to follow the course of the sun through the sky. Moreover, their design acumen is demonstrated by the fact that having the tail cast the shadow across the dial enhances the readability of the individual hours compared to a single perpendicular gnomon projecting its shadow down onto the grid from the upper left-hand corner.

This necessity to manipulate the instrument is what characterized the *viatoria pensilia* type: they were *pensilia*, hanging, because they had to be suspended and rotate freely around their axis to orient the gnomon toward the sun, and they were *viatoria*, traveling or portable, because they were not fixed in place but could be picked up and utilized where needed or where they could best capture the sun's rays, provided they remained within a reasonable distance of the design latitude. In the process of orienting the prosciutto's gnomon/tail to the sun, while the adjacent of the right triangle remains the shadow, the vertical gnomon constituting the opposite in a horizontal dial is transformed into a "diagonal gnomon" running from the tip of the gnomon/tail to the top of the vertical month line at its intersection with the upper line representing the horizon: different geometry, same result.

Producing the prosciutto sundial would have required scaling down a known dial to a size that would fit on the side of the prosciutto and setting the gnomon/tail to the correct height. Putting aside the difficulties of accurately positioning the gnomon and focusing on the grid alone, this process of scaling had magnified the potential that small errors in the position of the horary lines would yield incorrect readings. As Carcani and later commentators noted, even small discrepancies in the positioning of the lines can throw the readings off by many minutes. These errors, in turn, were amplified by the difficulty of transferring the grid onto the surface of the prosciutto in its beeswax form. Despite the best efforts of the artisan, the irregularities of the object's shape had resulted in small distortions to the width of the grid, the lengths of the month lines, and the undulating courses of the horary lines. Finally, all these errors are compounded further in modern studies by the heightened degree of precision introduced by advancements in science and mathematics that the ancient designer would not have sought in the first place nor been able to attain. The resulting errors in the ability of the prosciutto to clock the correct time likely would have been imperceptible or of little concern to its Roman user even if it was deployed at the latitude of the Bay of Naples rather than in Rome, as the Accademia had rightly noted.

Carcani's research that established the design latitude for the prosciutto as $41°\,39'\,45''$ ($41.66°$) was less a revelation than a confirmation of the most logical latitude for its design in the period of its construction: the "most known" latitude, as Carcani himself acknowledged. It did, however, rule out the alternatives by narrowing the geographical area in which the prosciutto could function accurately. Yet the Accademia's zeal to eclipse the French Académie had led them to introduce a nearly absurd degree of precision. Even La Condamine, the first professional astronomer to examine the prosciutto firsthand, had concluded its small scale would make it difficult to accurately determine its design latitude, as Calkoen, Drecker, and others would later affirm in their mathematical studies. Complicating the process of reconstructing how Carcani had conducted his research is the Accademia's failure to provide a complete account of their analysis, and this suggests they realized their haste had produced incomplete and problematic results.

Carcani had recognized that there were problems with the dial. He had concluded, for example, that a bulge in the prosciutto's surface near the meridiem line had caused the first and second vertical lines to be extended so the shadow could reach the meridiem line in those months. He observed that these longer lengths, in turn, would require a higher gnomon than was suitable for the dial as a whole, by which he meant that a higher gnomon would

Observations 287

change the shadow ratios for the entire dial. He did not, however, then go on to state what he believed the correct lengths of the lines should be. Because of this concern he did not use the length of the first vertical to derive the correct height for his reconstructed wax gnomon, even though this was the operative gnomon for the entire grid and, as a perpendicular measurement off the surface of the prosciutto, would have been substantially easier to measure with accuracy than the two measurements he did provide: the diagonal distances between the tip of his reconstructed wax gnomon and the points of intersection of the fourth and seventh vertical month lines along the horizon line. The exact end points of those diagonals on the grid would have been difficult to measure to Carcani's stated degree of precision—881 and 1,482 units, respectively—because of the undulating surface of the prosciutto, the uncertainty about where to end the diagonal along the horizon line, and the fact that the tip of his wax gnomon was suspended in space. While he could have measured these distances using precision calipers, such diagonals nevertheless would have been virtually impossible to establish so precisely and likely would have resulted in different numbers with each attempt. More importantly, he had fixed the position of his wax gnomon to take readings on the spring equinox in 1761 in Naples. He had himself established the latitude for Naples as 40° 50′ 15″, so any readings he obtained in 1761 for the equinoctial shadows with his reconstructed gnomon would have been based on a ratio of 0.851, not 0.881.[48]

It is therefore unlikely that he obtained these measurements from his reconstructed model, despite his statement that he had determined the lengths of the other lines and their tangents from the tip of his wax gnomon and taken his readings based on its position. More likely he established them mathematically, as have all subsequent scholars. The very precise nature of the exact value of 0.881 (41° 22′ 48″ = 41.38°) afforded the semblance of having been carefully derived from his empirical research. That it changed the value of the third decimal place of the 0.888 (= 41.63°) he could have arrived at mathematically from Vitruvius' or Pliny the Elder's 8:9 ratio or the 0.8897 of Ptolemy's 41° 40′ (= 41.6667°) latitude for Rome underscores the absurdity of Carcani's display of scientific rigor, especially since by subsequently factoring in for solar refraction, the semi-diameter of the sun, and solar parallax he wound up nevertheless raising it to a tangent of 0.889 (41° 39′ 45″ = 41.6625°).

---

48. The coordinates of the Chiesa di San Carlo alle Mortelle in Naples are latitude 40° 50′ 20″, longitude 14° 14′ 29″. The calculations for here in 1761 confirm an error of roughly six minutes in the second and third hours.

That this was sufficiently close to Ptolemy's latitude evidently had satisfied him that it had been designed to ancient standards, even though he acknowledged it was less than Rome's actual latitude, which he gave as 41° 54′ (= 41.9° = 0.897).

Carcani had not provided his measured length of the horizon line or the distance between the fourth and seventh lines, and this, coupled with the errors that had crept into the transfer of the grid onto the wax prototype and then the cast bronze, make it difficult to determine exactly where along the horizon line he had taken his measurements. The tolerances at this small scale are so fine that the slightest variation in where the measurement is taken changes the shadow to line length ratios, and consequently the dial's entire design, by shifting the placement of the vertical month lines. For example, the depressions in the bronze that should mark the points of intersection between the verticals and the horizon fell to the right and the left of the equinox and winter solstice lines, respectively. Circles centered on or near these points that use the length of Carcani's diagonals for their radii all intersect at varying distances to the left of the grid, rather than perpendicular to it, and yield relatively low gnomon heights (Figure 4.4). A lower gnomon would, in turn, correspond to a shorter line for the summer solstice, a fact that may have led Carcani to conclude the first two month lines were too long. The Accademia's

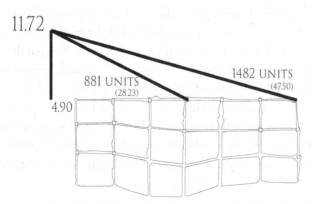

FIGURE 4.4 Probable height and location to the left of the grid of Carcani's reconstructed wax gnomon, based on dimensions of tip to equinox and tip to winter solstice lines provided in the Accademia's preface. Author's image.

Observations   289

scaled engraving of the prosciutto shows the tip of the wax gnomon ending roughly 7.50 mm to the left of the dial, an even more unlikely position for the gnomon in that it would pull all the vertical month lines substantially toward the left. Their engraving neatened up many of the irregularities found on the original but nevertheless closely reproduces the length and position of the grid's lines. Carcani likely based his calculations on the original drawing on which the Accademia's engraving was based rather than the prosciutto itself since he is not known to have frequented the museum.

Once Carcani had settled on the ratio of 0.889 for the equinoctial shadow, he could have taken two approaches to determine the ratio he provided for the winter solstice line. His stated approach was to divide the measurement he obtained from the tip of his gnomon to the top of the vertical representing the winter solstice along the horizon line by the length of that seventh line: $1482 \div 687 = 2.15720$, the tangent of $65° 7' 45''$ ($65.129°$).[49] On the other hand, he may have simply added the standard value of the solar declination in his own day ($23° 28' = 23.46°$) to his equinoctial latitude ($41° 39' 45''$; $41.66°$) to yield $65° 8' 3''$ ($65.1342°$), the tangent of which is 2.15726—a difference his calculations might not have revealed.[50] Multiplying this tangent by his length for the winter solstice line (687) would have generated his 1482 mathematically. He might have performed the same calculation with the summer solstice, but, in this case, he would have subtracted the solar declination from his equinoctial latitude to find $18° 11' 45''$ ($18.19°$), the tangent of which is 0.328. It may have been the value he obtained by multiplying this tangent by the length of the first vertical (1686 units) that caused him to conclude the height of the resulting gnomon (553 units or 17.83 mm) did not equate with his calculations for the equinox and winter solstice, much less the position of his reconstructed gnomon. That he evidently employed the value for solar declination in 1761 together with the observations he took with his reconstructed prosciutto on the spring equinox in Naples underscores the problematic nature of his analysis.

The Accademia must have considered Carcani's determination of the value of the obliquity of the ecliptic at the time of the prosciutto's manufacture and his method for determining the year in which it had been made to be the *pièce*

---

49. To this, Carcani similarly had added his values for the semi-diameter of the sun, refraction, and the solar parallax to increase the latitude of the winter solstice from $65° 7' 45''$ ($65.129°$) to $65° 26' 15''$ ($65.437°$), so the actual tangent was 2.187.

50. Carcani gave the value of the obliquity in his day as $23° 28' 18''$ so he would need to have dropped the seconds, as he had done with the latitude of Rome; this may have been common practice at the time since Calkoen had used $23° 46'$ and Drecker $23° 37'$.

de résistance of their response to the French Académie. Carcani had derived the obliquity of the ecliptic by subtracting the design latitude he had determined for the spring equinox from that of his winter solstice, that is, 41° 39′ 45″ (41.6625°) from 65° 26′ 15″ (65.437°) to give 23° 46′ 30″ (23.775°). De Louville's 1719 recalculation of the obliquity in Pytheas' day in the mid-fourth century BCE, which Carcani had used as his reference, had set it to 23° 49′ 23″ (23.82°), with a secular diminution of about 1′. Carcani's value for 28 CE was 2′ 53″ (0.048°) less, so not the full three minutes of diminution needed but close enough to Pytheas' to appear plausible. It also fit neatly with an 18′ 12″ rate of diminution to the value of 23° 28′ 18″ (23.4717°) assigned to it in Carcani's day. All of these values are substantially higher than the currently accepted ones. Today, De Louville's calculations of the rate in Pytheas' day would place him in 1080 BCE while Carcani's prosciutto would date to 660 BCE, though it was only in Calkoen's day that these discrepancies began to be realized. The true obliquity of the ecliptic in 28 CE was 23° 41′ 29″ (23.6914°), a 5′ 1″ (0.0836°) difference from Carcani's, while its true value in 1762, when Carcani conducted his analysis, differed only slightly from his, at 23° 28′ 13″ (23.4703°).[51] The current rate of secular diminution, moreover, is set to 46.8″ (0.013°), less than half of what De Louville had reckoned. Reversing Carcani's method for determining the obliquity by adding his equinoctial latitude to the actual obliquity in 28 CE generates 65° 21′ 14″ (65.3539°), tangent of 2.1795, for the prosciutto's design latitude for the winter solstice, rather than his 2.157.

Such modern recalibrations of the true value of the obliquity of the ecliptic in antiquity highlight the advances made in astronomical science from Chaerephon's twelve-foot dinner through the eighteenth century. The resulting enhanced precision, however, exaggerates the dial's inaccuracies and casts a poor light on what was a fairly sophisticated instrument at the time of its manufacture, but designed to less stringent standards based on substantially less astronomical knowledge. Vitruvius, for example, had set the value for the obliquity of the ecliptic in his day at 1/15 of a circle—an even 24°—a value first recorded in 320 BCE by Eudemus of Rhodes. Others, such as Ptolemy, set the value slightly lower, at 11/83 of a semicircle (23° 51′ 20″ = 23.85°).[52]

---

51. These values were calculated using PHP Science Labs' calculator for the Obliquity of the Ecliptic, Nutation in Obliquity and Latitudes of the Artic/Antarctic Circles, which can be set to specific dates, both BCE and CE: http://www.neoprogrammics.com/obliquity_of_the_ecliptic/.

52. Gibbs 1976, 9.

Observations 291

Most studies since the Accademia's have focused on the lengths of the vertical month lines, especially for the equinoxes and the solstices, and what they can reveal about the latitude for which the prosciutto was designed, by establishing the ratios between the length of the gnomon's shadow and the length of each vertical line (see Appendix A: Comparison of Grid Dimensions). This is in large part due to the Accademia having specified these dimensions alone. Relying on these longer values also reduces the potential for errors that can result from calculating the even smaller values for the ratios for the intersection of the shadows for each seasonal hour along each vertical month line. Both Delambre and Woepcke, although also largely theoretical in their overall approaches, had introduced the equations necessary for determining these ratios based on the solar altitude, or zenith distance, throughout the day. While their equations require knowing the value of the latitude, this more fine-grained level of detail can serve as a check on the wide range of combinations of potential design latitudes and vertical line lengths for it reveals how the length of the vertical month line affects at which point the shadow of each seasonal hour falls along it. The ratios derived from these hourly elevations are divided into the product of the tangent of the solar altitude at the meridiem multiplied by the length of the line, with the product providing the distance below the horizon line of each hour along the vertical month line. Performing these calculations across the entire zodiacal year allows the actual distances to be compared with those engraved on the prosciutto. The undulating horizontal lines on the prosciutto serve to connect the points of intersection of each of these shadow lengths along the vertical lines while also tracking the movement of the gnomon point at these hours in the days falling between each vertical. At the same time, such calculations further reveal the slight inaccuracies that arose from attempting to project these celestial phenomena onto a miniature prosciutto.[53] If Carcani had performed these calculations in arriving at his conclusion that the times were only off by a few minutes, he did not explicitly state so. Only Ferrari and Schaldach offered illustrations that compare the differences between the prosciutto's actual dial and the hour lines they had calculated, though neither had taken measurements directly off the prosciutto.[54] Ultimately, analysis of the prosciutto's design must take into consideration the height and position of the gnomon together with how the

---

53. These elevations can be determined with great precision through the Solar Position Calculator of the National Oceanic and Atmospheric Administration's Earth System Research Lab (https://gml.noaa.gov/grad/solcalc/azel.html). The calculator can be set to precise latitudes and longitudes for any hour in any year in the Common Era.

54. Schaldach 1998, 49, fig. 4; Ferrari 2008, figs. 15–20.

resulting shadows progressed up and down the vertical month lines to mark the seasonal hours.

Whatever the dial's designer may have intended for the prosciutto is unlikely to have been accurately realized in the final product. In an era when latitudes and shadow lengths were expressed primarily in ratios and events marked in hours, not minutes and seconds, a latitude as precise as the Accademia's 41° 39′ 45″ more likely would have been the product of manufacturing chance than intent.[55] Its proximity to Ptolemy's 41° 40′ latitude for Rome gave it the sheen of scientific probability, but it is absurdly precise.

This is confirmed by plotting the values of solar elevation at different latitudes to establish the ratio of shadow length to gnomon height for each seasonal hour across the zodiacal year (see Appendix B: Tables of Horary Readings). These calculations were performed for 41°; 42°; the Accademia's 41° 39′ 45″; and, for only the critical spring equinox and summer and winter solstices, for 40° 50′ 20″, the latitude of Naples at the Chiesa di San Carlo alle Mortelle, where Carcani conducted his analysis, and for 40° 48′ 27″, the Villa dei Papiri in Herculaneum. To ensure consistency, the values for the solar elevations were taken for the year 28 CE, the year of manufacture determined by the Accademia's analysis, even though this date falls in the twilight of Piso Pontifex's life.[56] To maintain consistency between grids the length of each vertical month line was rounded to a whole number—54, 48, 40, 31, 24, 21, 21, from left to right—even though their actual lengths on the prosciutto are fractional: 52.8, 47.30, 38.92, 31.23, 24.83, 21.97, and 21.60. One result of this rounding was that the deviation between the hour lengths at each meridiem was artificially exaggerated to be higher or lower than what is represented on the prosciutto. This resulted in greater inaccuracies at those times. Nevertheless, these whole values for line lengths yielded the smallest deviations for all seven month lines at each latitude: calculating with longer or shorter line lengths resulted in even higher discrepancies while calculations employing the actual fractional lengths on the prosciutto also did not provide satisfactory results. In two cases the lengths of the test lines had to be altered. The summer solstice line for the Villa dei Papiri had to be extended an additional two millimeters, to 56 mm or 3.52 mm longer than the engraved length, to yield ratios approximating those on the prosciutto. The second line

---

55. S. Remijsen, "The Postal Service and the Hour as a Unit of Time in Antiquity," *Historia: Zeitschrift für Alte Geschichte* 56.2 (2007) 127–40, presents the evidence for the heavy reliance on precise hours in the Ptolemaic and Roman postal services.

56. Woepcke had settled on 27 CE.

representing May/July at 42° yielded closer ratios by reducing the length of the line to 47 mm, bringing it closer to the actual length represented on the prosciutto. The latter correction provides some evidence that the intended design latitude was 42°, while the former, coupled with an overall tendency toward inaccuracy, indicated the prosciutto was not designed for use at the Villa, even if it could have functioned there, and so additional calculations were not performed at that latitude.

A different error between the astronomical values and those realized on the prosciutto appears when plotting the gnomon lengths for each line. This analysis determined that a height of 16.74 mm would have been required for the gnomon/tail for 41°; a 17.28 mm high gnomon for the latitude established by the Accademia; and a 17.82 mm gnomon for 42°. These, in turn, cast the diagonal shadows—in geometric terms the hypotenuse of the right angle created by the gnomon and the horizon line—that determined the position of the vertical month lines forming the grid. These too are calculated by multiplying the tangent of the sun's elevation at the meridiem by the length of the month line. As Calkoen, Drecker, and others have shown, the month lines should be spaced apart regularly along the horizon line in accordance with solar declination throughout the year. At all the plotted latitudes, however, the diagonals fail to intersect the top of the vertical month lines as they were engraved on the prosciutto. Contributing to the problem is the fact that the dots intended to mark the intersection of the month lines with the horizon line are not in their proper position. In most cases these differences can be attributed to the small scale, with errors rarely more than one millimeter enhanced by projecting precise geometry onto the imprecise surface of the prosciutto. The deviations are greatest in the case of 42°, however, with the second diagonal falling considerably short of its target and the seventh extending beyond the grid. Yet the grid engraved on the prosciutto at this latitude accords most closely not only with the calculated values but with real-world readings off the dial. This adds further corroboration to the idea that the craftsman had used an existing dial initially established, as Delambre had proposed, through empirical observation of the shadow ratios produced by a fixed gnomon over the course of the zodiacal year, rather than one designed purely mathematically by, for example, plotting an analemma.

As for the hourly grids produced mathematically at these various latitudes, the errors realized on the prosciutto would have been largely imperceptible to the naked eye: the scale is simply too small. Across the dial's face, the deviations between what is required astronomically for the individual hours and what is engraved on the prosciutto rarely exceed 1.00 mm, and most are substantially smaller, less than half a millimeter. Even the largest error, the length of the summer solstice line, amounts to only 1.52 mm, and some of this can be ascribed to the use of the standard length of 54 mm in the calculations. The convex surface of the prosciutto at this point may well have distorted the lines here. Most of the discrepancies highlighted through these calculations would be subsumed in the width of the engraved lines or masked by the width of the gnomon's shadow. The individual hours and the intervals between them would be hardest to discern along the winter solstice line because of the constricted distance between the horary lines and the proportionally larger hour dots made in the wax model and the size of the gnomon's shadow. The sharp drop-off in the curving outer edge of the prosciutto here also may have caused the vellum marked with the accurate grid to distort as it was pressed to conform to the surface, resulting in the hour dots and the vertical month line falling out of alignment. Even so, the errors average less than two minutes per hour. Deviations elsewhere in the grids average barely 0.25 mm between what should be represented and what is engraved, a detail revealed, and exaggerated, only thanks to modern methods of measurement and calculation. The horary grid generated by the Accademia's latitude is scarcely more accurate than that for 41°—an overall average error of 3.30 minutes per hour compared to 3.49 minutes—while the grid established mathematically for 42° corresponds most closely with the prosciutto's at a mere 2.84 minutes of error per hour. Dropping the decimals and rounding the numbers to more closely emulate the scale at which the ancient craftsman worked eliminates most of the incongruencies.[57]

Consequently, it is not possible to conclude with any greater accuracy than that the design latitude was between 41° and 42°, with this analysis slightly favoring 42°.[58] This is in keeping with later portable sundials on which the label "Rome," used to encompass a broad geographic band across the

---

57. Talbert 2017, 141–46, also noted that the differences between hours in both the same and in different latitudes would amount to no more than a few minutes and that the small scale of portable sundials would render these differences imperceptible. See also D. Savoie, *Recherches sur les Cadrans Solaires* (Turnhout 2014) 17–31 ("L'Exactitude des cadrans antiques"), and A. Jones, "Greco-Roman Sundials: Precision and Displacement," in K. Miller and S. Symons, eds., *Down to the Hour: Short Time in the Ancient Mediterranean and Near East* (Boston 2019) 125–57.

58. G. Houston, "Using Sundials," in B. Lee and D. Slootjes, eds., *Aspects of Ancient Institutions and Geography: Studies in Honor of Richard J. A. Talbert* (Leiden 2014) 302, notes the tendency for Roman sundials to employ different latitudes for the curves representing the winter and summer solstices.

central Italian peninsula, also varied their latitudes between 41° and 42°.[59] The fact that the latitude of Rome, at 41° 54' (41.90°), is closer to 42° than 41° adds further support to the conclusion that the prosciutto's grid was calculated through empirical observation.

The small deviations between the required length of the shadow at a specific hour and where the line actually is marked on the prosciutto's dial results in errors adding up to no more than a few minutes within each hour, as Carcani and Drecker had observed. In many cases it is substantially less. Determining these errors is to apply a standard of precision that exceeds the needs of the ancient operator.[60] Nevertheless, this analysis shows that at 42° on the spring equinox, the owner in antiquity relying on this device would have been little more than ten minutes late the entire day. Even given the problems with the engraving of the grid clearly visible to the eye, they would have crossed paths only with the most dilatory women exiting the baths at the eighth hour on the winter solstice. And on the summer solstice, this parasite would have encountered difficulties taking readings around the meridiem but otherwise would have arrived quite punctually for his *cena* at the ninth hour.

4.f) Slicing and Dicing: Recreating Carcani's Empirical Research in 3D

This evidence indicating the prosciutto sundial's ability to keep quite accurate time was largely confirmed by empirical research conducted as part of this study. The approach was patterned after Carcani's and fulfilled Lalande's desire for a study model of the prosciutto but modified through the application of more modern methodologies and technologies. This research was not scientifically rigorous: the specific questions and various approaches for interrogating the sundial were not conceived in advance but evolved over time, as the potential avenues of inquiry and the astronomical principles underlying its design, and its inherent limitations became clearer. A 3D digital model developed from digitized photographs of the original and reproduced in thermoplastic served as the basis for this research whose scope was limited from the outset by certain constraints. Obtaining an exact copy had not been a

---

59. Talbert 2017, 120.

60. For a summary of the evolution of the concept of the hour (*hora*) in ancient Rome and its relation to sundials, see A. Wolkenhauer, "Time, Punctuality, and Chronotopes: Concepts and Attitudes Concerning Short Time in Ancient Rome," in K. Miller and S. Symons, eds., *Down to the Hour: Short Time in the Ancient Mediterranean and Near East* (Boston 2019) 214–38.

principal goal initially, for example, and the available technologies employed at the time of the original image capture ultimately served as the main check on doing so: the model was a close approximation not an exact replica.[61] By coincidence, the latitude of Middletown, Connecticut, where this study was conducted, is close to that of the prosciutto's apparent design latitude, being 41° 31′ 52″ (= 41.53°), even if the longitude, which has no bearing on operating the prosciutto, is quite different.[62] This suggested a reconstructed gnomon could demonstrate the sundial in action and allow its accuracy to be evaluated. Several attempts to print a model with a fully reconstructed gnomon/tail in 3D failed because sufficient precision in the placement of the tip could not be attained through thermoplastic printing (Figure 4.5).[63] Ultimately, the best results were obtained with versions of the model employing a 20-gauge galvanized wire anchored to the prosciutto in an epoxy resin base. This allowed the gnomon to be set more precisely to the desired height and position in relation to the grid. The wire cast a sufficiently narrow shadow with a sharp terminus that facilitated accurate readings. This seems to confirm that the prosciutto's original gnomon/tail would have been similarly slim and without any curl. The shadows were most visible on models printed in white or grey thermoplastic, or with a scaled print of the grid pasted to the thermoplastic, adding further support to the conclusion that the silvering of the

61. One series of photographs in 360° of the original prosciutto was taken in the photographic studio of the MANN in 2018 by the author with a Sony Cyber-Shot 16.2 megapixel camera and another series with a LGE Nexis 5X 12 megapixel cellphone. These were uploaded to Autodesk ReCap Photo for conversion in the Cloud into a 3D digitized model. Autodesk Meshmixer was used to clean up the image and reconstruct the gnomon. Models based on the digital version were initially printed in resin, but subsequent ones used the "fused deposition modeling" (FDM) method on an Ultimaker 3 3D printer in the IDEAS lab of Wesleyan University. The final, fully reconstructed version weighed 28 grams, compared to the 285 grams of the original bronze, and this lighter form made it more susceptible to movement in the lightest breeze and therefore difficult at times to read. Reconstructing ancient sundials in 3D for both scientific and personal analysis has become popular. For a more scientific 3D reconstruction of a Roman sundial, see, e.g., M. Pistellato, A. Traviglia, and F. Bergamasco, "Geolocating Time: Digitization and Reverse Engineering of a Roman Sundial," in A. Bartoli and A Fusiello, eds., *Computer Visions—ECCV 2020 Workshops (Glascow, UK, August 23–28, 2020), Proceedings, Part II* (Switzerland 2020) 143–58. The Edition|Topoi website includes a repository of all known ancient Greek and Roman sundials, many in 3D, with detailed information (http://repository.edition-topoi.org/collection/BSDP). The prosciutto is inventoried as Dialface ID 544.

62. The longitude is 72° 39′ 31″ (= -72.65°), compared to Rome's 12° 29′ 44″ (= 12.49).

63. The method employed by this type of 3D printer consists of building up layers of material and requires the addition of temporary struts to support features projecting into space during printing. This and the tendency for the material to sag during printing rendered it impossible to reproduce the gnomon to anywhere near the degree of precision necessary to take readings.

Observations 297

FIGURE 4.5 Early version of the 3D model of the prosciutto sundial with reconstructed gnomon in operation on the spring equinox on the "Cardo III Inferiore" in Herculaneum in 2019, year 1940 from the eruption of Vesuvius. The jagged gnomon shadow highlights the difficulties presented by a 3D printed gnomon but underscores the need for a thin one. The time stamp on the photo is 10:00, the start of the fourth hour. Author's photo.

original had the practical effect of enhancing the grid's readability; readings were difficult on black thermoplastic. The 3D model included a fixed, rather than a moveable, printed suspension ring which was slightly distorted in shape and not as smooth as the original metal, so care had to be taken to ensure the prosciutto swung naturally into the correct vertical position via gravity, with the grid properly aligned with the horizon; a similar requirement of the original. The seasonal hours were calculated for each zodiacal month in the modern day at this latitude and the solar elevation corresponding with each hour plotted.[64] Readings were taken over the course of several years, between 2018 and 2023, though the most accurate findings were collected

---

64. These calculations, as elsewhere in this study, were made using https://www.suncalc.org and later refined with NOAA's Solar Position Calculator (https://gml.noaa.gov/grad/solcalc/azel.html).

only in the last year when what appeared to be the ideal gnomon height, established through observational analysis, was coupled with a more accurate hour grid.

Several different gnomon heights were used over the course of this research, ranging from 11.00 mm to 17.60 mm high.[65] The best results were realized with the tip resting above the upper left-hand corner of the dial; experiments conducted with the tip located to the left did not yield accurate readings. At this latitude, with the 52.48 mm engraved length of the summer solstice line, the gnomon height would need to be 17.32 mm above the grid. Indeed, the correspondence between the table of solar elevations and the engraved grid is remarkably close. This was confirmed by readings taken over the course of the zodiacal year that showed many of the same slight errors in the seasonal hour readings discernible in a comparison between the shadow lengths required by solar elevation and the hour points marked on the prosciutto (Figure 4.6 and Figure 4.7).[66] The shadow fell closest to, if not directly on, the hour points

FIGURE 4.6 The prosciutto at work at latitude 41° 31′ 52″ (41.53°). *From left to right*: hour 4.5 (10:58 EST), hour 5 (11:36 EST), and meridiem (12:52 EST) on the summer solstice. The increasing difficulty of reading the shadow around the meridiem is evident. Author's photo.

---

65. As a general rule, because of the difficulty of accurately measuring at this small scale, the gnomon heights were set to within a single decimal point, though accuracy at this scale with these materials is difficult.

66. Readings could not be taken consistently over the year or day, in many cases because of cloud cover, but enough to establish with some confidence that the dial functioned with a reasonable degree of accuracy at this latitude.

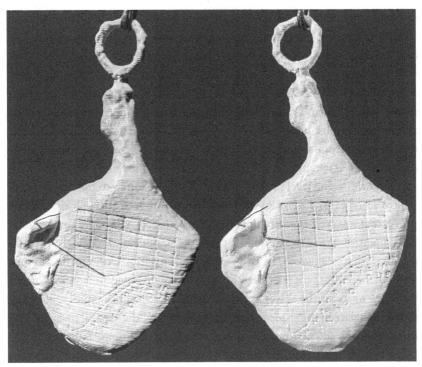

FIGURE 4.7 The prosciutto at work at latitude 41° 31′ 52″ (41.53°). *(Left)* Hour 4 on August 25 (sun entering Virgo) and *(right)* hour 3 on fall equinox (sun exiting Virgo). Author's photo.

within a couple of days before and after the verticals representing the start of each zodiacal sign before gradually progressing across the dial in the direction of the next sign. Moreover, it was possible to take quite accurate readings not only at the quarter- and half-hour points along the month lines but essentially at any time on any day, including all hours on the days falling between those marked by the grid's vertical month lines, though the accuracy in these transitional areas was reduced. Nevertheless, because of the complexity of determining where to situate the gnomon's shadow tip within these differing points of the grid, this empirical testing vividly illustrated that the readings serve to verify what the operator already knows rather than offering up any new information.

It should come as no surprise that both the astronomical data and the empirical evidence confirm that this portable sundial keeps accurate time, not precise but sufficient to meet the needs of its owner.[67] Despite its droll

---

67. Jones 2019, 143: "In any case, the ancient purchaser of an expensive, high-quality sundial could have taken comfort in the knowledge that, if for any reason he or she had to move to a

300   THE PROSCIUTTO SUNDIAL

outward appearance, the science underlying its design was sophisticated and sound for its times, while its form may have subtly and cleverly reflected its original owner's philosophical beliefs. In a period when Augustus' *horologium* still languished out of sync on the Campus Martius and ostensibly before Martial's daily life was strictly regimented into hourly tasks (*Ep.* 4.8), the prosciutto's precision and visual allure would have set it and its owner apart from all the other philosophers and timepieces, both viewed with suspicion in their day, to judge from a quip from Seneca: "I cannot tell you the fixed hour—philosophers agree more easily than sundials—still it was between the sixth and seventh."[68] The level of skill and knowledge required to operate the prosciutto to reveal the point in time of the sun's progress through the ecliptic proves that it served as much more than a mere object of prestige and personal adornment. If it was superfluous for telling the hour of the day, the explication of all that its shadow revealed—its synchronization with sunrise and sunset and the culmination of the sun in the sky, its vivid illustration of the varying lengths of the seasonal hours, the geometrical principles underlying the ratios between the shadow's lengths and those of the vertical lines; its relationship to the sun's entrance and exit of the zodiacal signs, and the correlation of those signs with the Roman calendar and the Julian and Augustan corrections—would have cast a bright light on the owner's scientific knowledge and elite social status. If Piso Pontifex had passed down the skills necessary to properly manipulate and decipher the prosciutto, the last patron of the Villa might have turned its tail for one more reading from its shimmering dial around the seventh hour, as the shadow traveled up the grid just to the right of the third vertical representing Virgo, and shortly before the sun was blotted out on that fateful day in August, 79 CE.[69]

The Prosciutto Sundial: Casting Light on an Ancient Roman Timepiece from the Villa dei Papiri in Herculaneum.
Christopher Parslow, Oxford University Press. © Oxford University Press 2024. DOI: 10.1093/9780197749418.003.0004

distant place, the sundial would still work reasonably well as a time-keeping device, if not as a calendar."

68. Seneca, *Apocolocyntosis* 2.2: *Horam non possum certam tibi dicere, facilius inter philosophos quam inter horologia conveniet, tamen inter sextam et septimam erat.*

69. Pliny the Younger (*Ep.* 6.16.4) set the start of the eruption of Vesuvius *hora fere septima* (around the seventh hour). For a detailed assessment of the evidence relating to the date, phases, and timing of the eruption, see Foss 2022, esp. 145–50.

APPENDIX A

Comparison of Grid Dimensions

These tables summarize the main calculations made by the principal commentators on the prosciutto, where these are obtainable, for the (**A**) length of each of the seven vertical lines representing the zodiacal months and (**B**) the mathematically calculated height of the gnomon required to cast a correct shadow on those lines. The quantities employ the original unit values established by Carcani's analysis on behalf of the Accademia Ercolanese, which set the value of the equinoctial line at 1,000 units. The values proposed by each author have been converted into millimeters based on the 31.23 mm length of the central equinoctial line (in italics) or using the value for the equinoctial line established independently by that author. Table **C** compares the length of the horizon line/width of the grid and the calculated values of the distance between the vertical month lines employed by each author with the actual values on the prosciutto.

A) Length of Vertical Month Lines

	CARCANI	DRECKER on Carcani	DRECKER @ 41.00°	FERRARI	ACTUAL
1	1686 = 52.65	1575 = 47.25 [49.19]	1677 = 50.26 [52.37]	1690 = 56.60 [54.11]	52.48 = 1680
2	1543 = 48.19	1489 = 44.67 [46.50]	1546 = 46.33 [48.28]	1540 = [49.31]	47.30 = 1514
3	1244 = 38.85	1242 = 37.26 [38.79]	1253 = 37.55 [39.13]	1240 = [39.70]	38.92 = 1246
4	1000 = **31.23**	1000 = 30.00 [**31.23**]	1001 = 30.00 [**31.23**]	1000 = 33.50 [**31.23**]	31.23 = 1000
5	804 = 25.11	806 = 24.18 [25.17]	810 = 24.27 [25.29]	790 = [25.30]	24.83 = 795
6	691 = 21.58	687 = 20.61 [21.45]	683 = 20.47 [21.33]	690 = [22.09]	21.97 = 703
7	687 = 21.46	687 = 20.61 [21.45]	686 = 20.56 [21.42]	690 = 23.10 [22.09]	21.60 = 691

B) Mathematical Height of Gnomon for Each Month Line

	CARCANI	CALKOEN	DRECKER on Carcani	DRECKER @ 41.00°	FERRARI	28 CE @ 41° 39' 45"	28 CE @ 42°	28 CE @ 41°
1		533 = 17.07	500 = 15.62	523 = 16.33	537 = 17.70*	17.28 = 553	17.82 = 570	16.74 = 536
2		582 = 18.64	570 = 17.80	579 = 18.08		18.24 = 584	18.19 = 582	17.76 = 569
3		686 = 21.96	710 = 22.17	700 = 21.86		22.40 = 717	22.76 = 728	22.00 = 704
4	881 = 27.51	831 = 26.60	881 = 27.51	870 = 27.17		27.22 = 872	27.56 = 882	26.60 = 852
5		1000 = 32.02	1070 = 33.42	1053 = 32.88		31.44 = 1006	31.92 = 1022	30.72 = 984
6		1182 = 37.85	1280 = 39.97	1269 = 39.63		39.48 = 1264	40.11 = 1284	38.43 = 1230
7	1482 = 46.28	1372 = 43.93	1482 = 46.28	1458 = 45.53		45.57 = 1459	46.65 = 1494	44.31 = 1419

* Ferrari calculated several heights for the gnomon depending on the length of the horizon line, from 1,220 units to 1,416; this is the value for the longer horizon line of 1,416 units = 44.22 mm

C) Length of Horizon Line and Distance between Vertical Month Lines

	CARCANI*	CALKOEN	DRECKER +	FERRARI 1 ++	FERRARI 2	ACTUAL
Horizon	40.00	1260 = *40.35*	1320 = 44.00 = [*41.22*]	1220 = 40.80 = [*38.10*]	1416 = *44.22*	41.35
1	6.53	210 = 6.72	220 = 7.33 = [*6.87*]	204 = 6.70 = [*6.37*]	[236 = *7.37*]	7.09
2	6.32	210 = 6.72	220 = 7.33 = [*6.87*]	204 = 6.70 = [*6.37*]	[236 = *7.37*]	6.81
3	6.44	210 = 6.72	220 = 7.33 = [*6.87*]	204 = 6.70 = [*6.37*]	[236 = *7.37*]	6.67
4	6.47	210 = 6.72	220 = 7.33 = [*6.87*]	204 = 6.70 = [*6.37*]	[236 = *7.37*]	6.78
5	6.18	210 = 6.72	220 = 7.33 = [*6.87*]	204 = 6.70 = [*6.37*]	[236 = *7.37*]	7.26
6	6.23	210 = 6.72	220 = 7.33 = [*6.87*]	204 = 6.70 = [*6.37*]	[236 = *7.37*]	6.74

* Carcani's values as measured off the engraving in the *Pitture di Ercolano*, vol. 3. The sum of the distances between verticals does not equal the total horizon line because of the difficulty of taking measurements at this small scale.

+ = Drecker's value for the equinoctial line was 30.00 mm rather than 31.23 mm.

++ = Ferrari 2008, fig. 8, though his conversions from the Accademia's dimensions regularly use 30.3 mm rather than the 33.5 mm length he obtained for the equinoctial line.

APPENDIX B

Tables of Horary Readings

The following tables show the data for the length of the shadow cast by the gnomon for the seasonal hours over the course of the zodiacal year compared with the corresponding horary lines on the prosciutto. For consistency's sake, the change in zodiacal sign was fixed at the twenty-second of each month; only the first six hours in seven months (in bold face) were calculated; the lengths of the vertical month lines were set at round values close to those on the prosciutto—54, 48, 40, 31, 24, 21, 21—with the exception of a few cases, especially in Table G, where slight adjustments (in bold face) brought the ratios closer to those on the prosciutto; only values for the spring equinox and the winter and summer solstices were calculated for Tables D & E, while only the equinox was calculated to illustrate the shadow lengths Carcani likely saw during his empirical tests in 1761 for Table F. Even though the real differences between them are quite small, full calculations were performed for 42°; for the Accademia's determination of 41° 39′ 45″; and for 41°; and partial calculations for Naples (40° 50′ 25″) and the Villa dei Papiri (40° 48′ 27″), all in the year 28 CE, while full calculations were performed for 41° 31′ 52″, the latitude of Middletown CT, in 2023 CE. The longitude throughout was set to that of Rome.

The calculations required determining the length in minutes of each seasonal hour in each zodiacal month; these are indicated at the upper right of each month chart. The solar elevation is determined for each seasonal hour and subtracted from 90° to produce the zenith angle. These values were obtained using the National Oceanic and Atmospheric Administration (NOAA) calculators available at https://gml.noaa.gov/grad/solcalc/sunrise.html and https://gml.noaa.gov/grad/solcalc/azel.html. The tangent of the zenith angle (e.g., 18.09 = 0.33) is the shadow length cast by a gnomon one meter in height at that hour. The tangent of the zenith angle at the meridiem is then multiplied by the full length of the month line to produce the "line tangent," which is equivalent to the height of the gnomon required to cast the shadows on that month line. The "line tangent" is then divided by the tangent of each of the seasonal hours, revealing where along each month line the gnomon shadow falls at each seasonal hour, expressed in millimeters. This is then compared to the actual intersections of the horary lines along the month lines measured off the prosciutto. The distance between each horary line is then divided by the total number of minutes in the hour to determine the dimensional length of each minute, and this, in turn, is multiplied by the difference between the astronomical length of the hour and the measured length of the hour on the prosciutto, and the products for each month averaged, regardless of negative or positive values, to yield the total deviation in minutes over the course of those first six seasonal hours of the day. Deviations greater than 5 mm are in bold face.

The resulting horary grids for 42°; 41° 39′ 45″; 41°; and 41° 31′ 52″ are superimposed on the prosciutto's grid in graphic form, with the height of the gnomon (the "line tangent") determined for each month represented across the top of the dial.

TABLE A
LATITUDE 42° IN 28 CE

Average minute errors per hour = 2.79

EQUINOX September/March [Vertical Line 4]
Line Tangent = 31 × 0.889 = 27.56   61 min/hour

A	B	C	D	E	F	G	H	I
1	7:13	10.56	79.44	5.36	5.14	5.20	+0.06	5.14/0.08/ +0.48
2	8:14	21.48	68.52	2.54	10.85	11.13	+0.28	5.71/0.09/ +2.52
3	9:15	31.61	58.39	1.62	17.01	16.70	-0.31	6.16/0.10/ -3.10
4	10:16	40.20	49.80	1.18	23.35	23.42	+0.07	6.34/0.10/ +0.70
5	11:17	46.19	43.81	0.96	28.71	28.67	-0.04	5.36/0.09/ -0.36
6	12:18	48.36	41.64	**0.889**	31.00	31.23	+0.23	2.29/0.04/ +0.92
							Avg. 0.17	10.56 Avg. 1.35

WINTER SOLSTICE [Vertical Line 7]
Line Tangent = 21 × 2.21 = 46.655   46 min/hour

A	B	C	D	E	F	G	H	I
1	8:27	6.24	83.76	9.14	5.07	5.09	+0.02	5.07/0.11/ +0.22
2	9:12	12.18	77.82	4.63	10.02	9.79	-0.23	4.95/0.11/ -2.53
3	9:57	17.20	72.80	3.23	14.37	14.34	-0.03	4.35/0.09/ -0.27
4	10:42	21.04	68.96	2.60	17.85	17.46	-0.39	3.48/0.08/ -3.12
5	11:27	23.48	66.52	2.30	20.18	19.72	-0.46	2.33/0.05/ -2.30
6	12:13	24.34	65.66	2.21	21.00	21.60	+0.60	0.82/0.02/ +1.20
							Avg. 0.29	9.64 Avg. 1.61

SUMMER SOLSTICE [Vertical Line 1]
Line Tangent = 54 × 0.33 = 17.82   76 min/hour

A	B	C	D	E	F	G	H	I
1	5:45	11.83	78.17	4.77	3.73	4.12	+0.39	3.73/0.05/ +1.95
2	7:01	25.37	64.63	2.11	8.45	8.84	+0.39	4.72/0.06/ +2.34
3	8:17	39.41	50.59	1.22	14.60	14.67	+0.07	6.15/0.08/ +0.56
4	9:33	53.28	36.72	0.75	23.76	23.51	-0.25	9.16/0.12/ -3.00
5	10:48	65.42	24.58	0.46	38.74	38.84	+0.10	14.98/0.20/ +2.00
6	12:06	71.69	18.31	0.33	54.00	52.48	-1.52	15.26/0.20/ **-30.40**
							Avg. 0.20	40.25 Avg. 6.71

MAY/JULY [Vertical Line 2]
Line Tangent = 47 × 0.387 = 18.19   74 min/hour

A	B	C	D	E	F	G	H	I
1	6:01	11.82	78.18	4.78	3.80	3.96	+0.16	3.80/0.05/ +0.80
2	7:15	25.21	64.79	2.12	8.58	8.99	+0.41	4.78/0.06/ +2.46
3	8:30	39.09	50.91	1.23	14.79	14.72	-0.07	6.21/0.08/ -0.56
4	9:44	52.37	37.63	0.77	23.62	23.96	+0.34	8.83/0.12/ +4.08
5	10:58	63.67	26.33	0.49	37.12	37.13	+0.01	13.50/0.18/ +0.18
6	12:12	68.83	21.17	0.387	**47.00**	47.30	+0.30	9.88/0.13/ +3.90
							Avg. 0.22	11.98 Avg. 2.00

APRIL/AUGUST [Vertical Line 3]
Line Tangent = 40 × 0.569 = 22.76   68 min/hour

A	B	C	D	E	F	G	H	I
1	6:29	11.64	78.36	4.85	4.69	4.80	+0.11	4.69/0.07/ +0.77
2	7:37	24.20	65.80	2.23	10.20	10.21	+0.01	5.51/0.08/ +-.08
3	8:46	36.72	53.28	1.34	16.98	16.64	-0.34	6.78/0.10/ -3.40
4	9:54	48.02	41.98	0.90	25.29	25.07	-0.22	8.31/0.12/ -2.64
5	11:02	56.77	33.23	0.66	34.48	34.12	-0.36	9.19/0.14/ **-5.04**
6	12:11	60.25	29.65	0.569	40.00	38.92	-1.08	5.52/0.08/ **-8.77**
							Avg.0.18	20.70 Avg. 3.45

FEBRUARY/OCTOBER [Vertical Line 5]
Line Tangent = 24 × 1.33 = 31.92   55 min/hour

A	B	C	D	E	F	G	H	I
1	7:25	8.52	81.48	6.68	4.78	5.17	+0.39	4.78/0.09/ +3.51
2	8:19	17.29	72.71	3.21	9.94	9.82	-0.12	5.16/0.09/ -1.08
3	9:13	25.05	64.95	2.14	14.92	14.87	-0.05	4.98/0.09/ -0.45
4	10:08	31.39	58.61	1.64	19.46	18.70	-0.76	4.54/0.08/ **-6.08**
5	11:02	35.47	54.53	1.40	22.80	22.24	-0.56	3.34/0.06/ -3.36
6	11:57	36.92	53.08	1.33	24.00	24.83	+0.83	1.20/0.02/ +1.66
							Avg.0.45	16.14 Avg. 2.69

JANUARY/NOVEMBER [Vertical Line 6]
Line Tangent = 21 × 1.91 = 40.11   48 min/hour

A	B	C	D	E	F	G	H	I
1	8:01	6.92	83.08	8.24	4.86	5.11	+0.25	4.86/0.10/ +2.50
2	8:47	13.43	76.57	4.19	9.57	9.37	-0.20	4.71/0.10/ -2.00
3	9:35	19.25	70.75	2.86	14.02	14.07	+0.05	4.45/0.09/ +0.45
4	10:23	23.75	66.25	2.277	17.66	17.38	-0.28	3.64/0.07/ -1.96
5	11:11	26.62	63.38	1.99	20.15	19.87	-0.27	2.56/0.05/ -1.46
6	11:59	27.63	62.37	1.91	21.00	21.97	+0.97	0.85/0.02/ +1.94
							Avg. 0.34	10.31 Avg. 1.72

A = Temporal Hour; B = Time; C = Solar Angle; D = Zenith Angle; E = Tangent of Zenith Angle = Shadow length of 1.00 m high gnomon; F = Corresponding Horary Lines (**Line Tangent ÷ E**); G = Horary Lines Measured on Prosciutto; H = Difference of G vs. F; I = Minute errors per hour (F ÷ min/hr × H)

SOURCES: https://gml.noaa.gov/grad/solcalc/sunrise.html and https://gml.noaa.gov/grad/solcalc/azel.html

LATITUDE 42° IN 28 CE

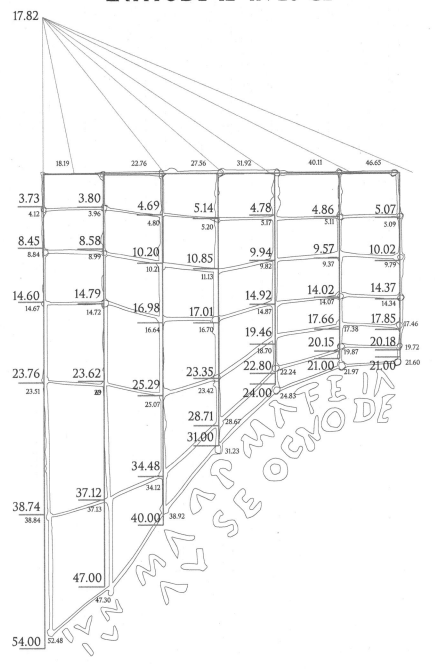

TABLE B
LATITUTE 41° 39′ 45″ (41.66°)

= Accademia's Findings

Average minute errors per hour = 3;30

EQUINOX September/March [Vertical Line 4]
Line Tangent = 31 × 0.878 = 27.22   —   61 min/hour

A	B	C	D	E	F	G	H	I
1	7:13	10.62	79.38	5.33	5.10	5.20	+0.10	5.10/0.08/ +0.80
2	8:14	21.60	68.40	2.53	10.76	11.13	+0.37	5.66/0.09/ +3.33
3	9:15	31.79	58.21	1.61	16.90	16.70	−0.20	6.14/0.10/ −2.00
4	10:16	40.45	49.55	1.17	23.26	23.42	+0.16	6.36/0.10/ +1.60
5	11:17	46.50	43.50	0.95	28.65	28.67	+0.02	5.39/0.09/ +0.18
6	12:18	48.70	41.30	0.878	31.00	31.23	+0.23	2.35/0.04/ +0.92
							Avg. 0.18	11.31 Avg. 1.47

WINTER SOLSTICE [Vertical Line 7]
Line Tangent = 21 × 2.17 = 45.57   —   46 min/hour

A	B	C	D	E	F	G	H	I
1	8:24	6.02	83.97	9.47	4.81	5.09	+0.28	4.81/0.10/ +2.80
2	9:09	12.06	77.93	4.67	9.76	9.79	+0.03	4.95/0.11/ +0.33
3	9:54	17.18	72.81	3.23	14.10	14.34	+0.24	4.34/0.09/ +2.26
4	10:41	21.28	68.72	2.57	17.73	17.46	−0.27	3.63/0.08/ −2.13
5	11:26	23.77	66.23	2.27	20.07	19.72	−0.35	2.34/0.05/ −1.78
6	12:13	24.68	65.32	2.17	21.00	21.60	+0.60	0.93/0.02/ +1.21
							Avg. 0.21	10.68 Avg. 1.78

SUMMER SOLSTICE [Vertical Line 1]
Line Tangent = 54 × 0.32 = 17.28   —   76 min/hour

A	B	C	D	E	F	G	H	I
1	5:47	12.06	77.94	4.68	3.69	4.12	+0.43	3.69/0.05/ +2.15
2	7:02	25.49	64.51	2.10	8.23	8.84	+0.61	4.54/0.06/ +3.66
3	8:18	39.61	50.39	1.21	14.28	14.67	+0.39	6.05/0.08/ +3.12
4	9:33	53.38	36.61	0.74	23.35	23.51	+0.16	9.07/0.12/ +1.92
5	10:48	65.65	24.35	0.45	38.40	38.84	+0.44	15.05/0.20/ **+8.80**
6	12:06	72.02	17.97	0.32	54.00	52.48	−1.52	15.60/0.21/ **−31.92**
							Avg. 0.59	51.57 Avg. **8.60**

A = Temporal Hour; B = Time; C = Solar Elevation; D = Zenith Angle; E = Tangent of Zenith Angle = Shadow length of 1.00 m high gnomon; F = Corresponding Horary Lines (Line Tangent ÷ E); G = Horary Lines Measured on Prosciutto; H = Difference of G vs. F; I = Minute errors per hour (F ÷ min/hr × H)

MAY/JULY [Vertical Line 2]
Line Tangent = 48 × 0.38 = 18.24   —   74 min/hour

A	B	C	D	E	F	G	H	I
1	6:02	11.90	78.10	4.75	3.84	3.96	+0.12	3.84/0.05/ +0.60
2	7:16	25.36	64.64	2.11	8.63	8.99	+0.33	4.79/0.06/ +1.98
3	8:31	39.31	50.68	1.22	14.93	14.72	−0.21	6.30/0.09/ +1.98
4	9:44	52.50	37.50	0.77	23.66	23.96	+0.30	8.73/0.12/ +3.60
5	10:58	63.92	26.09	0.49	37.22	37.13	−0.09	13.56/0.18/ −1.62
6	12:12	69.17	20.83	0.38	48.00	47.30	−0.70	10.78/0.15/ −10.50
							Avg. 0.29	20.28 Avg. 3.38

APRIL/AUGUST [Vertical Line 3]
Line Tangent = 40 × 0.56 = 22.40   —   68 min/hour

A	B	C	D	E	F	G	H	I
1	6:29	11.60	78.39	4.87	4.60	4.80	+0.20	4.60/0.06/ +1.20
2	7:37	24.23	65.76	2.22	10.09	10.21	+0.12	5.49/0.08/ +0.96
3	8:46	36.82	53.17	1.33	16.84	16.64	+0.20	6.75/0.10/ +2.00
4	9:54	48.21	41.79	0.89	25.17	25.07	−0.10	8.33/0.12/ −1.20
5	11:02	57.06	32.94	0.65	34.46	34.12	−0.34	9.29/0.14/ +4.76
6	12:11	60.69	29.31	0.56	40.00	38.92	−1.08	5.54/0.08/ **+8.64**
							Avg. 0.34	18.76 Avg. 3.13

FEBRUARY/OCTOBER [Vertical Line 5]
Line Tangent = 24 × 1.31 = 31.44   —   55 min/hour

A	B	C	D	E	F	G	H	I
1	7:25	8.65	81.35	6.57	4.79	5.17	+0.38	4.79/0.09/ +3.42
2	8:19	17.47	72.52	3.17	9.91	9.82	−0.09	5.12/0.09/ −0.81
3	9:13	25.29	64.71	2.11	14.90	14.87	−0.03	4.99/0.09/ −0.27
4	10:08	31.67	58.33	1.62	19.41	18.70	−0.71	4.51/0.08/ **−5.68**
5	11:02	35.79	54.20	1.38	22.78	22.24	−0.54	3.37/0.06/ −3.24
6	11:57	37.26	52.74	1.31	24.00	24.83	+0.83	1.22/0.02/ +1.66
							Avg. 0.43	15.08 Avg. 2.51

JANUARY/NOVEMBER [Vertical Line 6]
Line Tangent = 21 × 1.88 = 39.48   —   49 min/hour

A	B	C	D	E	F	G	H	I
1	7:58	6.67	83.33	8.55	4.61	5.11	+0.50	4.61/0.09/ +4.50
2	8:46	13.53	76.47	4.15	9.51	9.37	−0.14	4.90/0.10/ +1.40
3	9:34	19.41	70.58	2.84	13.90	14.07	+0.17	4.39/0.09/ +1.53
4	10:22	23.97	66.02	2.25	17.54	17.38	−0.16	3.64/0.07/ −1.12
5	11:10	26.91	63.09	1.97	20.04	19.87	−0.17	2.50/0.05/ −0.85
6	11:59	27.96	62.03	1.88	21.00	21.97	+0.97	1.96/0.04/ +3.88
							Avg. 0.35	13.28 Avg. 2.21

SOURCES: https://gml.noaa.gov/grad/solcalc/sunrise.html and https://gml.noaa.gov/grad/solcalc/azel.html

LATITUDE 41° 39' 45" (41.66°) IN 28 CE
ACCADEMIA'S LATITUDE

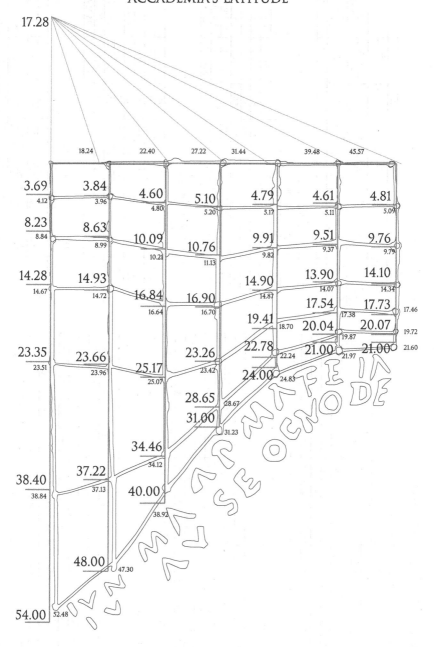

TABLE C
LATITUDE 41° IN 28 CE

Average minute errors per hour = 3.49

EQUINOX September/March [Vertical Line 4]
Line Tangent = 31 × 0.885 = 27.44
61 min/hour

A	B	C	D	E	F	G	H	I
1	7:13	10.72	79.28	5.28	5.13	5.20	+0.07	5.13/0.08 / +0.56
2	8:14	21.82	68.18	2.50	10.84	11.13	+0.29	5.71/0.09 / +2.61
3	9:15	32.15	57.85	1.59	16.94	16.70	−0.24	6.10/0.10 / −2.40
4	10:16	40.94	49.06	1.15	23.25	23.42	+0.17	6.31/0.10 / +1.70
5	11:17	47.11	42.89	0.93	28.88	28.67	−0.21	5.63/0.09 / −1.89
6	12:18	49.36	41.46	0.883	31.00	31.23	+0.23	2.12/0.03 / +0.69
					9.85		Avg. 0.20	Avg. 1.64

WINTER SOLSTICE [Vertical Line 7]
Line Tangent = 21 × 2.11 = 44.31
46 min/hour

A	B	C	D	E	F	G	H	I
1	8:27	6.30	83.70	9.06	4.89	5.09	+0.20	4.89/0.10 / +0.20
2	9:12	12.55	77.45	4.49	9.86	9.79	−0.07	4.97/0.11 / −0.77
3	9:57	17.84	72.16	3.11	14.25	14.34	+0.09	4.39/0.10 / +0.90
4	10:42	21.90	68.10	2.49	17.84	17.46	−0.38	3.21/0.07 / −2.66
5	11:27	24.46	65.54	2.20	20.13	19.72	−0.41	2.29/0.05 / −2.05
6	12:13	25.34	64.66	2.11	21.00	21.60	+0.60	0.87/0.02 / +1.20
					7.78		Avg. 0.29	Avg. 1.30

SUMMER SOLSTICE [Vertical Line 1]
Line Tangent = 54 × 0.31 = 16.74
76 min/hour

A	B	C	D	E	F	G	H	I
1	5:45	11.47	78.53	4.92	3.56	4.12	+0.56	3.56/0.05 / +2.80
2	7:01	25.19	64.81	2.13	8.08	8.84	+0.76	4.52/0.06 / +4.56
3	8:17	39.44	50.56	1.21	14.06	14.67	+0.61	5.98/0.08 / +4.88
4	9:33	53.58	36.42	0.74	23.25	23.51	+0.26	9.19/0.12 / +3.12
5	10:48	66.08	23.92	0.44	38.05	38.84	−0.09	15.68/0.20 / −1.80
6	12:06	72.69	17.31	0.31	54.00	52.48	−1.52	15.07/0.20 / −30.40
					47.56		Avg. 0.63	Avg. 7.93

MAY/JULY [Vertical Line 2]
Line Tangent = 48 × 0.37 = 17.76
74 min/hour

A	B	C	D	E	F	G	H	I
1	6:01	11.53	78.47	4.90	3.62	3.96	+0.34	3.62/0.05/ +1.70
2	7:15	25.10	64.90	2.13	8.33	8.99	+0.66	4.71/0.06/ +3.96
3	8:30	39.02	50.98	1.23	14.44	14.72	+0.28	6.11/0.08/ +2.24
4	9:44	52.57	37.73	0.77	23.06	23.96	+0.90	8.62/0.12/ +10.80
5	10:58	64.40	25.60	0.48	37.00	37.13	+0.13	13.94/0.18 / +2.34
6	12:12	69.84	20.16	0.37	48.00	47.30	−0.70	11.00/0.15/ −10.50
							Avg. 0.43	31.54 Avg. 5.26

APRIL/AUGUST [Vertical Line 3]
Line Tangent = 40 × 0.55 = 22.00
68 min/hour

A	B	C	D	E	F	G	H	I
1	6:29	11.91	78.09	4.74	4.64	4.80	+0.16	4.64/0.07 / +1.12
2	7:37	24.67	65.33	2.18	10.09	10.21	+0.12	5.45/0.08 / +0.96
3	8:46	37.21	52.79	1.32	16.66	16.64	−0.02	6.57/0.10/ −0.20
4	9:54	48.75	41.25	0.88	25.00	25.07	+0.07	8.34/0.12/ +0.84
5	11:02	57.73	32.27	0.63	34.92	34.12	−0.80	9.92/0.15/ −12.00
6	12:11	61.36	28.64	0.55	40.00	38.92	−1.08	5.08/0.07/ −7.56
							Avg. 0.42	22.68 Avg. 3.78

FEBRUARY/OCTOBER [Vertical Line 5]
Line Tangent = 24 × 1.28 = 30.72
55 min/hour

A	B	C	D	E	F	G	H	I
1	7:25	8.90	81.10	6.38	4.82	5.17	+0.35	4.82/0.09/ +3.15
2	8:19	17.83	72.17	3.11	9.88	9.82	−0.06	5.06/0.09/ −0.54
3	9:13	25.75	64.25	2.07	14.84	14.87	+0.03	4.96/0.09/ +0.99
4	10:08	32.23	57.77	1.58	19.44	18.70	−0.74	4.60/0.08/ −5.92
5	11:02	36.43	53.57	1.35	22.75	22.24	−0.51	3.31/0.06/ −3.06
6	11:57	37.92	52.08	1.28	24.00	24.83	+0.83	1.25/0.02/ +1.66
							Avg. 0.40	15.32 Avg. 2.55

JANUARY/NOVEMBER [Vertical Line 6]
Line Tangent = 21 × 1.83 = 38.43
49 min/hour

A	B	C	D	E	F	G	H	I
1	8:01	7.04	82.96	8.10	4.74	5.11	+0.37	4.74/0.10/ +3.70
2	8:47	13.99	76.01	4.01	9.58	9.37	−0.21	4.84/0.10/ −2.10
3	9:35	19.95	70.05	2.75	13.97	13.97	0.00	4.39/0.09/ 0.00
4	10:23	24.58	65.42	2.19	17.55	17.38	−0.17	3.58/0.07/ −1.19
5	11:11	27.55	62.45	1.92	20.01	19.87	−0.14	2.46/0.05/ −0.70
6	11:59	28.63	61.37	1.83	21.00	21.97	+0.97	0.99/0.02/ +1.94
							Avg. 0.53	9.63 Avg. 1.61

A = Temporal Hour; B = Time; C = Solar Elevation; D = Zenith Angle; E = Tangent of Zenith Angle =
Shadow length of 1.00 m high gnomon; F = Corresponding Horary Lines (Line Tangent ÷ E); G = Horary
Lines Measured on Prosciutto; H = Difference of G vs. F; I = Minute errors per hour (F ÷ min/hr × H)

SOURCES: https://gml.noaa.gov/grad/solcalc/sunrise.html and https://gml.noaa.gov/grad/solcalc/azel.html

LATITUDE 41° IN 28 CE

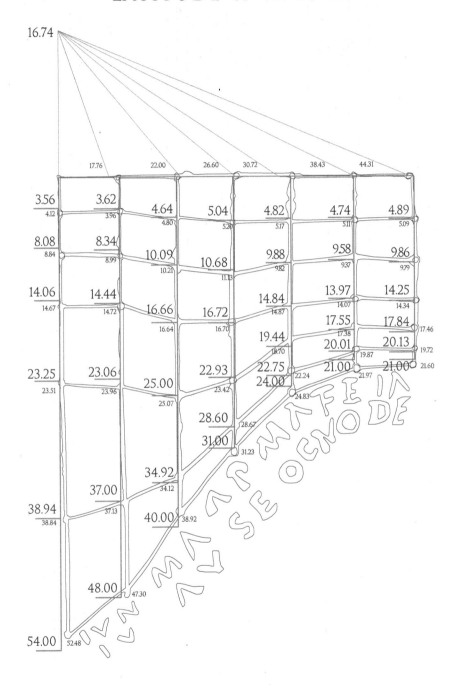

TABLE D
LATITUDE 40° 50′ 25″ (40.84°)

Naples in 28 CE

Average minute errors per hour = 6.73

EQUINOX September/**March** [**Vertical Line 4**]
Line Tangent = 31 × 0.85 = **26.35**   61 min/hour

A	B	C	D	E	F	G	H	I
1	7:06	10.76	79.24	5.26	5.00	5.20	+0.20	5.00/0.08 / +1.60
2	8:07	21.89	68.10	2.49	10.58	11.13	+0.55	5.58/0.09 / +4.95
3	9:08	32.24	57.76	1.58	16.67	16.70	+0.03	6.09/0.10 / +0.30
4	10:08	40.94	49.06	1.15	22.91	23.42	+0.51	6.24/0.10/ +5.10
5	11:09	47.19	42.80	0.93	28.33	28.67	+0.34	5.42/0.09 / +3.06
6	12:10	49.52	40.48	0.85	31.00	31.23	+0.23	2.67/0.04/ +0.92
							Avg. 0.28	15.93 Avg. 2.66

WINTER SOLSTICE [**Vertical Line 7**]
Line Tangent = 21 × 2.09 = **43.89**   46 min/hour

A	B	C	D	E	F	G	H	I
1	8:16	6.40	83.60	8.91	4.93	5.09	+0.15	4.93/0.11 / +1.65
2	9:02	12.67	77.33	4.45	9.86	9.79	−0.07	4.93/0.11 / −0.77
3	9:48	17.98	72.02	3.08	14.25	14.34	+0.09	4.39/0.09 / +0.81
4	10:34	22.05	67.95	2.46	17.84	17.46	−0.38	3.59/0.08 / −3.04
5	11:20	24.61	65.39	2.18	20.13	19.72	−0.41	2.29/0.05 / −2.05
6	12:06	25.50	64.50	2.09	21.00	21.60	+0.60	0.87/0.02/ +1.20
							Avg. 0.28	9.52 Avg. 1.58

SUMMER SOLSTICE [**Vertical Line 1**]
Line Tangent = **56** × 0.30 =**16.80**   76 min/hour

A	B	C	D	E	F	G	H	I
1	5:41	11.95	78.04	4.72	3.56	4.12	+0.56	3.56/0.05 / +2.80
2	6:57	25.73	64.26	2.07	8.12	8.84	+0.72	4.56/0.06 / +4.32
3	8:13	40.03	49.97	1.19	14.12	14.67	+0.55	6.00/0.08 / +4.40
4	9:29	54.18	35.83	0.72	23.33	23.51	+0.18	9.21/0.12 / +2.16
5	10:45	66.76	23.23	0.43	39.06	38.84	−0.22	15.73/0.21/ −4.62
6	12:00	72.85	17.15	0.30	**56.00**	52.48	**−3.52**	16.94/0.22/ −77.44
							Avg. 0.96	**95.74 Avg. 15.96**

TABLE E
LATITUDE 40° 48′ 27″ (40.81°)

Villa dei Papiri in 28 CE

Average minute errors per hour = 6.73

EQUINOX September/**March** [**Vertical Line 4**]
Line Tangent = 31 × 0.85 = **26.35**   61 min/hour

A	B	C	D	E	F	G	H	I
1	7:06	10.83	79.17	5.26	5.01	5.20	+0.19	5.01/0.08 / +1.52
2	8:07	21.97	68.03	2.49	10.58	11.13	+0.55	5.57/0.09 / +4.95
3	9:08	32.32	57.68	1.58	16.67	16.70	+0.03	6.09/0.10 / +0.30
4	10:08	41.02	48.98	1.15	22.91	23.42	+0.51	6.24/0.10/ **+5.10**
5	11:09	47.25	42.75	0.93	28.33	28.67	+0.34	5.42/0.09 / +3.09
6	12:10	49.55	40.45	0.85	31.00	31.23	+0.23	2.67/0.04/ +0.92
							Avg. 0.31	15.88 Avg. 2.65

WINTER SOLSTICE [**Vertical Line 7**]
Line Tangent = 21 × 2.09 = **43.89**   46 min/hour

A	B	C	D	E	F	G	H	I
1	8:16	6.40	83.60	8.91	4.93	5.09	+0.15	4.93/0.11 / +1.65
2	9:02	12.67	77.33	4.45	9.86	9.79	−0.07	4.93/0.11 / −0.77
3	9:48	17.98	72.02	3.08	14.25	14.34	+0.09	4.39/0.09 / +0.81
4	10:34	22.05	67.95	2.46	17.84	17.46	−0.38	3.59/0.08 / −3.04
5	11:20	24.61	65.39	2.18	20.13	19.72	−0.41	2.29/0.05 / −2.05
6	12:06	25.50	64.50	2.09	21.00	21.60	+0.60	0.87/0.02/ +1.20
							Avg. 0.28	9.52 Avg. 1.58

SUMMER SOLSTICE [**Vertical Line 1**]
Line Tangent = **56** × 0.30 =**16.80**   76 min/hour

A	B	C	D	E	F	G	H	I
1	5:41	11.95	78.04	4.72	3.56	4.12	+0.56	3.56/0.05 / +2.80
2	6:57	25.73	64.26	2.07	8.12	8.84	+0.72	4.56/0.06 / +4.32
3	8:13	40.03	49.97	1.19	14.12	14.67	+0.55	6.00/0.08 / +4.40
4	9:29	54.18	35.83	0.72	23.33	23.51	+0.18	9.21/0.12 / +2.16
5	10:45	66.76	23.23	0.43	39.06	38.84	−0.22	15.73/0.21/ −4.62
6	12:00	72.85	17.15	0.30	**56.00**	52.48	**−3.52**	16.94/0.22/ −77.44
							Avg. 0.96	**95.74 Avg. 15.95**

A = Temporal Hour; **B** = Time; **C** = Solar Elevation; **D** = Zenith Angle; **E** = Tangent of Zenith Angle = Shadow length of 1.00 m high gnomon; **F** = Corresponding Horary Lines (**Line Tangent** ÷ **E**); **G** = Horary Lines Measured on Prosciutto; **H** = Difference of **G** vs. **F**; **I** = Minute errors per hour

TABLE F
LATITUDE 40° 50′ 20 (40.84°)

Naples in 1761 CE at San Carlo alle Mortelle

EQUINOX September/**March** [**Vertical Line 4**]
Line Tangent = 31 × 0.851 = **26.38**   61 min/hour

A	B	C	D	E	F	G	H	I
1	7:05	10.60	79.40	5.34	4.94	5.20	+0.26	4.94/0.08/ +2.08
2	8:06	21.75	68.25	2.51	10.51	11.13	+0.62	5.57/0.09/ **+5.58**
3	9:07	32.12	57.88	1.60	16.49	16.70	+0.21	5.98/0.10/ +2.10
4	10:08	40.99	49.01	1.15	22.94	23.42	+0.48	6.45/0.10/ +4.80
5	11:09	47.26	42.74	0.92	28.67	28.67	0.00	5.73/0.09/ 0.00
6	12:10	49.59	40.41	**0.851**	31.00	31.23	+0.23	2.33/0.04/ +0.92
							Avg. 0.41	15.48 Avg. 2.58

A = Temporal Hour; **B** = Time; **C** = Solar Elevation; **D** = Zenith Angle; **E** = Tangent of Zenith Angle = Shadow length of 1.00 m high gnomon; **F** = Corresponding Horary Lines (**Line Tangent** ÷ E); **G** = Horary Lines Measured on Prosciutto; **H** = Difference of **G** vs. F; **I** = Minute errors per hour

TABLE G
LATITUDE 41° 31' 52" (41.53°) in 2023 CE

Average minute errors per hour = 2.35

EQUINOX September/March [Vertical Line 4]
Line Tangent = 31.20 × 0.875 = **27.30**   61 min/hour

A	B	C	D	E	F	G	H	I
1	7:53	10.52	79.48	5.38	5.07	5.20	+0.13	5.07/0.08/ +1.04
2	8:54	21.53	68.47	2.53	10.79	11.13	+0.34	5.72/0.09/ +3.06
3	9:55	31.75	58.25	1.62	16.85	16.70	−0.15	6.06/0.11/ −1.65
4	10:56	40.46	49.54	1.17	23.33	23.42	+0.09	6.48/0.11/ +0.99
5	11:57	46.56	43.44	0.95	28.74	28.67	−0.07	5.41/0.09/ −0.63
6	12:58	48.80	41.20	**0.875**	**31.20**	31.23	+0.03	2.46/0.04/ +0.12
							Avg. 0.14	7.49 Avg. 1.25

WINTER SOLSTICE [Vertical Line 7]
Line Tangent = 21.00 × 2.14 = **44.94**   45 min/hour

A	B	C	D	E	F	G	H	I
1	7:58	6.13	83.87	9.31	4.82	5.09	+0.27	4.82 /0.11/ +2.97
2	8:44	12.34	77.66	4.47	10.05	9.79	−0.16	5.23/0.12/ −1.92
3	9:29	17.50	72.50	3.17	14.17	14.34	+0.17	4.12/0.09/ +1.53
4	10:13	21.41	68.59	2.55	17.62	17.46	−0.16	3.45/0.08/ −1.28
5	11:02	24.18	65.82	2.23	20.15	19.72	−0.43	2.53/0.06/ −2.58
6	11:48	25.06	64.94	2.14	21.00	21.60	+0.60	0.85/0.02/ +1.20
							Avg. 0.30	11.48 Avg. 1.91

SUMMER SOLSTICE [Vertical Line 1]
Line Tangent = 52.48 × 0.33 = **17.32**   76 min/hour

A	B	C	D	E	F	G	H	I
1	6:32	11.80	78.20	4.71	3.66	4.12	+0.46	3.66/0.05/ +2.30
2	7:48	25.45	64.55	2.10	8.24	8.84	+0.60	4.58/0.06 +3.60
3	9:04	39.61	50.39	1.21	14.31	14.67	+0.36	6.07/0.08/ +2.88
4	10:20	53.58	36.42	0.74	23.40	23.51	+0.11	9.09/0.12/ +1.32
5	11:36	65.91	24.09	0.45	38.49	38.84	+0.35	15.09/0.20/ **+7.00**
6	12:52	71.90	18.01	0.33	**52.48**	52.48	0.00	13.99/0.18/ 0.00
							Avg. 0.53	17.10 Avg. 2.85

MAY/JULY [Vertical Line 2]
Line Tangent = 48 × 0.385 = **18.48**   74 min/hour

A	B	C	D	E	F	G	H	I
1	6:48	11.89	78.11	4.75	3.89	3.96	+0.07	3.89/0.05/ +0.35
2	8:02	25.40	64.60	2.11	8.76	8.99	+0.23	4.87/0.06/ +1.38
3	9:16	39.20	50.80	1.23	15.02	14.72	−0.30	6.26/0.08/ −2.40
4	10:30	52.56	37.44	0.77	24.00	23.96	−0.04	8.98/0.12/ −0.48
5	11:43	63.75	26.25	0.49	37.71	37.13	−0.58	13.71/0.18/ **+10.44**
6	12:57	68.94	21.06	0.385	48.00	47.30	−0.30	10.29/0.14/ +4.20
							Avg. 0.25	19.25 Avg. 3.21

APRIL/AUGUST [Vertical Line 3]
Line Tangent = 40 × 0.56 = **22.40**   68 min/hour

A	B	C	D	E	F	G	H	I
1	7:13	11.66	78.34	4.84	4.63	4.80	+0.17	4.63/0.07/ +1.19
2	8:21	24.32	65.68	2.21	10.14	10.21	+0.07	5.51/0.08/ +0.56
3	9:29	36.74	53.26	1.34	16.72	16.64	−0.08	6.58/0.10/ −0.80
4	10:37	48.12	41.88	0.90	24.88	25.07	+0.19	8.16/0.12/ +2.28
5	11:45	56.94	33.06	0.65	34.46	34.12	−0.34	9.58/0.14/ −4.76
6	12:53	60.51	29.49	0.56	40.00	38.92	−1.08	5.54/0.08/ **−8.64**
							Avg. 0.32	18.23 Avg. 3.04

FEBRUARY/OCTOBER [Vertical Line 5]
Line Tangent = 24 × 1.29 = **30.96**   55 min/hour

A	B	C	D	E	F	G	H	I
1	8:02	8.68	81.32	6.55	4.72	5.17	+0.45	4.72/0.08/ +3.60
2	8:56	17.57	72.43	3.16	9.79	9.82	+0.03	5.07/0.09/ +0.27
3	9:50	25.46	64.54	2.10	14.74	14.87	+0.13	4.95/0.09/ +1.17
4	10:45	31.94	58.06	1.60	19.35	18.70	−0.65	4.61/0.08/ −5.20
5	11:39	36.16	53.84	1.37	22.59	22.24	−0.35	3.24/0.06/ −2.10
6	12:35	37.72	52.28	1.29	24.00	24.83	+0.83	1.41/0.02/ +1.66
							Avg. 0.41	14.00 Avg. 2.33

JANUARY/NOVEMBER [Vertical Line 6]
Line Tangent = 21 × 1.83 = **38.43**   48 min/hour

A	B	C	D	E	F	G	H	I
1	7:33	6.87	83.13	8.30	4.63	5.11	+0.48	4.63/0.10/ +4.80
2	8:21	13.80	76.20	4.07	9.44	9.37	−0.07	4.81/0.10/ −0.70
3	9:10	19.87	70.13	2.77	13.87	14.07	+0.20	4.43/0.09/ +1.80
4	9:59	24.56	65.44	2.19	17.54	17.38	−0.16	3.67/0.07/ −1.12
5	10:47	27.50	62.50	1.92	20.01	19.87	−0.14	2.47/0.05/ −0.70
6	11:36	28.54	61.46	1.83	21.00	21.97	+0.97	0.99/0.02/ +1.94
							Avg. 0.34	11.06 Avg. 1.84

A = Temporal Hour; B = Time; C = Solar Elevation; D = Zenith Angle; E = Tangent of Zenith Angle = Shadow length of 1.00 m high gnomon; F = Corresponding Horary Lines (**Line Tangent** ÷ E); G = Horary Lines Measured on Prosciutto; H = Difference of G vs. F; I = Minute errors per hour (F ÷ min/hr × H)

LATITUDE 41° 31' 52" (41.53°) IN 2023 CE

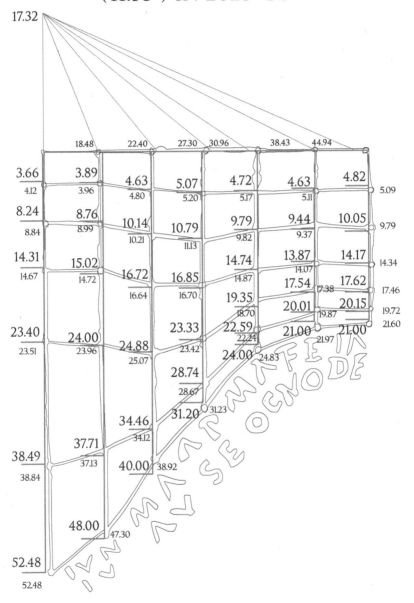

LATITUDE 41° 31' 52"
(41.53°) IN 2023 CE

Bibliography

BIOGRAPHICAL DICTIONARIES, ENCYCLOPEDIAS, AND MANUSCRIPTS

ASN, *Archivio Borbone, Inventari diversi*, Busta 304: "Inventario di tutto ciò che esiste in Palermo dei Reali Musei, Ercolanense, Capodimonte, Studi Vecchi ed altro di Napoli ordinato da farsi con R(eale) Dispaccio, 1807."

ASN, *Casa Reale Antica*, Primo Inventario 1529, 31/fol. 13: Letter of Pietro La Vega dated April 11, 1806.

ASN, *Maggiordomia Maggiore e Soprintendenza Generale di Casa Reale*, III Inventario, Categorie Diverse, vol. 46: *Diario de' Monumenti antichi, rinvenuti in Ercolano, Pompei e Stabia dal 1752 al 1799 formata dal Sig.r Camillo Paderni, Custode del Real Museo Ercolanese in Portici, e proseguito dal Signor Francesco La Vega* (= Paderni, *Diario*)

ASN, *Segreteria di Casa Reale detta casa reale antica* Fascio 1541/35: Letter of Camillo Paderni dated January 25, 1765, to Don Angelo Fernandez.

Biblioteca Nazionale di Napoli, Ms. IX.F.51: Letters of Antonio Piaggio to Giuseppe Vairo, (1) *I Piccioni*, addressed to Giuseppe Vairo, July 16, 1790; (2) *L'Iscrizione della Porta della Città di Pompeja*, October 16, 1791; (3) *Le Tavole di Eraclea, seque dall' Inscrizione della Porta della Città di Pompeja*, Resina, October 19, 1771; (4) *L'Orologio solare del Museo Ercolanese*; (5) *Candelabri, Caccia di Persano, Funerale*, Resina, November 21, 1793.

British Library Manuscripts and Archives, *Egerton Mss.* 3464, ff. 250–71, 303–18, 321–48: *Sir James Gray, 2nd Baronet, Diplomatist; Correspondence with Lord Holdernesse, 1749–1759*: (January 29, 1754, April 9, 1754, and October 29, 1754).

École Française de Rome, Bibliotèque Manuscript 26: Camillo Paderni to Charles III, *Monumenti antichi rinvenuti ne Reali Scavi di Ercolano e Pompei.*

Gori Archives BVII7.833r (Record 11953): Letter of Gasparo Cerati to Anton Francesco Gori dated January 18, 1754, Biblioteca Marucelliana in Florence (BMF): http://www.maru.firenze.sbn.it/gori/a.f.gori.htm.

Bibliography

Letters of Gori and Martorelli: Biblioteca Marucelliana in Florence (Mss. B-VII-18 and A-LIV; http://www.maru.firenze.sbn.it/gori/a.f.gori.htm).

Allgemeine Deutsche Biographie (Leipzig 1898) 44.209–10, s.v. *Woepcke, Franz* (W. Cantor).

Almanacco Reale del Regno delle Due Sicilie per l'Anno 1854 (Naples 1854).

Biografia Universale Antica e Moderna, Supplemento, vol. 4 (Venice 1839) 384–88, s.v. Pasquale Carcani.

Biographical Encyclopedia of Astronomers (New York 2014) 499–502, s.v. *d'Alembert, Jean-Le-Rond* (M. Chapront-Touze); 536–37, s.v. *de La Caille, Nicolas-Louis* (M. Murara); 542–43, s.v. *Delambre, Jean-Baptiste-Joseph* (A. Ten); 538–39, s.v. *de la Condamine, Charles-Marie* (A. Ten); 661–62, s.v. *Encke, Johann Franz* (M. Meo) 1264–66, s.v. *Lalande, Joseph-Jérôme* (S. Dumont).

Connaissance de Mouvemens Célestes pour l'Annee 1763 (Paris 1761).

Corpus Iuris Civilis, P. Krueger and T. Mommsen, eds. (Berlin 1893).

A Dictionary of British and Irish Travellers in Italy, 1701–1800 (New Haven 1997) 424–25, s.v. *Gray, Sir James*; 512–13, s.v. *Hollis, Thomas*.

Dictionary of National Biography (London 1908) 7.946–48, s.v. *Gatty, Margaret* (G. C. Boase).

Dictionary of National Biography, Second Supplement (London 1912) 1.634–37, s.v. *Evans, Sir John* (W. Wroth).

Dictionary of National Biography, 1941–1950 (Oxford 1959) 240–43, s.v. *Evans, Sir Arthur* (J. L. Myres).

Dictionnaire des Antiquites Grecques et Romaines d'apres les Textes et les Monuments, ed. C. Daremburg and E. Saglio, vol. 3.1 (Paris 1899) 260 n. 7 (s.v. *Horologium*).

Dizionario Biografico degli Italiani 80 (2015), s.v. *Paciaudi, Paolo Maria*; *Zanòtti, Francesco Maria*; and *Zanòtti, Eustachio. de la Condamine, Charles-Marie* (A. Ten). 32.564–67, s.v. *Danesi* (M. Miraglia).

Dorsten Lexicon (http://www.dorsten-lexikon.de/drecker-joseph/).

Edinburgh Encylopedia, 18 vols. (Edinburgh 1830) 7.691, s.v. *Dialling* (W. Wallace).

Encyclopedia Britannica: A dictionary of arts, sciences and general literature, 7th edition (Edinburgh 1842) 7.758, s.v. *Dialling* (H. Meickle).

Encyclopedia Britannica: A dictionary of arts, sciences and general literature, 9th edition (New York 1878) 7.154, s.v. *Dialling*.

Encylopedia Edinensis, or Dictionary of Arts Sciences and Literature, 6 vols. (Edinburgh 1827) 2.768–84, s.v. *Dialling*.

Encyclopedia Londinensis (London 1811) 10.287–369, s.v. *Horology*.

Encyclopédie, ou Dictionnaire Raisonné des Sciences, des Arts et des Métiers, par un societè de gens de lettres (Paris 1757) 7.725–26, s.v. *Gnomonique* (D'Alembert).

Encyclopédie, ou Dictionnaire Universel Raisonné des Connaissances Humaines, F. De Felice ed. (Yverdon 1773) 21.718–19, s.v. *Gnomon*.

Encyclopédie Méthodique: Mathematiques (Paris 1785) 2.145–47, s.v. *Gnomonique* (Lalande).

Bibliography

Lexicon Topographicum Urbis Romae, vols. 1 and 4 (Rome 1993, 1999).

Neue Deutsche Biographie, vol. 1 (Berlin 1953) 623, s.v. *Bassermann-Jordan, Ernst von* (E. Zinner)

Nouveau Dictionnaire pour server de Supplément aux Dictionnaires des Sciences des Arts et Des Metiers (Paris and Amsterdam 1777) 3.240–41, s.v. *Gnomon*.

Planches de l'Encyclopédie Méthodique: Mathematiques (Padua 1790).

Serie di vite e ritratti de' famosi personaggi degli ultimi tempi, 3 vols. (Milan 1818) vol. 3, s.v. *Giuseppe Girolamo di Lalande*.

Tables Astronomiques publiées par le Bureau des Longitudes de France (Paris 1806).

Thieme, U., and F. Becker, eds., *Allgemeines Lexikon der bildenden Künstler von der Antike bis zur Gegenwart*, vol. 22 (Leipzig 1978) 1, s.v. *Krügner, Johann Gottfried (1684–1769)*.

BOOKS AND JOURNALS

Acton, H. *The Bourbons of Naples, 1734–1825* (London 1956).

Aghion, I., M. Avisseau-Broustet, and Marc Fumaroli, eds. *Caylus, Mécène du Roi: Collectionner les antiquities au XVIIIᵉ siècle* (Paris 2002).

Aitken, R. G. "The James Arthur Foundation." *Astronomical Society of the Pacific Leaflet*s 5.238 (January 1949) 313–20.

Alonso Rodriguez, M. C. "La colección de antigüedades comprada por Camillo Paderni en Roma para el rey Carlos III." In J. Beltran Fortes et al., eds., *Illuminismo e Ilustración: Le antichità e i loro protagonisti in Spagna e in Italia nel XVIII secolo* (Rome 2003) 29–45.

Allroggen-Bedel, A. "Ein Malerei-fragment aus der Villa dei Papiri." *Cronache Ercolanese* 6 (1976) 85–88.

Allroggen-Bedel, A. "L'antico e la politica culturale dei Borbone." In R. Cantilena and A. Porzio, eds., *Herculanense Museum: Laboratorio sull' antico nella Reggia di Portici* (Milan 2008) 52–72.

Allroggen-Bedel, A., and H. Kammerer-Grothaus. "Il Museo ercolanese di Portici." In *La Villa dei Papiri*. (Cronache Ercolanese 13 Suppl. 2, Naples 1983) 83–127.

Amodeo, F., *Vita matematica napoletana: Studio storico, biografico, bibliografico*, 2 vols. (Naples 1905).

Anderson, M. A., and D. Robinson. *House of the Surgeon, Pompeii: Excavations in the Casa del Chirurgo (VI 1, 9–10.23)* (Oxford 2018).

Angeli, A. "La Villa dei Papiri e gli scavi sub divo fra archeologia, filologia e papirologia." *Studi di egittologia e di papirologia* 16 (2019) 9–70.

Anonymous. "Catalogo degli Antichi Monumenti de Ercolano." *Nova Acta Eruditorum*, August 1757 (Leipzig 1757) 443–45 [= Bayardi's *Catalogo*].

Anonymous. "Copy of a Letter from a Learned Gentleman of Naples, Dated February 25, 1755, Concerning the Books and Antient [*sic*] Writings Dug out of the Ruins of an Edifice Near the Site of the Old City of Herculaneum; To Monsignor Cerati,

320 *Bibliography*

of Pisa, F. R. S. Sent to Mr. Baker, F. R. S. and by Him Communicated; with a Translation by John Locke, Esq; F. R. S." *Philosophical Transactions, Royal Society of London* 49 (1754–55) 112–15.

Anonymous. "Herculaneum." *Bulletin des Sciences Historiques, Antiquités, Philologie* 1 (1824) 230.

Anonymous. "The Herculaneum Dial." *London Literary Gazette* 328 (May 3, 1823) 283–84.

Anonymous. "*Le Pitture Antiche d'Ercolano e Contorni*, etc. id est Picturae Antiquae Herculani, in aes incisae, cum explicationibus: Tomus I…" *Nova Acta Eruditorum*, January 1759 (Leipzig 1759) 1–11 [= PdE 1].

Anonymous. "*Le Pitture Antiche d'Ercolano e Contorni*, etc. id est Picturae Antiquae Herculani, in aes incisae, cum explicationibus: Tomus II…" *Nova Acta Eruditorum* (April–May 1764) (Leipzig 1764) 41–80 [= PdE 2].

Anonymous. "*Le Pitture Antiche d'Ercolano e Contorni*, etc. id est Picturae Antiquae Herculani, in aes incisae, cum explicationibus: Tomus III…" *Nova Acta Eruditorum* (March–April 1765) (Leipzig 1767) 280–335 [= PdE 3].

Anonymous. *Memorie per Servire all' Istoria Letteraria* 6.2 (August 1755) 3–7.

Anonymous. "Peintures antiques d'Herculanum etc, Tom. III." *Bibliotheque des Sciences et des Beaux Arts* 20.2 (The Hague 1763) 309–30.

Antonini, C. *Manuale di varj ornamenti tratti dalle fabbriche, e frammenti antichi: Per uso, e commodo de' pittori, scultori, architetti, scarpellini, stuccatori, intagliatori di pietre, e legni, argentieri, giojellieri, ricamatori, ebanisti, Volume 4 che contiene la Serie de' Candelabri, e orologgi Solari Antichi* (Roma 1790).

Arago, F. *Astronomie Populaire*, 4 vols. (Paris 1854).

Arnaldi, M., and K. Schaldach, "A Roman Cylinder Dial: Witness to a Forgotten Tradition." *Journal of the History of Astronomy* 28 (1997) 107–31.

Arthur, J. *Time and Its Measurement* (Chicago 1909). [Reprint of J. Arthur, "Time and Its Measurement." *Popular Mechanics* 12 (August 1909) 151–59, 294–303, 479–91, 627–38.]

Aspaas, P. P. "Maximilianus Hell (1720–1792) and the Eighteenth-Century Transits of Venus: A Study of Jesuit Science in Nordic and Central European Contexts." (PhD dissertation, University of Tromsø, 2012). http://hdl.handle.net/10037/4178.

Baldini, G. "Sopra un antica piastra di bronzo, che si suppone un' orologio da sole." *Saggi di dissertazioni accademiche pubblicamente lette nella nobile Accademia etrusca dell'antichissima citta di Cortona* 3 (1741) 185–94.

Barthélemy, J. J. *Explication de la mosaïque de Palestrine* (Paris 1760).

Barthélemy, J. J. *Memoires sur la vie et sur quelques-uns des ouvrage de JJ Barthelemy* (Paris 1830).

Barthélemy, J. J. *Oeuvres complètes: avec une notice sur la vie et les ouvrages de Barthélemy*, 4 vols. (Paris 1821).

Barthélemy, J. J. *Oeuvres diverses de J.J. Barthélemy* (Paris 1823).

Barthélemy, J. J. *Reflexions sur l'Alphabet et sur la Langue dont on se servoit autrefois a Palmyra* (Paris 1754).

Bibliography

Barthélemy, J. J. *Voyage en Italie de M. l'abbé Barthelemy...ecrites au Comte du Caylus*, ed. A. Serieys (Paris 1801).

Bassermann-Jordan, E. von. *Uhren: Ein Handbuch für Sammler und Liebhaber* (Berlin 1914).

Bassi, D. "Altre lettere inedite del P. Antonio Piaggio e spigolature dalle sue 'Memorie'." *Archivio Storico per le Province Napoletane* 33.1 (1908) 277–332.

Bassi, D. "Il P. Antonio Piaggio e i primi tentativi per lo svolgimento dei papiri ercolanesi." *Archivio Storico per le Province Napoletane* 32 (1907) 636–90.

Bayardi, O. A. *Catalogo degli antichi monumenti dissotterrati dalla discoperta città di Ercolano* (Naples 1755).

Bedos de Celles, F. *La Gnomonique Pratique, ou L'art de tracer les cadrans solaires avec la plus grande précision...* (Paris 1760).

Bell, W. J. *Patriot-Improvers: Biographical Sketches of Members of the American Philosophical Society* (American Philosophical Society 1997) 427–28.

Bernoulli, J. J. "Beschreibung des Herkulanischen Musäums nach den Zimmern." *Neue Bibliothek der schönen Wissenshaften und Freyen Künste* 17.1 (1775) 78–87.

Bianconi, G. L. *Lettere sopra A. Cornelio Celso al celebre abate Girolamo Tiraboschi* (Rome 1779).

Blank, D. "Philodemus." In E. Zalta ed., *The Stanford Encyclopedia of Philosophy* (Spring 2019). https://plato.stanford.edu/archives/spr2019/entries/philodemus/.

Bloch, H. "L. Calpurnius Piso Caesoninus in Samothrace and Herculaneum." *American Journal of Archaeology* 44.4 (1940) 485–93.

Bonnin, J. "*Horologia et memento mori*...Les Hommes, la mort et le temps dans l'Antiquité gréco-romaine." *Latomus* 72.2 (June 2013) 468–91.

Bonnin, J. *La Mesure du Temps dans l'Antiquité* (Paris 2015).

Borchard, L. *Die altägyptische Zeitmessung*. D. Wuensch and K. P. Sommer, eds. (Göttingen 2013).

Borchard, L. *Die altägyptische Zeitmessung*, Band 1, Lieferung B (E. von Bassermann-Jordan, ed., *Geschichte der Zeitmessung und der Uhren*, Berlin 1920).

Boscovich, R. J. "D'una antica villa scoperta sul dosso del Tusculo, d'un antico orologio a sole, e di alcune altre rarità, che si sono trà le rovine della medesima ritrovate (Art. XIV)." *Giornale de' Letterati di Roma* (1746) 115–35.

Britton, J. "Ptolemy's Determination of the Obliquity of the Ecliptic." *Centaurus* 14 (1969) 29–41.

Bruhns, C. *Johann Franz Encke: Seine Leben und Wirke* (Leipzig 1869).

Bryant, J. A., H. Plaisier, L. M. Irvine, A. McLean, M. Jones, and M. E. Spencer Jones. "Life and Work of Margaret Gatty (1809–1873) with Particular Reference to British Seaweeds (1863)." *Archives of Natural History* 43.1 (April 2016) 131–47.

Buchner, E. *Die Sonnenuhr des Augustus: Nachdruck aus RM 1976 und 1980 und Nachtrag über die Ausgrabung 1980/1981* (Mainz am Rhein 1982).

Burani, M. C. "Un direttore illuminato: Padre Paolo Maria Paciaudi (1763–1785)." In S. Pennestri, ed., *Complesso monumentale della Pilotta. Il Medagliere I. Storia e documentazione* (Rome 2018) 79–94.

Bibliography

Burlot, D. *Fabriquer l'antique: Les contrefaçons de peinture murale antique au XVIIIe siècle* (Naples 2012a).

Burlot, D. "Le Faux: Une tentation d'expert? Le cas de Camillo Paderni (1715–1781), artiste, restaurateur et antiquaire napolitain." In N. Étienne, ed., *L'Histoire à l'Atelier: Restaurer les oeuvres d'art, XVIIIe-XXIe siècles* (Lyon 2012b) 41–62.

Cadell, W. A. *A Journey in Carniola, Italy and France in the years 1817, 1818*, 2 vols. (Edinburgh 1820).

Cadell, W. A. "On the Lines that Divide Each Semidiurnal Arc into Six Equal Parts." *Transactions of the Royal Society of Edinburgh* 8 (1817) 61–82.

Calanca, R. *Il transito di Venere sul disco del sole* (Mestre 2004).

Calder III, W. "What Sort of Fellow Was Diels?" In W. Calder III and J. Mansfeld, *Hermann Diels (1848–1922) et la science de l' antiquité* (Geneva 1999) 1–28.

Calkoen, J. F. van Beeck. *De horologiis veterum sciothericis cui accedit theoria solariorum horam azimuthum et altitudinem solis* (Amsterdam 1797).

Cantilena, R. "Il 'medaglione' d'oro da Pompei." In C. Gasparri, G. Greco, and R. Pierobon Benoit, eds., *Dall'immagine alla storia: Studi per ricordare Stefania Adamo Muscettola* (Pozzuoli 2012) 459–76.

Cantilena, R., and A. Porzio, eds. *Herculanense Museum: Laboratorio sull' antico nella Reggia di Portici* (Milan 2008).

Capaccioli, M., G. Longo, and E. O. Cirella, *L'astronomia a Napoli dal Settecento ai nostri giorni: Storia di un' altra occasione perduta* (Naples 2009) 99–102.

Capasso, M. "Gli studi ercolanesi di Hermann Usener nel suo carteggio inedito con Hermann Diels." In M. Capasso et al., *Momenti della storia degli studi classici fra Ottocento e Novecento* (Naples 1987) 105–36.

Capasso, M. "Un omaggio dei Borboni al Padre Piaggio." In M. Gigante, ed., *Contributi alla Storia della Officina dei Papiri Ercolanesi*, vol. 1 (Rome 1980) 63–69.

Capasso, M. "Who Lived in the Villa dei Papiri?" In M. Zarmakoupi, ed., *The Villa of the Papyri at Herculaneum: Archaeology, Reception, and Digital Reconstruction* (Sozomena, Studies in the Recovery of Ancient Texts 1, Berlin 2010) 89–113.

Carcani, N. M. *Passaggio di Venere sotto il Sole, osservato in Napoli nel Real Collegio delle Scuole Pie, la mattina de' 6 giugno 1761* (Naples 1761).

Carcani, N. M. "Passaggio di Venere sotto il Sole, osservato in Napoli nel Real Collegio delle Scuole Pie, la mattina de' 6 giugno 1761." *Novelle Letterarie pubblicate in Firenze* (1761) col. 696–702 (= #44 of October 30, 1761), 714–20 (= #45 of November 6, 1761).

Carcani, N. M. "*Solaris defectus*, observatio habita Napoli die Octobris 1753." *Memorie per Servire all'istoria Letteraria* 2.4 (1753) 77–78.

Carpaci, C., and F. Zevi, eds. *Museo Archaeologico Nazionale di Napoli: La Collezione Epigrafica* (Verona 2017).

Casaubon, I. *Animadversiones in Athenaei Deipnosophistas libri* XV (1621).

Castaldi, G. *Della regale accademia ercolanese dalla sua fondazione sinora con un cenno biografico de' suoi soci ordinari* (Naples 1840).

Bibliography

Castorina, A., and F. Zevi, "*Antiquaria Napoletana* e cultura Toscana nel Settecento." In G. Cafasso et al., eds., *Il Vesuvio e le Città Vesuviane, 1730–1860: In ricordo di Georges Vallet* (Naples 1998) 115–32.

Castrén, P. *Ordo populusque Pompeianus: Polity and society in Roman Pompeii* (Acta Instituti Romani Finlandiae 8, Rome 1975).

Casaubon, I. *Animadversiones in Athenaei Deipnosophistas libri* XV (Lyon 1600).

Caylus, A. C. P. *Recueil d'Antiquités Égyptiennes, Étrusques, Grecques, Romaines et Galouises* (Paris 1752–1767).

Caylus, A. C. P., et al. *Correspondance inédite avec le P. Paciaudi, théatin (1757–1765) suivie de celles de l'abbé Barthélemy et da P. Mariette avec la même* (Paris 1877).

Ceserani, G. "The Antiquary Alessio Simmaco Mazzocchi: Oriental Origins and the Rediscovery of Magna Graecia in Eighteenth-Century Naples." *Journal of the History of Collections* 19.11 (2007) 249–59.

Chalmers, A. *The General Biographical Dictionary*, vol. 26 (London 1816) 166–67, s.v. *Rezzonico, Anthony Joseph.*

Chamorro, S. E., and M. C. Sánchez. "A Note on a New Graffito from Pompeii: *Porculus* (Casa della Diana Arcaizzante)." *Zeitschrift für Papyrologie und Epigraphik* 206 (2018) 234–35.

Chandler, B., and C. Vincent. "A Sure Reckoning: Sundials of the 17th and 18th Centuries." *Metropolitan Museum of Art Bulletin* 26.4 (December 1967) 154–69.

Choisi, E. " 'Humanitates' e Scienze: La Reale Accademia napoletana di Ferdinando IV, Storia di un progetto." *Studi Storici* 30.2 (1989) 435–56.

Choisi, E. "La Reale Accademia Ercolanese." In R. Ajello and M. D'Addio, eds., *Bernardo Tanucci: Statista, letterator, giurista* (Naples 1986) 495–517.

Cirillo, S. *Herculanensium Voluminum quae supersunt: Collectio prior* (Napoli 1844).

Clarke, J. T. "A Doric Shaft and Base Found at Assos." *American Journal of Archaeology and of the History of the Fine Arts* 2.3 (July–September 1886) 267–87.

Comparetti, D. "La Villa dei Pisoni in Ercolano e la sua biblioteca." In *Pompei e la regione sotterrata dal Vesuvio nell' anno 79* (Naples 1879) 159–76.

Comparetti, D., and G. De Petra, *La Villa Ercolanese dei Pisoni* (Turin 1883).

Coyer, F.G. *Voyage d'Italie et Hollande*, 2 vols. (Paris 1775) 1.230: Letter 29 from Naples (February 4, 1764).

Cramer, H. *Nachrichten zur Geschichte der Herkulanischen Entdeckungen* (Halle 1773).

Cust, L. *History of the Society of Dilettanti* (London 1914).

D'Alconzo, P. "Everyone Should Be Wary Whenever They Hear Boasts about Paintings Said To Have Emerged from the Excavations at Herculaneum: Bourbon Excavations and the Diffusion of Forged Pompeian Paintings." In M. Grimaldi, ed., *The Painters of Pompeii: Roman frescoes from the National Archaeological Museum of Naples* (Rome 2021) 82–87.

D'Alconzo, P. "La *Memoria dell' Osservazioni fatte sopra gl'antichi monumenti d'Ercolano l'anno 1769* di Camillo Paderni: Un istantanea del Museo Ercolanese di Portici." *Cronache Ercolanese* 49 (2019) 245–86.

Bibliography

D'Alconzo, P. "La tutela del patrimonio archeologico nel Regno di Napoli tra Sette e Ottocento." *Mélanges de l'École française de Rome—Antiquité* 113.2 (2001) 507–37.

D'Alembert, J. "Résponse a un article du Mémoire de M. L'Abbé de la Caille, sur la théorie du soleil." *Mémoires*, in *Histoire de L'Académie Royale des Sciences avec les Mémoires de Mathématique et de Physique, 1757* (Paris 1762) 145–79.

D'Aloé, S. *Nouveau guide du Musée royal bourbon* (Naples 1854).

D'Amore, M. "The Politics of Learned Travel: The Royal Society and the Grand Tour of the South of Italy 1740–1790." In L. De Michelis, L. Guerra, and F. O'Gorman, eds., *Politics and Culture in 18th Century Anglo-Italian Encounters: Entangled Histories* (Newcastle upon Tyne 2019) 102–22.

D'Arms, J. *Romans on the Bay of Naples* (Cambridge 1970).

Débarbat, S. "The French Savants, and the Earth-Sun Distance: A Résumé." In C. Sterken and P. P. Aspaas, eds., *Meeting Venus (The Journal of Astronomical Data 19.1, 2013)* 109–19.

De Caro, S. *Il santuario di Iside a Pompei e nel Museo archeologico nazionale* (Naples 2006).

De Clercq, P. "Lewis Evans and the White City Exhibitions." *Sphaera* 11 (Spring 2000). https://www.mhs.ox.ac.uk/sphaera/index.htm.

Delambre, J-B. J. "Cadran antique trouvé dans l'Ile de Délos, et par occasion de la Gnomonique des Anciens…" *Magazin Encyclopédique, ou Journal des Sciences, des Lettres et des Arts* (Septembre 1814) 361–91.

Delambre, J-B. J. *Histoire de l'Astronomie Ancienne*, 2 vols. (Paris 1817).

Della Corte, M. "Groma." *Monumenti Antichi*, vol. 28 (Milan 1922) 85–88.

Della Torre, G. M. *Storia e Fenomini del Vesuvio* (Naples 1755).

De Simone, A. "Rediscovering the Villa of the Papyri." In M. Zarmakoupi, ed., *The Villa of the Papyri at Herculaneum: Archaeology, reception, and digital reconstruction* (Sozomena: Studies in the Recovery of Ancient Texts 1, Berlin 2010) 1–20.

De Simone, A., and F. Ruffo. "Ercolano e la Villa dei Papiri alla luce dei nuovi scavi." *Cronache Ercolanese* 33 (2003) 279–312.

De Solla Price, D. "Gears from the Greeks: The Antikythera Mechanism, a Calendar Computer from c. 80 BCE." *Transactions of the American Philosophical Society* 64.7 (1974) 1–70.

De Solla Price, D. "Portable Sundials in Antiquity, Including an Account of a New Example from Aphrodisias." *Centaurus* 14.1 (1969) 242–66.

De Solla Price, D. "Precision Instruments: To 1500." In C. Singer et al., eds., *History of Technology*, 8 vols. (Oxford 1957) 3.594–96.

De Vos, M. "Camillo Paderni, la tradizione antiquaria romana e i collezionisti inglesi." In L. Franchi Dall'Orto, ed., *Ercolano 1739–1988: 250 anni di ricerca archeologica* (Rome 1993) 99–116.

Di Castiglione, R. *La Massoneria nelle Due Sicilie e i "Fratelli" Meridionale del '700* (Naples 2008).

Diels, H. *Antike Technik*[2] (Leipzig 1920).

Diels, H. "Stichometrisches." *Hermes* 17 (1882) 383–84.

Di Gerio, M., and A. Genovese. "Studio zoo-archeologico sui reperti provenienti dalla Regio IX, Insulae 11/12, saggi B e C." *Rivista di Studi Pompeiani* 22 (2011) 143–49.

Disney, J. *Memoirs of Thomas Brand-Hollis* (London 1808).

Dodero, E. *Ancient Marbles in Naples in the Eighteenth Century: Findings, Collections* (Leiden 2019).

Drecker, J. *Die Theorie der Sonnenuhren*, Band 1, Lieferung E (E. von Bassermann-Jordan, ed., *Geschichte der Zeitmessung und der Uhren*, Berlin 1925).

Drummond, W., and R. Walpole. *Herculanesia: or archaeological and philological dissertations containing a manuscript found among the ruins of Herculaneum* (London 1810).

Dumont, S., and M. Gros. "The Important Role of the Two French Astronomers J.-N. Delisle and J.-J. Lalande in the Choice of Observing Places during the Transits of Venus in 1761 and 1769." In C. Sterken and P. P. Aspaas, eds., *Meeting Venus (The Journal of Astronomical Data 19.1, 2013)* 131–44.

Dunbabin, K. M. D. *The Roman Banquet: Images of Conviviality* (Cambridge 2003).

Dunbabin, K. M. D. "*Sic erimus cuncti . . .* the Skeleton in Graeco-Roman Art." *Jahrbuch des Deutschen Archäologischen Instituts* 101 (1986) 185–255.

Dunbabin, K. M. D., I. Adak Adıbelli, M. Çavuş, D. Alper, and W. Slater, "A Sundial, a Raven, and a Missed Dinner Party on a Mosaic at Tarsus." *JRA* 32 (2019) 329–58.

Echlin, A. "Dynasty, Archaeology and Conservation: The Bourbon Rediscovery of Pompeii and Herculaneum in Eighteenth-Xentury Naples." *Journal of the History of Collections* 26.2 (2014) 145–59.

Eristov, H. "Les Antiquités d'Herculanum dans la correspondance de Caylus." In G. Cafasso et al., eds., *Il Vesuvio e le Città Vesuviane, 1730–1860: In ricordo di Georges Vallet* (Naples 1998) 145–61.

Esposito, M. R. "La collezione dei rami incisi." In P. G. Guzzo et al., eds., *Città vesuviane: antichità e fortuna: il suburbio e l'agro di Pompei, Ercolano, Oplontis e Stabiae* (Rome 2013) 42–49.

Falconet, M. "Dissertation sur Jacques de Dondis, auteur d'une horloge singulière; et à cette occasion sur les anciennes horloges." *Memoires de Litterature, tires des registres de L'Académie Royale des Inscriptions et Belles-Lettres* 20 (1753) 440–58.

Farninelli, L. ed. *Paolo Maria Paciudi e i suoi corrispondenti* (Parma 1985).

Ferrari, G. "L'Orologio romano conosciuto come 'prosciutto di Portici." In *XV Seminario Nazionale di Gnomonica (Monclassico, Trento, May 30–June 1, 2008)* 1–18. https://archive.org/details/LorologioRomanoConosciutoComeprosciutto DiPortici.

Ferrari, G. "The Roman sundial known as the Ham of Portici." *The Compendium* 26.2 (June 2019) 19–32.

Bibliography

Filangieri di Candida, A. "Monumenti ed oggetti d'arte trasportati da Napoli a Palermo nel 1806." *Napoli Nobilissima* 10 (1901) 13–15.

Finati, G. B. *Le Musée Royal-Bourbon* (Naples 1843).

Fino, E. "Il carteggio tra Anton Francesco Gori e Paolo Maria Paciaudi: Ricostruzione e studio." In C. De Benedictis and M. G. Marzi, eds., *L'Epistolario di Anton Francesco Gori: Saggi critici, antologia delle lettere e indice dei mittenti* (Florence 2004) 85–99.

Fiorelli, G. *Giornale degli Scavi di Pompei* (Naples 1865).

Fiquet du Boccage, A.-M. *Letters Concerning England, Holland and Italy* (London 1770).

Fish, F. "Lucius Calpurnius Piso Caesoninus: A Philosophical Statesman of the Late Roman Republic." In K. Lapatin, ed., *Buried by Vesuvius: The Villa dei Papiri in Herculaneum* (Los Angeles 2019) 5–9.

Flammarion, C. *Astronomie Populaire* (Paris 1881).

Forcellino, M. *Camillo Paderni Romano e l'immagine storica degli scavi di Pompei, Ercolano e Stabia* (Rome 1999).

Forcellino, M. "La Formulazione e il metodo di Camillo Paderni." *Eutopia* 2.2 (1993) 49–64.

Foss, P. *Pliny and the Eruption of Vesuvius* (New York 2022).

Fougeroux de Bondaroy, A. D. *Recherches sur les Ruines d'Herculanum et ses lumieres qui peuvent en résulter relativement à l'état présent des Sciences et des Arts* (Paris 1770).

Froehner, R. "Schinken aus Herkulanum: Ein kleiner Beitrag zur Geschichte der animalischen Nahrungsmittel." *Deutsche Tieraerztliche Wochenshrift* 16 (January 4, 1908) 12.

Frost, F. "Sausage and Meat Preservation in Antiquity." *Greek Roman and Byzantine Studies* 40 (1999) 241–52.

Fulford, M., et al. "Towards a History of Pre-Roman Pompeii: Excavations beneath the House of Amarantus (I.9.11–12), 1995–8." *Papers of the British School at Rome* 67 (1999) 37–144.

Gandellini, G. G. *Notizie Istoriche degli Intagliatori*, 3 vols. (Siena 1808), s.v. *Piaggio, Antonio*.

Gargano, M. "The Development of Astronomy in Naples: The Tale of Two Large Telescopes Made by William Herchel." *Journal of Astronomical History and Heritage* 15.1 (2012) 30–41.

Gargano, M. "The Status of Astronomy in Naples before the Foundation of the Capodimonte Observatory." In S. Esposito, ed., *Società Italiana degli Storici della Fisica e dell' Astronomia, Atti del XXXVI Convegno Annuale* (Naples 2016) 205–14.

Gasparini, D., and M. Peloso. *Le istituzioni scolastiche a Genova nel Settecento* (Genova 1995).

Gatty, A. *The Book of Sun-Dials*, enlarged edition (London 1889). Enlarged and re-edited by H. K. F. Eden and Eleanor Lloyd (London 1900).

Bibliography

Gell, W., and J. Gandy. *Pompeiana: The Topography, edifices, and ornaments* (London 1817–1819).

Gerhard, E., and T. Panofka. *Neapels antike Bildwerke* (Stuttgart 1828).

Gibbs, S. *Greek and Roman sundials* (New Haven 1976).

Gigante, M. *Philodemus in Italy* (Ann Arbor 1995).

Giustiniani, L. *Guida per lo Real Museo Borbonico*² (Naples 1824).

Goethe, J. W. *Travels in Italy, France, and Switzerland*, trans. R. D. Boylan (Boston 1883).

Goldstein, B. "The Obliquity of the Ecliptic in Ancient Greek Astronomy." *Archives Internationales d'Histoire des Sciences* 33.3 (1983) 3–14.

Goold, G. P. *Manilius: Astronomica* (Harvard 1977).

Gordon, A. E., and J. A. Gordon. *Contributions to the Paleography of Latin Inscriptions* (University of California Publications in Classical Archaeology 3.3, Berkeley 1957).

Gordon, A. R. "Subverting the Secret of Herculaneum: Archaeological Espionage in the Kingdom of Naples." In V. C. Gardner Coates and J. Sedyl, eds., *Antiquity Recovered: The Legacy of Pompeii and Herculaneum* (Los Angeles 2007) 37–58.

Gori, A. F. *Le Gemme Antiche di Anton-Maria Zanetti di Girolamo* (Venice 1750).

Gori, A. F. *Notizie del memorabile scoprimento dell' antica città di Ercolano* (Florence 1748).

Gow, A. S. F., and D. L. Page. *The Greek Anthology: The Garland of Philip, and Some Contemporary Epigrams* (London 1968).

Graevius, J. G. *Thesaurus antiquitatum Romanarum: in quo continentur lectissimi quique scriptores, qui superiori aut nostro seculo Romanae reipublicae rationem, disciplinam, leges, instituta, sacra, artesque togatas ac sagatas explicarunt & illustrarunt*, 11 vols. (Leiden 1694–1699).

Grafton, A. *Joseph Scaliger: A Study in the History of Classical Scholarship* (Oxford 1983).

Grattan-Guinness, I. "Tailpiece: The History of Mathematics and Its Own History." In *Companion Encyclopedia of the History and Philosophy of the Mathematical Sciences*, 2 vols. (New York 1994) 1665–75.

Gray, J. "Extract of a Letter from Sir James Gray, Bart. His Majesty's Envoy to the King of Naples, to the Right Honorable Sir Thomas Robinson knight of the Bath... Relating to the Same Discoveries at Herculaneum." *Philosophical Transactions, Royal Society of London* 48 (1754) 825–26.

Griggs, T. "Ancient Art and the Antiquarian: The Forgery of Giuseppe Guerra, 1755–1765." *Huntington Library Quarterly* 74.3 (2011) 471–503.

Gruterus, J. *Inscriptiones antiquae totius orbis Romani* (Amsterdam 1707).

Guadagno, G. "Note prosopografiche ercolanesi: I Mammii e L. Mammius Maximus." *Cronache Ercolanese* 14 (1984) 149–56.

Guidobaldi, M. "L'Impronta epicurea nella Villa dei Papiri alla luce delle recenti indagini archeologiche." In M. Beretta, F. Citti, and A. Iannucci, eds., *Il Culto di Epicuo: Testi, Iconografia e Paesaggio* (Florence 2014) 151–61.

Bibliography

Guidobaldi, M. P., and D. Esposito. "New Archaeological Research at the Villa dei Papiri in Herculaneum." In M. Zarmakoupi, ed., *The Villa of the Papyri at Herculaneum: Archaeology, Reception and Digital Reconstruction* (Sozomena: Studies in the Recovery of Ancient Texts 1, New York 2010) 21–62.

Guidobaldi, M. P., et al. "L'*Insula* I, l'*Insula* Nord-Occidentale e la Villa dei Papiri di Ercolano: Una sintesi delle conoscenze alla luce delle recenti indagini archeologiche." *Vesuviana* 1 (2009) 43–180.

Gunter, E., *The Works of Edmund Gunter* (London 1673).

Hall, M. B. "The Royal Society and Italy, 1667–1795." *Notes and Records of the Royal Society of London* 37.1 (August 1982) 63–81.

Hankins, T. *Jean D'Alembert: Science and the Enlightenment* (Oxford 1970).

Hannah, R. *Greek and Roman Calendars: Constructions of Time in the Classical World* (London 2005).

Hannah, R. *Time in Antiquity* (New York 2009).

Hannah, R. "Time Telling Devices." In G. Irby, ed., *A Companion to Science, Technology, and Medicine in Ancient Greece and Rome* (London 2016) 1094–112.

Harderwijk, K. J. R. ed. *Biographisch woordenboek der Nederlanden* (Haarlem 1858) 3.7–8, s.v. *Calkoen (Jan Frederik van Beeck)*.

Hardesty, K. *The Supplément to the Encyclopédie* (The Hague 1977).

Hayter, J. *A Report upon the Herculaneum Manuscripts in a Second Letter Sddressed, by Permission, to His Royal Highness, the Prince Regent* (London 1811).

Héron de Villefosse, A. *L'Argenterie et les bijoux d'or du Trésor de Boscoreale: Description des pièces conservées au Musée du Louvre* (Paris 1903).

Héron de Villefosse, A. *Le trésor de Boscoreale* (*Fondation Eugène Piot* 5, Paris 1899).

Heslin, P. "Augustus, Domitian and the So-called Horologium Augusti." *Journal of Roman Studies* 97 (2007) 1–20.

Hey, J. D. "On an Ancient Sundial Unearthed at Pompeii." *Transactions of the Royal Society of South Africa* 73.1 (2018) 56–72.

Hislop, W. "What Is Horology?" *Horological Journal* 1.1 (September 1858) 28–29.

Hoefer, M. *Histoire de l'astronomie depuis ses origins jusqu'à nos jours* (Paris 1873).

Hollis, T. "Thomas Hollis, Esq. to Professor Ward. Copies of letters from Camillo Paderni, Respecting the Various Articles Found at Herculaneum; and the Establishment of the Museum at Portici. Monseigneur Baiardi's Work on Herculaneum. Stuart and Revett Issue New Proposals. Incloses [*sic*] the Copy of an Inscription Recently Found Near Rome.." In H. Ellis, ed., *Original Letters of Eminent Literary Men of the Sixteenth, Seventeenth, and Eighteenth Centuries* (London 1843) 389–94.

Housman, A. E. *M. Manilii Astronomica* (Cantabrigae 1932).

Houston, G. "Using Sundials." In B. Lee and D. Slootjes, eds., *Aspects of Ancient Institutions and Geography: Studies in Honor of Richard J. A. Talbert* (Leiden 2014) 298–313.

Houzeau, J-C. and A. Lancaster, *Bibliographie générale de l'astronomie* (Brussels 1882).

Jacquier, R. P. "Lettre du R.P. Jacquier aux auteurs de la Gazette Littéraire." *Gazette Littéraire de l'Europe* (Paris 1764–1766) 295–99.

Jamineau, I. "An Extract of the Substance of Three Letters from Isaac Jamineau, Esq; His Majesty's Consul at Naples, to Sir Francis Hoskins Eyles Stiles, Bart. and F. R. S. concerning the Late Eruption of Mount Vesuvius." *Philosophical Transactions, Royal Society of London* 49 (1755–1756) 24–28.

Jones, A. "Eratosthenes, Hipparchus, and the Obliquity of the Ecliptic." *Journal of the History of Astronomy* 33.1 (2002) 15–19.

Jones, A., "Greco-Roman Sundials: Precision and displacement." In K. Miller and S. Symons, eds., *Down to the Hour: Short Time in the Ancient Mediterranean and Near East* (Boston 2019) 125–57.

Jones, A. *A Portable Cosmos: Revealing the Antikythera Mechanism, Scientific Wonder of the Ancient World* (Oxford 2017).

Jones, A. ed., *Time and Cosmos in Greco-Roman Antiquity* (New York and Princeton 2016).

Justi, K. *Winckelmann und seine Zeitgenossen*, 3 vols. (Leipzig 1923).

Kaibel, G. *Philodemi Gadarensis Epigrammata* (Greifswald 1885).

Kelly, J. "The Portraits of Sir James Gray (c.1708–73)." *The British Art Journal* 8.1 (2007) 15–19.

Kelly, J. *The Society of Dilettanti: Archaeology and Identity in the British enlightenment* (New Haven 2009).

Kenner, F. "Römische Sonnenuhren aus Aquileia." *Mittheilungen der k.k. Central-Commission zur Erforschung und Erhaltung der Kunst- und historischen Denkmale* 6 (1880) 18–20.

Kilian, Georg Christoph. *Li corntorni delle pitture antiche d'Ercolano, con le spiegazioni incise d'appresso l'originale*, 7 vols. (Augsburg 1778).

Klima, S., ed. *Joseph Spence: Letters from the Grand Tour* (Montreal 1975).

Knight, C. "Le lettere di Camillo Paderni alla Royal Society di Londra sulle scoperte di Ercolano (1739–1758)." *Rendiconti dell'Accademia di Archeologia Lettere e Belle Arti di Napoli* 66 (1997) 13–58.

Knight, C. "Un inedito di Padre Piaggio: Il Diario Vesuviano (1779–1795)." *Rendiconti dell'Accademia di Archeologia Lettere e Belle Arti di Napoli* 61 (1989–90), 59–131.

Konstan, D. "Epicurean Happiness: A Pig's Life." *Journal of Ancient Philosophy* 6.1 (2012) 1–24.

La Caille, N. L. *Astronomiae fundamenta novissimis solis et stellarum observationibus stabilita lutetiae in collegio mazarinaeo et in Africa ad caput Bonae Spei* (Paris 1757).

La Caille, N. L. "Diverses observations faites pendant le cours de trois differentes traversées pour un voyage au Cap du Bonne-Esperance et aux Isles de France et de Bourbon." *Memoires de l'Académie Royale des Sciences, Année 1754* (Paris 1759) 94–130.

La Caille, N. L. "Mémoire sur la Théorie du Soleil." *Memoires de l'Académie Royale des Sciences, Année 1757* (Paris 1762) 108–44.

Bibliography

La Caille, N. L. "Recherches sur les réfractions astronomiques, et sur la hauteur du pôle à Paris; avec une nouvelle table de réfractions." *Mémoires de l'Académie Royale des Sciences, Année 1755* (1761) 547–93.

La Caille, N. L. "Sur la parallaxe de soleil...faite en l'annee 1751..." and "Memoire sur la parallaxe de soleil." *Histoire de L'Académie Royale des Sciences avec les Mémoires de Mathématique et de Physique* (1761) 108–10 (*Histoire*), and 90–92 (*Mémoires*).

La Caille, N. L. *Tabulae Solares* (Paris 1758).

La Condamine, C. M. "Extract of a Letter from Mons. la Condamine, F.R.S. to Dr. Maty, F.R.S," *Philosophical Transactions, Royal Society of London* 49 (1755–1756) 622–24.

La Condamine, C. M. *Journal du voyage fait par ordre du roi, a l'équateur: Servant d'introduction historique a la Mesure des trois premiers degrés du méridien* (Paris 1751).

La Condamine, C. M. d. "Extrait d'un journal de voyage en Italie (April 10, 1757)." *Mémoires de l'Académie royale des sciences de l'Institut de France: Histoire de L'Académie royale des sciences* (1762) 336–410.

Lalande, J. J. *Astronomie*, 3 vols. (Paris 1792).

Lalande, J. J. *Bibliographie astronomique, avec l'histoire de l'astronomie depuis 1781 jusqu'a 1802* (Paris 1802).

Lalande, J. J. *Tables Astronomiques: Calculées sur les observations les plus nouvelles, pour servir à la troisieme édition de l' Astronomie* (Paris 1792).

Lalande, J. J. *Voyage d'un François en Italie, fait dans les années 1765 & 1766* (Paris 1769).

La Niece, S. "Silver Plating on Copper, Bronze and Brass." *The Antiquaries Journal* 70 (1990) 102–14.

Lapatin, K. ed. *Buried by Vesuvius: The Villa dei Papiri in Herculaneum* (Los Angeles 2019).

Lawrence, L. "Stuart and Revett: Their Literary and Architectural Careers." *Journal of the Warburg Institute* 2.2 (1938) 128–46.

Leach, E. "[Review of] M. R. Wojcik, *La Villa dei Papiri ad Ercolano.*" *American Journal of Archaeology* 92 (1988) 145–46.

Le Roy, J-D. *Les ruines des plus beaux monuments de la Grece* (Paris 1758).

Leospo, E. *La Mensa Isiaca di Torino* (Leiden 1978).

Lewis, M. W. *The Official Priests of Rome under the Julio-Claudians: A Study of the Nobility from 44 BC to 68 AD* (Rome 1955).

Leybourn, W. *The Art of Dialling* (London 1669).

Longo Auricchio, F. "La figura del P. Antonio Piaggio nel Carteggio Martorelli-Vargas." In M. Gigante, ed., *Contributi alla Storia della Officina dei Papiri Ercolanesi*, vol. 2 (Rome 1986) 17–23.

Longo Auricchio, F. and M. Capasso, "Nuove accessioni al dossier Piaggio." In *Contributi alla storia della Officina dei Papiri Ercolanesi* (Naples 1980) 17–59.

MacKinnon, M. "High on the Hog: Linking Zooarchaeological, Literary, and Artistic Data for Pig Breeds in Roman Italy." *American Journal of Archaeology* 105.4 (2001) 649–73.

Mansi, M. G. "Di alcune fonti settecentesche per 'l'archeologia e l'antiquaria in epoca borbonica nei fondi manoscritti della Biblioteca Nazionale di Napoli." In P. Giulierini, A. Coralini, and E. Calandra, eds., *Miniere della Memoria: Scavi in archivi, depositi, e biblioteche* (Sesto Fiorentino 2020) 235–43.

Mansi, M. G. "La Stamperia Reale di Napoli." In M. R. Nappi, ed., *Imagini per il Grand Tour: L'attività della Stamperia Reale Borbonica* (Naples 2015) 21–48.

Mansi, M. G. "Libri del re: *Le Antichita di Ercolano* esposte." In R. Cantilena and A. Porzio, eds. *Herculanense Museum: Laboratorio sull' antico nella Reggia di Portici* (Naples 2008) 115–45.

Mansi, M. G. "Per un profilo di Camillo Paderni." *Papyrologica Lupiensia, Bollettino annuale del Centro di Studi Papirologici dell' Università degli Studi di Lecce* V (1997) 79–108.

Mansi, M. G. "Per un profilo di Nicola Ignarra." In M. Capasso, ed., *Contributi alla Storia della Officina dei Papiri Ercolanesi* vol. 3 (Naples 2003) 11–85.

Maréchal, S., and F. David, *Les Antiquités d'Herculanum: avec leurs explications en François*, 10 vols. (Paris 1780–89).

Martini, G. H. *Abhandlung von den Sonnenuhren der Alten* (Leipzig 1777).

Martini, G. H. *Das gleichsam auflebende Pompeji* (Leipzig 1779).

Martorelli, J. *De Regia Theca Calamaria* (Naples 1756).

Martuscelli, D. *Biografia degli uomini illustri del regno di Napoli*, vol. 1 (Naples 1813), s.v. *Pasquale Carcani.*

Marx, C. "Analysis of the latitudinal data of Eratosthenes and Hipparchus." *Mathematics and Mechanics of Complex Systems* 3 (2015) 309–39.

Mattusch, C. C. *Johann Joachim Winckelmann: Letter and Report on the Discoveries at Herculaneum* (Los Angeles 2011).

Mattusch, C. C., and H. Lie. *The Villa dei Papiri at Herculaneum: Life and afterlife of a sculpture collection* (Los Angeles 2005).

Mau, A. *Pompeii: Its Life and Art* (New York 1899).

Mazzocchi, A. S. *Commentarii in Regii Herculanensis Musei aeneas tabulas Heracleenses*, 2 vols. (Naples 1754–55).

Mazzocchi, A. S. *In mutilum Campani amphitheatri titulum aliasque nonnullas Campanas inscriptiones commentarius* (Naples 1727).

Mener, H., and J. Schulze eds., *Winckelmanns Werke*, 11 vols. (Berlin 1824).

Mersman, J. "Moving Shadows, Moving Sun: Early Modern Sundials Restaging Miracles." *Nuncius* 30 (2015) 96–123.

Milanese, A. "Exhibition and Experiment: A History of the Real Museo Borbonico." In E. Risser and D. Saunders, eds., *The Restoration of Ancient Bronzes: Naples and beyond* (Los Angeles 2013) 13–29.

Milanese, A. "Il Museo Reale di Napoli al tempo di Giuseppe Bonaparte e di Gioacchino Murat: La prima sistemazione del 'museo delle statue' e delle altre raccolte (1806–1815)." *Rivista dell' Istituto Nazionale d'Archeologia e Storia dell'Arte*, S. 3, XIX–XX (1996–1997) 345–405.

Miller, A. R. *Letters from Italy, Describing the Manners, Customs, Antiquities, Paintings, &c. of that Country, in the Years MDCCLXX and MDCCLXXI, to a Friend Residing in France*, 2 vols. (London 1777).

Mills, A. "Altitude Sundials for Seasonal and Equal Hours." *Annals of Science* 53 (1996) 75–84.

Mommsen, T. *C. Iulii Solini Collectanea rerum memorabilium* (Berlin 1895).

Mommsen, T. "Inschriftbüsten I: Aus Herculaneum." *Antike Zeit* 38 (1880) 32–36.

Monaco, D. *A Complete Guide to the Small Bronzes and Gems in the Naples Museum According to the New Arrangement* (Naples 1889).

Montfaucon, B. *L'Antiquité expliquée et représentée en figures*, 5 vols. (Paris 1719).

Montucla, J.-E. *Histoire des Mathématiques dans Laquelle on Rend Compte de leurs Progrès Depuis leur Origine jusqu'à nos Jours: où l'on Expose le Tableau et le Développement des Principales Découvertes dans Toutes les Parties des Mathématiques, les Contestations qui se Sont Élevées Entre les Mathématiciens, et les Principaux Traits de la vie des Plus Célèbres* (Paris 1799–1802).

Moormann, E. "Wall paintings in the Villa of the Papyri: Old and new finds." In M. Zarmakoupi, ed., *Sozomena: The Villa of the Papyri at Herculaneum: Archaeology, Reception, and Digital Reconstruction* (Berlin 2010) 63–78.

Moormann, E. M. "Le pitture della Villa dei Papiri ad Ercolano." In *Atti del XVII Congresso Internazionale di Papirologia* (Naples 1983) 637–75.

Morrison, J. "The Sundial of Ahaz," *Popular Astronomy* 6 (1898) 537–49.

Mulholland, J. D. "Measures of Time in Antiquity." *Proceedings of the Astronomical Society of the Pacific* 84.499 (June 1972) 357–64.

Munn, M. *The Mother of the Gods, Athens, and the Tyranny of Asia: A Study of Sovereignty in Ancient Religion* (Berkeley 2006).

Murr, C. G. *Abbildungen der Gemälde und Alterthümer in dem königlich Neapolitanischen Museo zu Portici*... (Augsburg 1793).

Nani, G. *Collezione di tutte le antichita che si conservano nel museo Naniano di Venezia* (Venice 1815).

Newcomb, S. "Discussion of the observations of the transits of Venus in 1761 and 1769," *Astronomical Papers prepared for the use of the American Ephemeris and Nautical Almanac* (1890) 263–405.

Nisard, C. *Correspondance inédite du Comte du Caylus avec le P. Paciaudi, Théatin* (Paris 1877).

Olivier, jr., A. "The Changing Fashions of Roman Silver." *Record of the Art Museum Princeton University* 63 (2004) 2–27.

Olivieri, A. *Philodemi peri tou kath Omeron agathou basileos libellus* (Leipzig 1909).

Ottley, W. Y. *Notices of Engravers, and Their Works*...(London 1831), s.v. *Carlo Antonini*.

Owen, T. *The Fourteen Books of Palladius Rutilius Taurus Aemilianus on Agriculture* (London 1807).

Paciaudi, P. M. *Monumenta Peloponnesia* (Rome 1761).

Paciaudi, P. M., et al. *Lettres de Paciaudi...au comte de Caylus* (Paris 1802).

Paderni, C. "An Account of the Late Discoveries of Antiquities at Herculaneum, etc., in Two Letters from Camillo Paderni, Keeper of the Museum Herculanei to Thomas Hollis, esq.: An Extract of a Letter from C. Paderni dated at Naples June 28, 1755." *Philosophical Transactions, Royal Society of London* 49 (1755–56) 490–98.

Paderni, C. "Extract of a Letter from Camillo Paderni Dated at Naples, January 1755." *Philosophical Transactions, Royal Society of London* 49 (1755) 110–12.

Paderni, C. "Extracts of Two Letters from Sig. Camillo Paderni at Rome to Allan Ramsay, Painter...Concerning Some Ancient Statues, Pictures, and Other Curiosities Found in a Subterraneous Town Lately Discovered Near Naples...Rome, November 20, 1739." *Philosophical Transactions, Royal Society of London* 41 (1739–1741) 484–89.

Pagano, M., and R. Prisciandaro. *Studio sulle provenienze degli oggetti rinvenuti negli scavi borbonici del Regno di Napoli: Una lettura integrata, coordinata e commentata della documentazione* (Studi e Ricerche della Soprintendenza per i Beni Archeologici del Molise, I (Castellammare di Stabia [NA] 2006).

Page, D. L. *Further Greek Epigrams: Epigrams before AD 50 from the Greek Anthology and Other Sources* (New York 1991).

Pagliano, A. *Le Ore del Sole: Geometria e astronomia negli antichi orologi solari Romani* (Rome 2018).

Pamir, H., and N. Sezgin. "The Sundial and Convivium Scene on the Mosaic from the Rescue Excavation in a Late Antique House of Antioch." *Andalya* 19 (2016) 251–80.

Pandermalis, D. "Sul programma della decorazione scultorea." In *La Villa dei Papiri* (Cronache Ercolanese Suppl. 2, Naples 1983) 19–50.

Pannuti, U. *Il 'Giornale degli scavi' di Ercolano, 1738–1756* (Memorie della Accademia nazionale dei Lincei 8.26.3, Rome 1983).

Pannuti, U. "Incisori e disegnatori della stamperia reale di Napoli nel secolo XVIII.La pubblicazione 'Delle Antichità di Ercolano.'" *Xenia Antiqua* 9 (2000) 151–78.

Paolini, R. *Memorie sui monumenti di antichità e di belle arti ch'esistono in Miseno, in Baoli, in Baja, in Cuma, in Pozzuoli, in Napoli, in Capua antica, in Ercolano, in Pompei, ed in Pesto* (Naples 1812).

Parente, A. R. "Caylus e l'archeologie en Italie au XVIIIe siecle." *Les Nouvelles de l'Archeologie* 110 (2007) 17–23.

Parslow, C. "Camillo Paderni's *Monumenti Antichi* and Archaeology in Pompeii in the 1760s." *Rivista di Studi Pompeiani* 18 (2007) 7–22.

Parslow, C. *Rediscovering Antiquity: Karl Weber and the excavation of Herculaneum, Pompeii, and Stabiae* (New York 1995).

Parslow, C. "The *Sacrarium* of Isis in the *Praedia* of Julia Felix in Pompeii in Its Archaeological and Historical Contexts." In C. C. Mattusch, ed., *Rediscovering the*

Ancient World on the Bay of Naples, 1710–1890 (Studies in the History of Art 79, Washington 2013) 47–72.

Pattenden, P. "Sundials in Cetius Faventinus." *Classical Quarterly* 29.1 (1979) 203–12.

Petau, D., et al. *Dionysii Petavii... opus De doctrina temporum: auctius in hac nova editione notis & emendationibus quamplurimis, quas manu sua codici adscripserat Dionysius Petavius* (Antwerp 1703).

Piano, V. "Il *PHerc* 1067 latino: Il rotolo, il testo, l'autore," *Cronache Ercolanese* 47 (2017) 163–250.

Pigatto, L. "The 1761 Transit of Venus Dispute between Audiffredi and Pingré." In D. W. Kurtz, ed., *Transits of Venus: New Views of the Solar System and Galaxy* (Proceedings of the International Astronomical Union, Colloquium No. 196, 2004) 74–86.

Pistellato, M., A. Traviglia, and F. Bergamasco. "Geolocating Time: Digitization and Reverse Engineering of a Roman Sundial." In A. Bartoli and A Fusiello, eds., *Computer Visions—ECCV 2020 Workshops (Glascow, UK, August 23–28, 2020), Proceedings, Part II* (Switzerland 2020) 143–58.

Piva, C. "Custodi: Da curatori a guardian, un ruolo professionale per i musei italiani e le sue definizioni storiche." *Venezia Arti* 25 (December 2016) 89–106.

Proctor, R. *The Transits of Venus* (London 1871).

Quaranta, B. *L'Orologio a sole di Beroso scoperto in Pompei addì XXIII di settembre MDCCCLIV* (Naples 1854).

Ratcliff, J. *The Transit of Venus Enterprise in Victorian Britain* (Pittsburgh 2016).

Remijsen, S. "The postal service and the hour as a unit of time in antiquity," *Historia: Zeitschrift für Alte Geschichte* 56.2 (2007) 127–40

Richardson jr, L. *A New Topographical Dictionary of Ancient Rome* (Baltimore 1992).

Ridley, R. T. "A Pioneer Art-Historian and Archaeologist of the Eighteenth Century: The Comte de Caylus and his *Recueil.*" *Storia dell'Arte* 76 (1992) 362–75.

Roberts, C. "Living with the Ancient Romans: Past and Present in Eighteenth-Century Encounters with Herculaneum and Pompeii." *Huntington Library Quarterly* 78.1 (2015) 61–85.

Rohr, R. *Sundials: History, Theory, Practice* (Toronto 1970).

Roland De La Platière, J.-M. *Lettres ecrites de Suisse, d'Italie, de Sicile, et de Malthe,* 4 vols. (Amsterdam 1780).

Romanelli, D. *Antica topografia istorica del regno di Napoli* (Naples 1815).

Romanelli, D. *Isola di Capri: Manoscritti inediti* (Naples 1816).

Romanelli, D. *Viaggio a Pompei, a Pesto e di Ritorno ad Ercolano ed a Pozzuoli* (Naples 1817).

Ruesch, A., ed. *Guida Illustrata del Museo Nazionale di Napoli* (Naples 1908).

Salamand, G. "Paul-Joseph Vallet, Lieutenant de Police, et 'L'Encyclopédie d'Yverdon.'" *Les Affiches de Grenoble et du Dauphiné: Hebdomadaire de chroniques juridiques, commerciales, économiques, financières et d'échos régionaux* 4701 (Octobre 2014) 39.

Salmasius, C. *Cl. Salmasii Plinianæ Exercitationes in Solini Polyhistora: Item Caii Julii Solini Polyhistor ex veteribus libris emendatus* (Paris1629).

Salomies, O. "The Roman Republic." In C. Bruun and J. Edmondson, eds., *The Oxford Handbook of Roman Epigraphy* (Oxford 2015) 153–77.

Salzman, M. *On Roman Time: The Codex-Calendar of 354 and the Rhythms of Urban Life in Late Antiquity* (Berkeley 1990).

Santoro, L. "Two Neapolitan scientists in the XVIII Century: Felice Sabatelli and Nicola Maria Carcani." *Società Italiana degli Storici della Fisica e dell'Astronomia. Atti del XXXVIII Convegno annuale/Proceedings of the 38th Annual Conference* (Pavia 2018) 17–22.

Sarton, G. "Montucla (1725–1799): His Life and Works." *Osiris* 1 (January 1936) 519–67.

Savoie, D. *Recherches sur les Cadrans Solaires* (Turnhout 2014).

Scaliger, J. *M. Manilii Astronomicon libri quinque* (Strasbourg 1655).

Scatozza Horicht, L. A., and F. Longo Auricchio. "Dopo il Comparetti-De Petra." *Cronache Ercolanese* (1987) 157–67.

Schade, P. "Food and Ancient Philosophy." In J. Wilkins and R. Nadeau, eds., *A Companion to Food in the Ancient World* (Chichester 2015) 72–75.

Scaldach, K. "A Plea for a New Look at Roman Portable Dials." *Bulletin of the British Sundial Society* 98.1 (1998) 48–50.

Schaldach, K. "Measuring the Hours: Sundials, Water Clocks, and Portable Sundials." In A. Jones, ed., *Time and Cosmos in Greco-Roman Antiquity* (Princeton 2016) 63–93.

Schoy, C. *Die Gnomonik der Araber*, Band 1, Lieferung F (E. von Bassermann-Jordan, ed., *Geschichte der Zeitmessung und der Uhren*, Berlin 1923).

Schütrumpf, E. E. "Hermann Diels." In W. W. Briggs and W. C. M. Calder III, eds., *Classical Scholarship: A Biographical Encyclopedia* (New York 1990) 52–60.

Schutz, M. "The Horologium on the Campus Martius." *Journal of Roman Archaeology* 24 (2011) 78–86.

Sedillot, J. J. *Traité des Instruments Astronomiques des Arabes*, 2 vols. (Paris 1835).

Seigneux de Correvon, F. *Lettres sur la decouverte de l'ancienne ville d'Herculane et de ses principals antiquities*, 2 vols. (Yverdon 1770).

Sena Chiesa, G., and M. P. Lavizzari Pedrazzini, eds., *Tesori dei Postumia* (Milan 1998).

Severino, N. *De Monumentis Gnomonicis apud Graecos et Romanos: Catalogo degli orologi solari Greco-Romani* (Rome 2011).

Severino, N. "Quando Venere passa davanti al Sole." *Astronomia: La rivista dell' Unione Astrofili Italiani* 3 (1999) 8–12.

Shcheglov, D. "Hipparchus' Table of Climata and Ptolemy's *Geography*." *Orbis Terrarum: International Journal for the Historical Geography of the Ancient World* 9 (2007) 159–91.

Short, J. "The Observations of the Internal Contact of Venus with the Sun's Limb, in the Late Transit, Made in Different Places of Europe Compared with the Time of the Same Contact Observed at the Cape of Good Hope, and the Parallax of the Sun from thence Determined." *Philosophical Transactions, Royal Society of London* 52 (1761–1762) 611–28.

Short, J. "Second Paper Concerning the Parallax of the Sun Determined from Observations of the Late Transit of Venus in which This Subject Is Treated More at Length and the Quantity of the Parallax More Fully Ascertained." In *Philosophical Transactions, Royal Society of London* 53 (1763) 300–45.

Sider, D. *The Library of the Villa dei Papiri at Herculaneum* (Los Angeles 2005).

Smith, D. "The Early Contributions of Carl Schoy." *The American Mathematical Monthly* 33.1 (January 1926) 28–31.

Souchon, A. *La Construction des Cadrans Solaires: Ses principes, sa pratique précédé d' une histoire de la gnomonique* (Paris 1905).

Souchon, A. *Traité d'astronomie théorique contenant l'exposition du calcul des perturbations planétaires et lunaires et son application à l'explication et à la formation des tables astronomiques avec une introduction historique et de nombreux exemples numériques* (Paris 1891).

Spence, J. "Extract of a Letter of the Reverend Mr. Joseph Spence, Professor of Modern History in the University of Oxford, to Dr. Mead, F. R. S, (December 7, 1753)." *Philosophical Transactions, Royal Society of London* 48.48 (1754) 486.

Spence, J. *Polymetis, or, An Enquiry Concerning the Agreement between the Works of the Roman Poets, and the Remains of the Antient Artists, Being an Attempt to Illustrate Them Mutually from One Another: In Ten Books* (London 1747).

Spon, J. *Voyage d'Italie, de Dalmatie, de Grece et du Levant* (The Hague 1724).

Stebbins, F. A. "A Roman Sundial." *Journal of the Roman Astronomical Society of Canada* 52 (1958) 250–54.

Stolzenberg, D. *Egyptian Oedipus: Athanasius Kircher and the Secrets of Antiquity* (Chicago 2013).

Strazzullo, F. *Il carteggio Martorelli—Vargas Macciucca*. (Collana napoletana di studi e documenti in memoria del conte Giuseppe Matarazzo di Licosa, Naples 1984).

Stuart, J., and N. Revett. *The Antiquities of Athens*, 5 vols. (London 1762–1816).

Sudhaus, S. "Philodemum." *Rheinisches Museum für Philology* 64 (1909) 475–76.

Talbert, R. "A Lost Sundial Found and the Role of the Hour in Roman Daily Life." *Indoevropeĭskoe ĭazykoznanie i klassicheskaĭa filologiĭa* 23.2 (2019) 971–88.

Talbert, R. *Roman Portable Sundials: The Empire in Your Hand* (New York 2017).

Tanucci, B. *Epistolario*, ed. M. G. Maiorini, vol. 8 (Rome 1985).

Tanucci, B. *Epistolario*, ed. M. G. Maiorini, vol. 10 (Rome 1988).

Tanucci, B. *Epistolario*, ed. S. Lollini, vol. 11 (Rome 1990).

Taylor, L. R. *Voting Districts of the Roman Republic* (Rome 1960).

Tran Tam Tinh, V. *Le culte des divinités orientales à Herculanum* (Leiden 1971).

Travaglione, A. "Padre Antonio Piaggio: Frammenti biografici." In M. Capasso, ed., *Bicentennario della Morte di Antionio Piaggio* (Naples 1997) 15–48.

Travaglione, A. "Testimonianze su Padre Piaggio." In *Epicuro e l'Epicureismo nei Papiri Ercolanesi* (Naples 1993) 53–80.

Bibliography

Trombetta, V. "La biblioteca napoletana della 'Croce di Palazzo' nel Piano di Organizzazione dell' abate Romanelli." *La Specola: Annuario di Bibliologia e Bibliofilia*, anno II (1992–1993) 169–98.

Trombetta, V. *L'Editoria a Napoli nel Decennio francese: Produzione libraria e stampa periodica tra Stato e imprenditoria privata (1806–1815)* (Milan 2011).

Turnbull, G. *A Treatise on Ancient Painting* (London 1740).

Turner, A. J. "Sundials: History and Classification," *History of Science* (1989) 303–18.

Turner, G. L'E. "James Short, FSR, and His Contribution to the Construction of Reflecting Telescopes." *The Royal Society Journal of the History of Science* 24.1 (June 1969) 91–108.

Udias, A. *Jesuit Contribution to Science: A History* (Springer 2015).

Vazquez-Gestal, P. "Printing Antiquities: Herculaneum and the cultural politics of the Two Sicilies." In K. D. S. Lapatin, ed., *Buried By Vesuvius: The Villa dei Papiri in Herculaneum* (Los Angeles 2019) 37–45.

Volkmann, J. J. *Historisch Kritische Nachrichten von Italien²*, 3 vols. (Leipzig 1778).

Walstein, C. *Herculaneum: Past, Present and Future* (London 1908).

Warren, J. *Epicurus and Democritean Ethics: An archaeology of* ataraxia (Cambridge 2002).

Webb, P. C. *An Account of a Copper Table: Containing Two Inscriptions, in the Greek and Latin Tongues, Discovered in the Year 1732 near Heraclea, in the Bay of Tarentum, in Magna Grecia* (London 1760).

Whistler, C. "Venezia e l'Inghilterra: Artisti, collezionisti e mercato dell' arte, 1700–1750." In L. Borean and S. Mason, *Il Collezionismo d'Arte a Venezia: Il Settecento* (Vicenza 2009) 89–102.

Wilcox, D. J. *The Measure of Times Past: Pre-Newtonian Chronologies and the Rhetoric of Relative Time* (Chicago 1987).

Wilson, C. "Perturbations and Solar Tables from Lacaille to Delambre: the Rapprochement of Observation and Theory, Part 1." *Archive for History of Exact Sciences* 22.1/2 (1980) 53–188.

Winckelmann, J. J. *Briefe*, 4 vols. (Berlin 1952–1957).

Winckelmann, J. J. *Sendschreiben von den herculanischen Entdeckungen an den hochgebohrnen Herrn Heinrich Reichsgrafen von Brühl* (Dresden 1762).

Woepcke, F. *Disquisitiones archaeologico-mathematicae circa solaria veterum* (Berlin 1842).

Wojcik, M. R. *La Villa dei Papiri ad Ercolano: Contributo alla ricostruzione dell'ideologia della nobilitas tardorepubblicana* (Rome 1986).

Wolkenhauer, A. "Time, Punctuality, and Chronotopes: Concepts and Attitudes Concerning Short Time in Ancient Rome." In K. Miller and S. Symons, eds., *Down to the Hour: Short Time in the Ancient Mediterranean and Near East* (Boston 2019) 214–38.

Woolf, H. "Eighteenth-Century observations of the transits of Venus." *Annals of Science*, 9:2 (1953) 176–90.

Woolf, H. *The Transits of Venus: A Study of Eighteenth-Century Science* (Princeton 1959).

Wright, M. T. "Greek and Roman Sundials: An Essay in Approximation." *Archive for History of Exact Sciences* 55.2 (December 2000) 177–87.

Ximenes, L. *Del vecchio e nuovo gnomone fiorentino e delle osservazioni astronomiche, fisiche et architettoniche fatte nel verificarne la costruzione, libri IV* (Florence 1757).

Zaccaria, F. A. *Storia Letteraria d'Italia*, vol. 7 (September 1752–June 1753) 137–38.

Zanker, P. *The Mask of Socrates: The Image of the Intellectual in Antiquity*, trans. A. Shapiro (Berkeley 1996).

Zanni, N. "Lettere di Camillo Paderni ad Allan Ramsey: 1739–1740." *Eutopia* 2.2 (1993) 65–77.

Zuzzeri, G. L. *D'Una antica villa scoperta sul dosso del Tuscolo, e d'un antico orologio a sole tra le rovine della medissima trovato: Dissertazioni due* (Venice 1746).

Index

For the benefit of digital users, indexed terms that span two pages (e.g., 52–53) may, on occasion, appear on only one of those pages.

Accademia Ercolanese, Reale 15–16, 46–47, 64–66, 76–78, 81–98, 100–148, 154, 156–163, 166–169, 171–174, 176–178, 181, 184–189, 192–196, 203–211, 217, 220–222, 233, 239–246, 248–255, 258–264, 278–279, 286, 288–294
  Le Antichità di Ercolano Esposte 44–46, 81–86, 89–90, 98, 135–136, 139, 153–157, 161, 166–167, 176–179, 213–214, 217–220, 233
  Le Pitture Antiche di Ercolano e Contorni 81, 83–85, 100–103, 144, 157–159, 175
Achaz, sundial of 216, 229
Alcubierre, Roque Joaquin de 2–4, 7, 13–14, 36–37, 64–66
altitude or shepherd's or cylinder or pillar dial 63, 169–171, 176, 209–210, 245–249, 251, 254–257, 261–262
analemma 203, 243–244, 261–262, 284–285
Antonini, Carlo 181–183
arachnen sundial of Apollonius 283–284
Aristophanes 116–117, 122–123

Arnaldi, Mario 169–171, 246–247, 254–255, 258–259
Arthur, James 229–232
Askew, Anthony (*aka* Milord Acri) 47–49, 67–70
ataraxia (tranquility) 271–272
Athenaeus of Naucratis 38–40, 106–108, 110–111, 118–119, 121–122, 156–157
Audiffredi, Giovanni Battista 135–136
Augustus, Gaius Julius Caesar 14–15, 30–31, 82, 111–112, 129–130, 138–142, 161, 193, 208–209, 217–220, 258–261, 268–270, 276–277, 299–300
Ausonius 112–113

Baldini, Gianfrancesco 106–108, 176–177, 205
Barthélemy, Jean-Jacques 13, 32–34, 47–57, 80–89
Bassermann-Jordan, Ernst von 236–237, 240–243, 249–250
Bato 38–40, 43, 106–108, 156–157, 185, 195–196, 213

340

Index

Bayardi, Ottavio Antonio 13–17, 36–37, 47–49, 54–56, 62, 64–66, 76–78, 100–103, 129–130

    Catalogo degli antichi monumenti dissotterrati dalla discoperta città di Ercolano 13–15, 17–18, 64–66, 98, 156–157

    Prodromo delle antichità di Ercolano 13

Bernoulli, Jean (Johann III) 161–163, 176–177

Berosus the Chaldean 106–108, 146, 200–201, 203, 214, 221–222, 227–228

Besselian years 208–209

Boscoreale silver cups 272–274

Boscovich, Roger Joseph (Ruđer Bošković) 106–108, 146–148

boustrophedon 18–19, 22–24, 30–31, 37–38, 56–57, 94–98, 111–112, 142–144, 168–169, 175, 197, 222–223, 235, 252–253

Brand, Thomas 67–68

Cadell, William Archibald 210

Caesar, Gaius Julius 13–15, 111–112, 129–130, 258–259, 264–265, 270

Calkoen, Jan Frederik van Beeck 184–193, 203–206, 208–210, 240–243, 257–259, 289–290, 293

Campana, Ferdinando 81–83, 85–86, 88, 139–140

Carcani, Nicola 90–93, 124–144, 146, 151–152, 164–166, 192–193, 221, 260–261, 286–293, 295–298

Carcani, Pasquale 44–46, 92–93

Casaubon, Isaac 40, 118–119

Cassini, Giovanni Domenico 56–57, 90–92, 132

Cato, Marcus Porcius 108–109, 270–271

Caylus, Anne-Claude-Philippe de Tubieres de Grimoard de Pestels de Lévis de, Marquis d'Esternay, Baron

de Branzac 32, 44, 46–49, 51–54, 83–90, 98–105

Censorinus 19–21, 111–116, 130, 138–139

Chaerephon 110–111, 116, 118–122, 290

Charles of Bourbon, King of Naples, and Charles III, King of Spain 2–4, 13, 15–16, 24–27, 44, 46, 57–59, 64–67, 83–93, 100–103, 126, 152–153, 161–163

Chrysippus of Soli 270–271

Cicero, Marcus Tullius 106–108, 196, 264–267

Civita (ancient Pompeii) 168–169

clepsydra 19–20, 39, 106–108, 110–111, 119–122, 156–157, 176–177, 180–181, 185, 195–196, 200

Collegio Reale delle Scuole Pie, San Carlo alle Mortelle, Naples 90–92, 164–166

Colonna Nania 22–23, 27–28, 32

Columna Dianiae 32, 181–183

Comparetti, Domenico 235–237, 258–259, 265–268

conum sundial of Dionysodorus 283–284

Correvon, Francois Seigneux de 157–160

Coyer, Abbe Francois Gabriel 161

Cramer, Heinrich 163

D'Alembert, Jean Le Rond 28–34, 38, 43–44, 47–49, 53–54, 86–89, 92–93, 106, 118, 154, 167–169, 173–175, 178–179, 194–197

D'Aloé, Stanislao 220–221

Danesi, Michele 235–236

De Louville, Jacques-Eugene d'Allonville, Chevalier 131, 136–139, 155–156, 192–193, 289–290

De Petra, Giulio 7–13, 235–237, 252, 258–259

Index

De Solla Price, Derek 246–252, 263

Delambre, Jean-Baptiste Joseph 198–206, 209–210, 214, 227–228, 257–259, 293

Delisle, Joseph-Nicolas 90–92, 133–136, 164–166

Di Martino, Pietro 90–92

Diderot, Denis 28–29, 53–54, 167, 178–179, 194–195

Diels, Hermann Alexander 237–240, 266–267

Dioskurides of Samos 100–103, 149
   mosaic *emblemma*, Villa di Cicerone, Pompeii (MANN 9985) 100–103, 149

Drecker, Joseph 240–246, 257–259, 286, 293, 295

Drunken satyr, Villa dei Papiri (MANN 5628) 161, 217–220

ecliptic 106, 112, 117, 124–125, 130–132, 136–139, 147–148, 155–157, 176–177, 179–180, 186–189, 192–193, 204, 206–210, 216–217, 233, 240–243, 252, 255–257, 260–261, 289–290, 299–300

Egyptianizing base from Herculaneum (MANN 1107/76384) 83–89, 135, 161–163

Encke, Johann Franz 204–205

Encyclopédie, ou dictionnaire raisonné des sciences, des arts et des métiers 28–36, 44, 53–54, 86–89, 93–97, 154–159, 161, 167–169, 173–175, 178–181, 185–186, 194–195, 213, 227–228, 260–261
   Gnomonique 28–29, 93, 155, 161, 167, 174, 178–179

Epicurean/Epicureanism 264–266, 268–276

Epicurus 195–196, 271–274

Eratosthenes of Cyrene 136–138, 284–285

Este cylinder dial 169–171, 254–257

Eubulus 110–111, 121–122

Eudemus of Rhodes 290

Evans, Lewis 224–227

Ferdinand IV, King of Naples 217–220

Ferrari, Gianni 258–262

Fiorelli, Giuseppe 252

Fiquet du Boccage, Anne-Marie 80

Fougeroux de Bondaroy, Auguste-Denis 161, 164–166, 178–179

Froehner, Reinhard 232–233

Galiani, Ferdinando 46–47, 83–85, 135

Gatty, Margaret 222–227

Gell, William 233–234

Gellius, Aulus 109–110, 115–116, 276–277

Gold coin of Augustus and Diana (MANN 3692) 81–83, 161, 217–220

Gori, Anton Francesco 27–28, 32–34, 76

Gray, Sir James 24–28, 34, 47–49, 64–68

Guerra, Giuseppe 52, 81–83, 100–103

Gunter, Edmund 70–72

Halley, Edmund 133–134

heliotropion of Meton 119–120

hemicyclium or hemispherical sundial 106–108, 110–111, 146, 181–183, 185, 200–201, 221–222, 247–248, 276–277
   Casa di Diana, Pompeii 144–148
   Villa di Cicero, Pompeii 149–153
   Tusculum, Villa Rusinella 106–108, 146–148, 168, 176–177, 196, 201, 205, 227–228

Herculanense Museum, Portici 54–56, 69–70, 76–78, 81–83, 100–103, 149, 161–166, 178–179, 217–220

Index

Hesychius of Miletus 116–117

Hipparchus of Nicaea 136–138, 284–285

Hislop, William 216–218

Hollis, Thomas 4–5, 13, 15–17, 67–69

Horace 271–272, 275–276

Horapollo 106–108, 195–196, 240–243

horizontal or planar sundial 23–24, 42, 62–63, 70–72, 74–75, 80, 94, 106–108, 149–153, 203–204, 206–207

horologia Brumalia 119–120

Ignarra, Nicola 62, 64–66, 72–76

inscription, calendrical (prosciutto) 1–2, 22–23, 27–28, 30–31, 34–36, 51–52, 66, 94–96, 129–130, 138–139, 142–144, 156–160, 168–169, 175–178, 181, 186–187, 193, 197–200, 204–206, 208–211, 214, 223–224, 232–235, 248–249, 252–253, 258, 260–261, 288

Jamineau, Isaac 24

Kilian, Georg Christoph 177–179

Kircher, Athanasius 99–100, 106–108

Krügner, Johann Gottfried 156–158

La Caille, Nicolaus-Louis 90–92, 128, 131–134, 138–139, 192

La Condamine, Charles Marie de 51, 53–57, 61, 63–64, 79–81, 92, 100–103, 164–167, 286

La Vega, Francesco 2–4, 149

Lalande, Joseph Jérôme Lefrançais de 53–54, 132, 156–157, 164–176, 178–181, 184–185, 192–195, 198–200, 209–210, 227–228, 248–249, 295–298

Lambecius (Peter Lambeck) 112–113

Landscape painting, Villa dei Papiri (MANN 9423) 12–13

Lex Pacuvia de mense sextilii (8 BCE) 138–139, 270

Lloyd, Eleanor 222–227

Lucretius 237–239, 264–265

lunar calendar 31, 38, 43, 94, 111–112, 168–169, 175, 258–259

Macrobius 109, 258–259

Maiuri, Amedeo 252

Manilius, Marcus 118–120

Maréchal, Pierre-Sylvain 178–182, 197

Martial 46, 66, 108–109, 113–115, 299–300

Martini, Georg Heinrich 176–178, 195–196, 201, 203–204, 206, 214–217

Martorelli, Giacomo 32, 57–59, 64–66, 75–80, 152

 De Regia Theca Calamaria 76–78

Mazzocchi, Alessio Simmacho 22, 24–28, 43, 57–59, 62, 64–66, 72–76, 78–79, 86–89

Memorie per Servire all' Istoria Letteraria 17–24, 27–29, 32–37, 44, 76, 129–130, 154–155

Menander 110–111, 116–117, 119–122

Meton of Athens 119–120

Miller, Lady Anna 163–164

Milord Acri (*aka* Anthony Askew) 67–69

Mommsen, Theodore 233–235, 237–239, 266–267

Montucla, Jean-Etienne 194–198, 200, 214, 227–228

Morghen, Giovanni Elia 139–140

Nani, Bernardo and Giacomo 22–23, 27–28, 32–34

Novius Facundus 276–277

Index

obliquity of the ecliptic 130–132, 136–139, 147–148, 155–157, 180, 186–188, 192–193, 204, 206–209, 243, 252, 255–257, 289–290

Octavius, Marcus 237–239, 266–267

Paciaudi, Paolo Maria 32–49, 52, 74–75, 83–85, 89–90, 92–93, 97–98, 100–106, 118, 149, 154, 156–157, 166–169, 181–183, 217, 223–224, 227–228, 233, 249–250
  Monumenta Peloponnesia 32–44, 46, 97–98, 155, 181–183, 217

Paderni, Camillo 2–10, 13–18, 23–27, 32–34, 36–37, 44, 47–49, 51, 54–56, 57–60, 61–72, 74, 75–76, 76–80, 83–85, 85–89, 100–103, 135–136, 144–153, 161–164, 222, 264–265
  Diario de' Monumenti antichi, rinvenuti in Ercolano, Pompei e Stabia dal 1752 al 1799 2–4, 83–85, 87, 144–145, 149–150
  Monumenti antichi rinvenuti ne Reali Scavi di Ercolano e Pompei 85–87, 144–145, 150–151

Palladius Rutilius 37–38, 117–118, 121–124, 185

Pancrazi, Giuseppe Maria 17

pelicinum (axe-shaped) sundial of Patrocles 110–111

perna 108–111, 156–157, 234, 263–264, 283–284

Persius 37–38, 121–122

petaso 108–109, 156–157

Petavius, Dionysius (aka Denis Pétau) 118–119, 122–124, 185

Petronius 21–22, 176–177, 272–274
  horologium in triclinio 21–22, 110–111, 272–274
  Trimalchio 21–22, 110–111, 115–116, 272–274

Phaidros, son of Zoilos 200, 214–216

pharetra (quiver) sundial of Apollonius 110–111

Philocrates 110–111, 121–122

Philodemus of Gadara 264–269, 276

Piaggio, Antonio 6–7, 57–83, 86–92, 100–103, 125, 142–144, 150–153, 161–163, 191
  L'Orologio Solare del Museo Ercolanese 59–80, 152–153

piglet, inscribed (MANN 4905) 139–143

piglet, leaping (MANN 4893) 275–276

Pingré, Alexandre Guy 135–136

Piso Caesoninus, Lucius Calpurnius 264–269

Piso Pontifex, Lucius Calpurnius 268–270

Plautus 109–110, 113–115, 180–181, 195–196, 213, 276–277

Pliny the Elder 19–21, 115–116, 118–119, 258–259, 268–270

Pliny the Younger 115

Pollux, Julius 116–117, 122, 210–211

Pompeii 47, 54–56, 61, 67–68, 81–83, 100–103, 144–153

Pontifex/pontifices 269–270

porcius 263–264

Porcius, Marcus (Pompeian) 263–264

Portici 2–4, 24–27, 44, 49–50, 54–57, 61, 74, 76–78, 80–83, 155–157, 161, 168–169, 173, 175, 179–180, 196–197, 202, 213–214, 216–220, 222, 228–229, 236–240, 253, 258–259

Praedia Iuliae Felicis, Pompeii (Regio II.iv.1–12) 15–16, 54–56, 67–68, 102
  sacrarium of Isis 15–16, 54–56
  satyr tripod, ithyphallic (MANN 27874) 54–56, 61, 64–68, 74–75, 100–103, 264–265

344   *Index*

Ptolemy, Claudius 119–120, 136–138, 147–148, 188–189, 202, 287–288, 290, 292

Pytheas of Massilia 136–139, 289–290

Quaranta, Bernardo 221–223

Real Museo Borbonico, Naples 217–220

Rezzonico, Count Gastone della Torre 210–211

Richardson, J. Wigham 224–225, 227–229

Rohr, Rene 253

Roland de la Platière, Jean-Marie 161–163

Romanelli, Domenico 210–212

Royal Society of London 4–5, 15–16, 24, 27, 32–34, 47–49, 53–56, 63–64, 67–72, 164–166, 210, 264–265
Philosophical Transactions 4–5, 24, 47–49, 67–68

Sabatelli, Felice 90–92

Salmasius, Claudius (Claude de Saumaise) 118–124, 176–177, 185

San Carlo alle Mortelle, Naples 90–92, 286–287, 292–293

Sauli, Padre Andrea 61–63, 70

Scaliger, Joseph 118–122

Schaldach, Karlheinz 254–257, 291–292

seasonal or temporary or temporal hour 114, 206, 227, 243, 251, 276–278, 282, 284–285, 291–293, 295–300

semi-diameter of the sun 128, 132–133, 136–138, 147–148, 164–166, 287–288

Seneca the Younger 227, 268–269, 299–300

Severino, Nicola 253–254, 258–259

Short, James 69–70, 135–136

Sidonius Apollinaris 110–111, 115

silver plating 1–2, 93–94, 168, 282–283, 295–298

Society of Dilettanti 24–27

Sogliano, Antonio 236–237

solar altitude 31, 291–292

solar declination 243–245, 251, 289, 293

solar parallax 90–92, 128, 132–136, 164–166, 204–205, 240–245, 287–288

solar refraction 128, 131–132, 136–138, 147–148, 287–288

Solinus, Gaius Julius 118–121

Souchon, Abel 227–228

Starke, Mariana 161–163

Statius, Publius Papinius 115

stoicheion 116–117

Strabo 108–109, 136–138

Suetonius Tranquillus, Gaius 139–142, 268–269

Suillus (*nomen*) 263–264

Tabulae Heracleenses (MANN 2481) 24–27, 64–66

Tanucci, Berardo 15–16, 32–34, 44–47, 83–90, 92–93, 100–103, 105–106, 126, 135

Tiberius (Julius Caesar Augustus) 138–139, 268–270

Toelken, Ernst Heinrich 204–205

transit of Mercury (1786) 198–200

transit of Venus (1761 and 1769) 90–92, 133–136, 164–166

Tremellius (Macrobius *Sat.* 1.6) 109

Twelve Tables of 450 BCE 115–116

twelve-foot dinner 110–111, 116–119, 121–124, 185, 290

Vallet, Paul-Joseph (*aka* V.A.L.) 173–175

Varro, Marcus Terentius 19–20, 108–109, 113–115, 263–264, 271–272

Vergil 113–115

viatoria pensilia (Vitr. 9.8.1) 1–2, 40, 106–108, 176–177, 185, 204–205, 237–243, 246–247, 255–257, 283–285

Villa dei Papiri, Herculaneum 1–2, 6–12, 24–27, 57–59, 106, 161, 168–169, 217–220, 235–239, 252–253, 258–259, 263–270, 275–276, 292–293, 299–300

Vitruvius 1–2, 19, 21, 40, 106–108, 110–111, 119–120, 146, 176–177, 185, 203–204, 221–222, 227–228, 240–243, 255–257, 261–263, 283–285, 287–288, 290

Volkmann, Johann Jacob 163–164

Von Murr, Christoph Gottlieb 177–179, 181

Waldstein, Charles 236–237

wax (beeswax) 70, 72, 125–126, 142–144, 278–283, 286–289, 293–294

Weber, Karl Jacob 7–10, 12–13, 57–59, 252

Winckelmann, Johann Joachim 32, 47–49, 57–59, 105, 145–146, 149

Woepcke, Franz 204–210, 237–243, 258–259, 291–292

Wojcik, Maria Rita 252–253, 267

Ximenes, Leonardo 56–57, 135–136, 138–139

Yverdon *Encyclopédie* 167, 173–174

Zanetti, Girolamo F. and Anton-Maria 17, 22–23, 27–28, 32–34

Zanotti, Francesco Maria 42, 90–92, 135–136

zenith distance 203–204, 206, 291–292

zodiacal calendar 1–2, 31, 38, 43, 94–97, 112–113, 118–120, 124–127, 159, 168–169, 175–176, 180, 188–189, 227, 230–231, 243–245, 255–257, 277–278, 284–285, 291–293, 295–300

Zuzzeri, Giovanni Luca 106–108, 146, 167, 176–177, 196, 227–228